Alan Turing's Automatic Computing Engine

Alan Turing's Automatic Computing Engine

The Master Codebreaker's Struggle to Build the Modern Computer

Edited by

B. Jack Copeland

OXFORD

UNIVERSITY PRESS

OXFORD
UNIVERSITY PRESS

Great Clarendon Street, Oxford OX2 6DP

Oxford University Press is a department of the University of Oxford.
It furthers the University's objective of excellence in research, scholarship,
and education by publishing worldwide in

Oxford New York

Auckland Cape Town Dar es Salaam Hong Kong Karachi
Kuala Lumpur Madrid Melbourne Mexico City Nairobi
New Delhi Shanghai Taipei Toronto

With offices in

Argentina Austria Brazil Chile Czech Republic France Greece
Guatemala Hungary Italy Japan South Korea Poland Portugal
Singapore Switzerland Thailand Turkey Ukraine Vietnam

Oxford is a registered trade mark of Oxford University Press
in the UK and in certain other countries

Published in the United States
by Oxford University Press Inc., New York

British Library Cataloguing in Publication Data

Data available

Library of Congress Cataloging in Publication Data

Data available

Typeset by Newgen Imaging Systems (P) Ltd., Chennai, India
Printed in Great Britain
on acid-free paper by
Biddles Ltd., King's Lynn, Norfolk

ISBN 0 19 856593 3 (Hbk)

10 9 8 7 6 5 4 3 2

This is only a foretaste of what is to come, and only the shadow of what is going to be. We have to have some experience with the machine before we really know its capabilities . . . I do not see why it should not enter any one of the fields normally covered by the human intellect, and eventually compete on equal terms.

Alan Turing

(Quoted in *The Times*, 11 June 1949:

'The Mechanical Brain')

Foreword

Donald W. Davies

It was on May 10 1950 that the Pilot Model of the Automatic Computing Engine (ACE) ran its first program. This lit the lamps along the top of the control desk, one at a time, at a rate that could be controlled by the input keys. It was a great event for those who had been building the machine, simple though the program was. That small beginning culminated in the National Physical Laboratory's commercial computing service and led on to several ranges of computers.

The Second World War saw scientific research projects of a size and complexity that reached new levels. Underlying much of the work were complex mathematical models, and the only way to get working solutions was to use numerical mathematics on a large scale. In the Tube Alloys project, for example, which became the UK part of the Manhattan Project to make a fission bomb, we had to determine the critical size of a shape of enriched uranium and then estimate mathematically what would happen when it exploded. For this problem we used about a dozen 'computers'—young men and women equipped with hand calculators (such as the Brunsviga). These human computers were 'programmed' by physicists like myself. The same story, with different physics and different mathematics, was repeated in many centres across the United Kingdom.

In meetings which began in 1943 it was decided that a centre of excellence would be formed, as soon as possible after the war ended, in order to develop numerical mathematics for peaceful applications of complex mathematical models. This centre became the Mathematics Division of the National Physical Laboratory. The Division began life in 1945 under its first superintendent, J. R. Womersley.

At the same time the idea was born of building a large, fast, programmed digital computer which the new Division could use to exploit its numerical expertise. The ENIAC electronic calculator had illuminated the way forward, and there were two other major influences on this plan. One was the extraordinary paper written by Alan Turing in 1936, where, in the course of resolving a fundamental problem in mathematical logic, Turing had described the design of a computer that could calculate anything capable of being calculated by means of an algorithm. Turing's design was, to an extent, idealized, but it became the basis for all the following work on programmed computation. The second major influence was the ultra-secret work on codebreaking machines at Bletchley Park, in particular the 'Colossus'. A large, fast electronic machine, Colossus was not itself a general-purpose programmed computer, but it demonstrated the technology needed to convert Turing's ideas of 1936 into reality.

Putting all these things together, it is not surprising that in the NPL Mathematics Division a computer project soon began, with Turing at its head, and involving two of the builders of Colossus. Turing rapidly wrote his report 'Proposed Electronic Calculator', which set out a detailed design study for the ACE. This brilliant report covered all aspects of the computer, from the physics of the 'delay lines' that formed its memory to the principles of programming. Alan Turing is celebrated as a genius, both for his mathematical work in 1936 and later, and for his codebreaking skill, exemplified by the now famous 'Bombe', the machine used to break the German Enigma code. Turing's brilliance showed again in the design of the ACE. The Pilot Model of the ACE could calculate faster than any computer of its generation and many of the next generation too.

Turing was joined in 1946 by Jim Wilkinson, who would later become a world expert in matrix inversion, eigenvalue calculations, and related numerical processes. (His breakthroughs in these areas were only possible with a fast computer—the Pilot ACE—on which to test his work.) Then Mike Woodger joined, followed by Gerald Alway, myself, Betty Curtis, Henry Norton, and others. Turing's innovative design was a brilliant start. Moreover, we had the mathematicians who were to make the ACE into a great working machine, expertise from Colossus, the major electronic digital project of the Second World War, and the resources of the National Physical Laboratory to fund and build the computer.

But then followed an unfortunate part of the story—delays that cost us two years. The chance to be the builders of the first stored-program computer escaped us. Several things rescued us from the log-jam. A key figure was Harry Huskey, who joined us in 1947 after working on the ENIAC. Huskey got things done and without him another year would have been lost. He proposed to build a small-scale version of the ACE, called the Test Assembly, to test the electronics and construction. I'm not sure how much Turing approved of this particular effort! The next important development was the recruiting of two expert electronic engineers from EMI, Edward (Ted) Newman and David (Tubsy) Clayden. Clayden's chapter 'Circuit Design of the Pilot ACE and the Big ACE' (Chapter 19) describes the revolutionary style of logic device they brought with them, employing what was called 'current-steering'. When Huskey left us to return to the United States his Test Assembly was partly built. A battle ensued between those who had been designing it and Newman, for whom its style of circuits, based on the ENIAC precedent, was anathema. In the end Huskey's design work was largely abandoned, but not entirely so, for we kept to the logical design of the Test Assembly in building the Pilot Model ACE.

Once the Pilot Model was working reliably, the whole machine was moved across the NPL site from Bushy House, where it had been constructed, into a new room where it became a working computing service. The machine was running again a very short time after the move, a tribute to its sound engineering. Not only was it the fastest computer of its day (and the hardest to program!) but with 1000 electronic valves it was by far the most compact. The Pilot Model ACE enabled Mathematics Division to make great advances in numerical analysis, and provided industry with the first commercial electronic computing service.

The main honours for the design and construction of the Pilot Model ACE are shared by Alan M. Turing, James H. Wilkinson, and Edward A. Newman, aided by a sharp nudge from Harry D. Huskey, who really started the Pilot Model idea rolling.

Acknowledgements

This book began its life in 2000 at the Dibner Institute for the History of Science and Technology, Massachusetts Institute of Technology, was continued at the University of Canterbury, New Zealand, and completed at the University of Melbourne, Australia. I am grateful to these institutions for support; also to Michael Woodger for encouragement, to Diane Proudfoot for inspiration and criticism, and to Margaret Jones and Susan Osborne at the National Physical Laboratory for their generous assistance.

B.J.C.

December 2003

Contents

Part III The ACE Computers

Part IV Electronics

Contents

Part V Technical Reports and Lectures on the ACE
1945–47

List of Photographs

Contributors

Martin Campbell-Kelly is Professor of Computer Science at the University of Warwick. He has written extensively on the history of computing; his books include *From Airline Reservations to Sonic the Hedgehog: A History of the Software Industry* (Cambridge, Mass: MIT Press, 2003), *ICL: A Business and Technical History* (Oxford: Oxford University Press, 1989) and, with W. Aspray, *Computer: A History of the Information Machine* (New York: Basic Books, 1996). He is the editor of *Works of Charles Babbage* (London: Pickering & Chatto, 1989).

David Clayden joined the National Physical Laboratory (NPL) in 1947 and was a designer of the Pilot Model ACE and of the Big ACE. He subsequently worked on optical character recognition, the RSA encryption algorithm, and various other projects. He retired from the NPL in 1981.

Jack Copeland was taught at the University of Oxford by Turing's student and friend Robin Gandy. Copeland is Professor of Philosophy at the University of Canterbury, New Zealand and Co-Director of The Turing Archive for the History of Computing. His books include *Artificial Intelligence* (Oxford: Blackwell, 1993) and he is the editor of a number of volumes on Turing's work.

Mary Croarken is author of *Early Scientific Computing in Britain* (Oxford: Oxford University Press, 1990). She is currently Visiting Fellow at the University of Warwick where she conducts research in the history of computing.

Donald Davies FRS heard of the planned Automatic Computing Engine while in his final year at university and immediately applied to join the NPL, becoming in 1947 a member of the small team surrounding Turing. Davies was a designer of the Pilot Model ACE and of the Big ACE. From 1966 he headed the NPL's Autonomics Division. Davies invented 'packet switching', used in the ARPANET (forerunner of the Internet), and from 1979 worked on data security and public key cryptosystems. His books include, with D. Barber, *Communication Networks for Computers* (Chichester: Wiley, 1973) and, with W. Price, *Security for Computer Networks* (Chichester: Wiley, 1984). He died in 2000.

Bob Doran is Professor Emeritus of Computer Science at the University of Auckland, where he was Head of Computer Science. He also worked for Amdahl Corporation in California as a computer architect. He edited, with B. Carpenter, *A. M. Turing's ACE Report of 1946 and Other Papers* (Cambridge, Mass: MIT Press, 1986).

Geoff Hayes joined the NPL in 1947 after wartime work in armaments research. He was among the first to program the Pilot Model ACE and claims the title 'World's First Ex-Programmer'. He gained international recognition for his work in numerical analysis. Hayes retired from the NPL in 1983 and died in 2001.

Harry Huskey was teaching mathematics at the University of Pennsylvania when by chance he applied for part-time work on the ENIAC. Thereafter he was an important figure in a number of the early computer projects, including the EDVAC, the ACE, and the SEAC. He designed the SWAC (Standards Western Automatic Computer) and the Bendix G15 computer. Huskey is Professor Emeritus at the University of California.

Eileen Magnello is Research Associate at the Wellcome Trust Centre for the History of Medicine at University College, London. Her publications include *An Illustrated History of the National Physical Laboratory: A Century of Measurement* (Bath: Canopus, 2000).

Henry John Norton joined Turing's group at the NPL in 1947 and was involved in programming the Pilot Model ACE. He now teaches computer architecture at King's College, London.

Teresa Numerico lectures at the University of Bologna and is writing a book on Turing and machine intelligence. She co-edited, with A. Vespignani, *Informatica per le scienze umanistiche* [Computers in the Humanities] (Bologna: Mulino, 2003).

Diane Proudfoot is Senior Lecturer in Philosophy at the University of Canterbury, New Zealand and Co-Director of The Turing Archive for the History of Computing. She is the author of a number of academic and popular articles on Turing's work.

Alan Turing FRS invented the 'universal computing machine', on which all electronic stored-program digital computers are based, in 1936. Following his crucial wartime work breaking German codes, Turing joined the NPL in 1945 to design the Automatic Computing Engine. He left in 1948 to take up a Readership at the University of Manchester. Turing pioneered Artificial Intelligence and Artificial Life. He died in 1954.

Tom Vickers joined the NPL in 1946 after wartime work in armaments research. Once the Pilot Model ACE was operational, Vickers became the manager of the computing service based around it—the first external scientific computing service in the world. He subsequently became Head of the NPL's General Computing Group. After retiring from the NPL in 1977, he worked as a member of the Computing Board of the Council for National Academic Awards, taking a wide interest in computer education.

Robin Vowels was from 1961 to 1964 engineer-in-charge and programmer of an English Electric DEUCE at the New South Wales University of Technology. Vowels was thereafter Senior Lecturer in Computer Science at the Royal Melbourne Institute of Technology, retiring in 1998. He is the author of several textbooks on computer programming.

Benjamin Wells is Professor of Mathematics and Computer Science at the University of San Francisco. The last doctoral student of noted mathematical logician Alfred Tarski, Wells works in the intersection of logic, algebra, and computing. He also works in the areas of computer graphics and visual communication.

Maurice Wilkes FRS designed and built the University of Cambridge EDSAC. The EDSAC, which first ran in 1949, was the second electronic stored-program computer to work. Wilkes was Professor of Computer Technology at Cambridge and Director of the Computer Laboratory there until 1980, when he became an Adjunct Professor at MIT and a Senior Consulting Engineer at Digital Equipment Corporation, later working with Olivetti Research. He was knighted in 2000. His books include *Memoirs of a Computer Pioneer* (Cambridge, Mass: MIT Press, 1985).

Jim Wilkinson FRS joined the NPL in 1946 after wartime work in armaments research. Wilkinson worked first as Turing's assistant and then, after Turing left the NPL, headed the ACE Section. Wilkinson contributed to the design of the ACE Test Assembly and played a key role in the design and construction of the Pilot Model ACE; he also contributed to the design of the Big ACE. Once the Pilot Model ACE was operational, Wilkinson concentrated on the development of numerical analysis. He is the author of several standard works in the field. In 1970 he received the A. M. Turing Award of the Association for Computing Machinery for his unique contributions to computational mathematics. Wilkinson retired from the NPL in 1980. He died in 1986.

Mike Woodger joined Turing and Wilkinson at the NPL in 1946. He contributed to the design of the ACE Test Assembly, the Pilot Model ACE, and the Big ACE. Woodger is co-author of Algol 60, the first international programming language. He also played a part in developing the real-time programming language Ada, co-editing the 1980 Reference Manual for the language and the 1983 ANSI standard. He retired from the NPL in 1983.

Alan M. Turing

The Pilot Model of the Automatic Computing Engine

Introduction

B. Jack Copeland

A s anyone who can operate a personal computer knows, the way to make the machine perform some desired task is to open the appropriate program stored in the computer's memory. Life was not always so simple. The earliest large-scale electronic digital computers, the British Colossus (1943) and the American ENIAC (1945), did not store programs in memory. To set up these computers for a fresh task, it was necessary to modify some of the machine's wiring, re-routing cables by hand and setting switches. The basic principle of the modern computer—the idea of controlling the machine's operations by means of a program of coded instructions stored in the computer's memory—was conceived by Alan Turing. His abstract 'universal computing machine' of 1936, soon known simply as the universal Turing machine, consists of a limitless memory, in which both data and instructions are stored, and a scanner that moves back and forth through the memory, symbol by symbol, reading what it finds and writing further symbols.[1] By inserting different programs into the memory, the machine is made to carry out different computations. It was a fabulous idea—a single machine of fixed structure that, by making use of coded instructions stored in memory, could change itself, chameleon-like, from a machine dedicated to one task into a machine dedicated to a quite different one. Turing showed that his universal machine is able to accomplish *any* task that can be carried out by means of a rote method (hence the characterization 'universal'). Nowadays, when so many people possess a physical realization of the universal Turing machine, Turing's idea of a one-stop-shop computing machine might seem as obvious as the wheel. But in 1936, when engineers thought in terms of building different machines for different purposes, Turing's concept was revolutionary.

By the end of 1945, thanks to wartime developments in digital electronics, groups in Britain and in the United States had embarked on creating a universal Turing machine in hardware. Turing headed a group situated at the National Physical Laboratory (NPL) in Teddington, London. His technical report 'Proposed Electronic Calculator', dating from the end of 1945 and containing his design for the Automatic Computing Engine (ACE), was the first relatively complete specification of an electronic stored-program digital computer. Turing saw that speed and memory were the keys to computing (in the words of his assistant, Jim Wilkinson, Turing 'was obsessed with the idea of speed on the machine'[2]). Turing's design for the ACE had much in common with today's RISC (Reduced Instruction Set Computer) architectures and called for a high-speed memory of roughly the same capacity as an early Apple Macintosh computer (enormous by the standards of his day).

In the United States the Hungarian American mathematician John von Neumann shared Turing's dream of building an electronic universal stored-program computing machine. Von Neumann had learned of the universal Turing machine before the war—he and Turing came to know each other during 1936–8, when both were at Princeton University. Like Turing, von Neumann became aware of the potential of high speed digital electronics as a result of wartime work. Von Neumann's 'First Draft of a Report on the EDVAC', completed in the spring of 1945, also set out a design for an electronic stored-program digital computer ('EDVAC' stood for 'Electronic Discrete Variable Computer'). Von Neumann's report, to which Turing referred in 'Proposed Electronic Calculator', was more abstract than Turing's, saying little about programming or electronics. Harry Huskey, the electronic engineer who subsequently drew up the first detailed hardware designs for the EDVAC, said that the information in von Neumann's report was of no help to him in this.[3] Turing, in contrast, supplied detailed circuit designs, full specifications of hardware units, specimen programs in machine code, and even an estimate of the cost of building the ACE.

Turing's ACE and the EDVAC differed fundamentally in design. The ACE was a 'low level' machine (a point taken up in the chapter 'Computer Architecture and the ACE Computers')—programs were made up entirely of instructions like 'Transfer the contents of Temporary Store 15 to Temporary Store 16'. The EDVAC had (what is now called) a central processing unit or CPU, whereas in the ACE the different Temporary Stores and other memory locations had specific logical or numerical functions associated with them. For example, if two numbers were transferred to a certain destination in memory their sum would be formed there, ready to be transferred elsewhere by a subsequent instruction. Instead of writing mathematically significant instructions such as

> Multiply x by y and store the result in z,

the programmer composed a series of low-level transfer instructions producing that effect. A related difference was that, in Turing's design, complex behaviour was to be achieved by complex programming rather than by complex equipment: his philosophy was to dispense with additional hardware (such as a multiplier, divider, and hardware for floating-point arithmetic) in favour of software, and he spoke disparagingly of 'the American tradition of solving one's difficulties by means of much equipment rather than thought'.[4]

In order to increase the speed of a program's execution, Turing proposed that instructions be stored, not consecutively, but at carefully chosen positions in memory, with each instruction containing a reference to the position of the next. Also with a view to speed, he included a small fast-access memory for the temporary storage of whichever numbers were used most frequently at a given stage of a computation. According to Wilkinson in 1955, Turing 'was the first to realise that it was possible to overcome access time difficulties with . . . mercury lines . . . or drum stores by providing a comparatively small amount of fast access store. Many of the commercial machines in the USA and . . . in this country make great use of this principle.'[5]

The delays mentioned by Davies in the Foreword (and described more fully in the chapter 'The Origins and Development of the ACE Project') meant that it was several years after the completion of 'Proposed Electronic Calculator' before any significant progress was made on the physical construction of the ACE. While waiting for the hardware to be built, Turing and his group pioneered the science of computer programming, writing a library of sophisticated mathematical programs for the planned machine. The result of these delays, which were not of Turing's making, was that the NPL lost the race to build the world's first stored-program electronic digital computer—an honour that went to the University of Manchester, where the 'Manchester Baby' ran its first program on 21 June 1948. As its name implies, the Baby was a very small computer, and the news that it had run what was only a tiny program—just 17 instructions long—for a mathematically trivial task was 'greeted with hilarity' by Turing's group.[6]

The Manchester computer project was the brainchild of Turing's friend and colleague Max Newman, whose section at Britain's wartime codebreaking headquarters, Bletchley Park, had contained 10 Colossus computers working around the clock to break German codes. Newman, like von Neumann in the United States, was profoundly influenced by Turing's pre-war conception of a universal computing machine. Frustrated by the delays at the NPL, and eager to get his hands at last on a stored-program computer, Turing accepted his friend's offer of a job and left London for Manchester.

Had Turing's ACE been built as planned, it would have been in a different league from the other early computers, but his colleagues at the NPL thought the engineering work too ambitious and a considerably smaller machine was built. Known as the Pilot Model ACE, this machine ran its first program on 10 May 1950. With a clock speed of 1 MHz it was for some time the fastest computer in the world. Despite having only a few per cent of the memory capacity that Turing had specified, the Pilot ACE in other respects adhered closely to what Turing called 'Version V' of his ACE design.

The Pilot ACE was preceded by several other electronic stored-program computers. The EDSAC, built by Maurice Wilkes at the University of Cambridge Mathematical Laboratory, was the second to run, in May 1949. Later in 1949 came the BINAC, built by the creators of the ENIAC, Presper Eckert and John Mauchly, at their Electronic Control Company, Philadelphia (opinions differ as to whether the BINAC ever actually worked, however), the CSIR Mark 1, built by Trevor Pearcey at the Commonwealth Scientific and Industrial Research Organisation Division of Radiophysics, Sydney, Australia, and Whirlwind I, built by Jay Forrester at the Digital Computer Laboratory, Massachusetts Institute of Technology. The SEAC, built by Samuel Alexander and Ralph Slutz at the US Bureau of Standards Eastern Division, Washington DC, first ran in April 1950. The EDVAC itself was not completed until 1952 but most of the computers just mentioned were influenced by the EDVAC design.

The English Electric Company built a production version of the Pilot Model ACE called the 'DEUCE' (Digital Electronic Universal Computing Engine). The first DEUCE was delivered in March 1955 (to the NPL). The DEUCE was a huge success and more than 30 were sold—confounding the suggestion, made in 1946 by the Director of the NPL, Sir Charles Darwin, that 'it is very possible that . . . one machine would suffice to solve all the problems that are demanded of it from the whole country'.[7] The last DEUCE went out of service in about 1970.

The basic principles of Turing's ACE design were used in the G15 computer, built and marketed by the Detroit-based Bendix Corporation. The G15 was designed by Huskey, who spent 1947 at the NPL, working in the ACE Section.

The G15 was arguably the first personal computer. By following Turing's philosophy of minimizing hardware in favour of software, Huskey was able to make the G15 small enough (it was the size of a large domestic refrigerator) and cheap enough to be marketed as a single-user computer. Yet thanks to the ACE-like design, the G15 was as fast as computers many times its size. The first G15 ran in 1954.[8] Over 400 were sold worldwide and the G15 remained in use until about 1970.

Other derivatives of Turing's ACE design include the MOSAIC (Ministry of Supply Automatic Integrator and Computer), which played a role in Britain's air defences during the Cold War period, the EMI Business Machine, a relatively slow electronic computer with a large memory, designed for the shallow processing of large quantities of data that is typically demanded by business applications, the low-cost transistorized Packard-Bell PB250, and the 'Big ACE', constructed at the NPL and fully operational in 1960.[9]

This book tells the story of the ACE computers. Much of it is in the words of the pioneers who designed or programmed these machines: Clayden, Davies, Hayes, Huskey, Norton, Vowels, Vickers, Wilkinson, Woodger, and, of course, Alan Turing himself. Wilkes compares the electronic techniques that he used in the EDSAC with those adopted in the Pilot ACE. Chapters by Magnello, Croarken, and Copeland explain how Britain's first attempt to build an electronic stored-program computer came to take place at the NPL and describe the ups and downs of the ACE project. Chapters by Copeland and Proudfoot, Campbell-Kelly, Numerico, and Doran assess the ACE computers, evaluate their impact, and investigate the claim—based upon the influence of his pre-war work—that Turing is the father of the modern computer. Turing's work on Artificial Intelligence and Artificial Life is also described.[10]

BRITAIN TO MAKE A RADIO BRAIN

---◆---

"Ace" Superior To U.S. Model

BIGGER MEMORY STORE

Britain is to make a radio "brain" which will be called "Ace," at a cost of between £100,000 and £125,000, it was announced by the Department of Scientific and Industrial Research last night Only one will probably be made.

Ace stands for automatic computing engine. The machine will work at least as fast as the American invention called Eniac (electronic numerical integrator and computor).

The invention of Eniac was disclosed by Viscount Mountbatten when he spoke at the dinner of the British Institution of Radio Engineers last Thursday. The machine, which cost £100,000, used 18,000 valves and 5,000 switches and consumed as much power as 100 electric radiators.

The memory storage of Ace will be higher than the American invention—75,000 decimal digits compared with 200, and by means of an exhaustive library of prefabricated instructions contained on specially punched cards, the English machine will be able to deal with more complex instructions.

TIME SAVED

The organisation of these prefabricated instructions will obviate the laborious system of plugs and switches employed in Eniac, that British scientists of the mathematics division of the National Physical Laboratory feel that they have made an important new contribution.

Instructions may take a couple of minutes compared to two hours on Eniac.

Numbers are represented by a series of 1's and 0's, and answers will be given in the decimal system. The machine will multiply two 10-figure numbers in 2,000ths of a second.

Well within its scope will be the class of problem which, by its extreme complexity and the enormous length of time needed to solve it, is almost an impossibility for the pencil and paper worker. It will, for instance, be able to tackle simultaneous equations with 50 or 100 unknowns.

THREE YEARS TO BUILD

It will be able to cope by itself with all the abstruse problems for which it is designed. Further advances will probably enable production of machines designed to do even more than Ace. It will take two or three years to build.

Leading the team working on the "brain" are Sir CHARLES DARWIN, Director of the laboratory; Dr. A. M. TURING, who is 34 years old and conceived the idea of Ace; Dr. J. R. WOMERSLEY, superintendent of the division, and Prof. D. HARTREE, of Cambridge University, the only man in Britain who has worked Eniac in the United States.

The Daily Telegraph, 7 November 1946.

Source: By permission of Telegraph Group Ltd.

'ACE' WILL SPEED JET FLYING

◆

SOLVING PROBLEMS IN AERODYNAMICS

DAILY TELEGRAPH REPORTER

Revolutionary developments in aerodynamics, which will enable jet-planes to fly at speeds vastly in excess of that of sound, are expected to follow the British invention of " Ace," which has been commonly labelled the electronic " brain."

The machine, as reported in THE DAILY TELEGRAPH yesterday, will solve in seconds abstruse mathematical problems which baffle the human brain for weeks. It will be able to determine the momentum of the airflow around aircraft at speeds greater than that of sound, which experts in aerodynamics would otherwise have little hope of obtaining.

At the National Physical Laboratory at Teddington yesterday the three scientists chiefly responsible for Ace — automatic computing machine—told me about their invention.

They are Dr. A. M. TURING, Dr. J. R. WOMERSLEY, superintendent of the mathematics division of the laboratory, and Prof. D. HARTREE, who worked on Eniac, the American equivalent of Ace.

1,000 TIMES FASTER

Prof. Hartree said: " The implications of the machine are so vast that even we cannot conceive how they will affect our civilisation. Here you have something which is making one field of human activity 1,000 times faster.

In the field of transport, the equivalent of Ace would be the ability to travel from London to Cambridge—about 52 miles—in five seconds as a regular thing. It is almost unimaginable."

Prof. Hartree disclosed for the first time that a third electronic computer —at the Institute for Advanced Studies at Princeton, New Jersey— was in the course of preparation.

Although mathematical plans were almost complete, he said, it was not likely that Ace would be in full operation for two years. Experiments on parts of the machine would, however, be carried on in the meantime.

LIKE TELEPHONE EXCHANGE

New facts given to me about the machine were:

It will use about 5,000 radio valves, as compared with the 18,000 required by Eniac.

Like Eniac, it will resemble the racks in a telephone exchange.

It will do 150,000,000 multiplications of 10 digit numbers in a week and write the results. One man engaged merely in copying down the results would take, Prof. Hartree estimated, 500 years.

Dr. Turing, who conceived the idea of Ace, said that he foresaw the time, possibly in 30 years, when it would be as easy to ask the machine a question as to ask a man.

Dr. Hartree, however, thought that the machine would always require a great deal of thought on the part of the operator.

He deprecated, he said, any notion that Ace could ever be a complete substitute for the human brain, adding:

" The fashion which has sprung up in the last 20 years to decry human reason is a path which leads straight to Nazism." ●

═══════

The Daily Telegraph, 8 November 1946.

Source: By permission of Telegraph Group Ltd.

"ACE" MAY BE FASTEST BRAIN

---◆---

BRITISH ROBOT ON DISPLAY

DAILY TELEGRAPH REPORTER

An electronic "brain," which is expected to outshine all rivals by its speed in working out mathematical problems, is being developed by the National Physical Laboratory. It is known as "Ace" (automatic computing engine).

One of Ace's 43 "brain cells," 6ft high, was displayed in the library of the Royal Society, Burlington House, yesterday. It was an exhibit in a collection illustrating the development of the National Physical Laboratory, which celebrates its jubilee this year.

Dr. E. C. BULLARD, director of the laboratory told me he hoped that Ace would be completed, with "memory" built in, by the summer. It would then tackle calculations a thousand times as quickly as a girl with a desk computor, and would be able to "remember" 256 10-digit numbers at a time.

Ace should surpass the world's most advanced electronic calculator, completed at Cambridge University mathematical laboratory last summer. It should prove invaluable to scientists engaged on research into atomic energy or aero-dynamics.

Young demonstrators operated yesterday a test panel as easily as if it had been a cricket score-board. But they admitted that Ace could not test Prof. Einstein's latest formulæ. "Ace does not deal with theories—only with practicalities."

ANSWER BY CARD

Instructions are fed into the machine in the form of figures punched into cards. Ace automatically converts these into "decimal binary tables." This is a code of ones and noughts, in which the figure 56, for example, is represented by 111000.

It then turns them into pulse patterns, seen as a green line on a screen, juggles with them, decodes the result into numbers, and hands out the answer in punched cards again.

The Daily Telegraph, 31 January 1950.

Source: By permission of Telegraph Group Ltd.

MONTH'S WORK IN A MINUTE

ACE CALCULATOR

FROM OUR SPECIAL CORRESPONDENT

TEDDINGTON, Nov. 29

With the completion of the pilot model of the National Physical Laboratory's automatic computing engine, known as Ace, the Department of Scientific and Industrial Research will be glad to hear of problems the solving of which requires long and intricate arithmetical calculations. The Ace itself will be built later, but the model demonstrated here to-day is none the less a complete electronic calculating machine, claimed as one of the fastest and most powerful computing devices in the world.

Its function is to satisfy the ever-increasing need in science, industry, and administration, for rapid mathematical calculation which in the past, by traditional methods, would have been physically impossible or required more time than the problems justified. The speed at which this new engine works, said Dr. E. C. Bullard, F.R.S., director of the laboratory, could perhaps be grasped from the fact that it could provide the correct answer in one minute to a problem that would occupy a mathematician for a month. In a quarter of an hour it can produce a calculation that by hand (if it were possible) would fill half a million sheets of foolscap paper.

The automatic computing engine uses pulses of electricity, generated at a rate of a million a second, to solve all calculations which resolve themselves into addition, subtraction, multiplication, and division; so that for practical purposes there is no limit to what Ace can do.

HOLED CARDS

On the machine the pulses are used to indicate the figure 1, while gaps represent the figure 0. All calculations are done with only these two digits in what is known as the binary scale. When a sum is put into the machine the numbers are first translated into the binary scale and coded; instructions are also given to the machine by coding them as holes in cards. To carry out long sequences of operations the engine must be endowed with a " memory."

This " memory " section is highly complicated. It depends upon the slower time of travel of supersonic waves, into which the electric pulses are converted, through a column of mercury. One thousand pulses—representing digits—can be stored in this way and extracted at the precise moment when they are needed by the " arithmetic section," which handling pulses of electricity, is working 100,000 times faster than the supersonic section. The completed calculation appears in code as a holed card, representing the answer in the binary scale, which is translated back into ordinary numbers.

When experience has been gained some improvements will doubtless be made to the pilot Ace and embodied in the first standard prototype model. The cost of development and construction of the pilot model, which uses some 800 thermionic valves, was about £40,000. Now it is ready to " do business " and is expected to more than earn its keep.

The Times, 30 November 1950.

The DEUCE, the production model of the Pilot ACE, manufactured by the English Electric Company. This DEUCE, photographed in 1956, was installed at the National Physical Laboratory in 1955.

Source: National Physical Laboratory. © Crown copyright; reproduced by permission of the Controller of HMSO.

Notes

1. Turing, A. M. (1936) 'On computable numbers, with an application to the Entscheidungsproblem', *Proceedings of the London Mathematical Society*, Series 2, 42 (1936–37), 230–65.

2. Wilkinson in interview with Christopher Evans in 1976 (*The Pioneers of Computing: An Oral History of Computing*. London: Science Museum).

3. Letter from Huskey to Copeland (4 February 2002).

4. Memo from Turing to Womersley, *c.*December 1946 (in the Woodger Papers, National Museum of Science and Industry, Kensington, London (catalogue reference M15/77); a digital facsimile is in the Turing Archive for the History of Computing <www.AlanTuring.net/turing_womersley_cdec46>).

5. Letter from Wilkinson to Newman, 10 June 1955 (among the Turing Papers in the Modern Archive Centre, King's College, Cambridge (catalogue reference A.7)).

6. Woodger in interview with Copeland (June 1998).

7. Sir Charles Darwin, 'Automatic Computing Engine (ACE)', 17 April 1946 (in the British Public Record Office (PRO), Kew, Richmond, Surrey (document reference DSIR 10/385); a digital facsimile is in The Turing Archive for the History of Computing <www.AlanTuring.net/darwin_ace>). A leading British expert on automatic computation, Douglas Hartree, appears to have thought that on the contrary a total of three digital computers would probably be adequate for the country's computing needs (Hartree's opinion is reported by Vivian Bowden (1975) in 'The 25th anniversary of the stored program computer', *The Radio and Electronic Engineer*, 45, 326.

8. Letter from Huskey to Copeland (20 December 2001).

9. Coombs, A. W. M. (1954) 'MOSAIC', in *Automatic Digital Computation: Proceedings of a Symposium Held at the National Physical Laboratory*. London: Her Majesty's Stationery Office; Froggatt, R. J. (1957) 'Logical design of a computer for business use', *Journal of the British Institution of Radio Engineers*, 17, 681–96; Bell, C. G. and Newell, A. (1971) *Computer Structures: Readings and Examples*. New York: McGraw-Hill, pp. 44, 74; Yates, D. M. (1997) *Turing's Legacy: A History of Computing at the National Physical Laboratory 1945–1995*. London: Science Museum.

10. This book grew out of the ACE 2000 Conference, a joint meeting of the British Society for the History of Science and the Computer Conservation Society, organized by Copeland in order to mark the 50th Anniversary of the Pilot Model ACE. ACE 2000 was held on 18–19 May 2000 at the London Science Museum and the National Physical Laboratory. A total of 103 people attended and 18 papers were presented. Details of the conference, including the programme and a list of the attendees, is at <www.AlanTuring.net/ace2000>. Thirteen chapters of this volume are derived from papers presented at ACE 2000.

Part I
The National Physical Laboratory and the ACE Project

1 The National Physical Laboratory

Eileen Magnello

The campaign for the endowment of science

The National Physical Laboratory (NPL), one of the world's great national standards laboratories, has its origins in the campaign for the endowment of science in the latter part of the nineteenth century. The first serious debate about the role of science in British industry took place in response to the Paris International Exhibition of 1867, at which continental manufacturers won a significantly larger proportion of the prizes than at previous exhibitions. Although the finest examples of British industry (such as the 1866 Atlantic telegraph cable) were not represented, many agreed on the basis of the Paris Exhibition that unless scientific education was increased, Britain was bound eventually to be eclipsed by competitor nations.[1] In response to this threat the Liberal Government of 1870 enacted a scheme of universal primary education, in order to promote a minimum standard of literacy and numeracy in the workforce. More tellingly, industrial communities across the country (notably in Leeds, Birmingham, and Sheffield) sponsored the creation of their own new civic colleges, later universities, with little support from the government. These soon developed their own facilities for laboratory research.

The campaign for the endowment of science found a strong ally in the journal *Nature*, established in 1869. Even in its early issues it lobbied for an enquiry into the state of science in Britain. The journal's editor, Norman Lockyer, drew attention to the involvement of the German government in promoting science, noting that Britain had resources and talent of the same order as Germany, but lacked government help.[2]

In February 1870 the Liberal prime minister, William Gladstone, agreed to appoint a Royal Commission under William Cavendish, the seventh Duke of

Devonshire, to study the existing national provision for scientific instruction and for the advancement of science. Lockyer's persistence, and the testimony of sympathetic witnesses from the scientific community, succeeded in persuading the Devonshire Commission that government aid, especially in the form of funds for the establishment of public laboratories, was essential to the future of science. In 1874 the Commission recommended that a national technical laboratory and physical observatory be built.[3]

A national standards laboratory

Although *Nature* conceded that the recent establishment of the Davy–Faraday Laboratory at the Royal Institution met some existing industrial needs, it argued that much more was required. In his presidential address to the British Association in 1895, Sir Douglas Galton uttered a plea for the foundation of a national physical laboratory supported by government funding. He proposed that the new laboratory be managed by the Royal Society, with a substantial sum of money allotted by the government for an extension to the Kew Observatory, along the lines of the *Reichsanstalt* in Berlin (the German state repository for determining standards).[4] The Kew Observatory had long been the national centre for calibrating magnetic and meteorological instruments.

It was resolved that a public institution should be established to determine and verify instruments, test materials, determine physical constants, and undertake investigations into the strength and durability of materials. The first NPL Executive Committee meeting was held on 16 May 1899, with Lord Rayleigh (brother-in-law of Arthur Balfour, First Lord of the Treasury) in the Chair. Members of the General Board of the NPL were named. At a second meeting on 5 July it was recommended that Richard Tetley Glazebrook be appointed director of the Laboratory, from 1 January 1900.[5] Glazebrook had established his reputation as a manager while Senior Bursar of Trinity College, Cambridge (a position he held from 1895).[6]

Finding a site for the NPL

The Executive Committee still had to determine the most suitable location for building the Laboratory. They visited sites at Eltham in Kent, Oxshott in Surrey, and Hainault Forest in Essex, all of which were deemed unsatisfactory. Hints reached Glazebrook that Bushy House in Teddington, near London,

Bushy House today. The Pilot Model of the Automatic Computing Engine was built here in what was originally the butler's pantry.

might be offered as an alternative to Kew.[7] He and his Executive Committee decided that Bushy House suited their requirements. The Laboratory was established in the existing building and opportunities for work opened in many directions (including the standards and verification work previously carried out at the Kew Observatory).

In May 1901 the Finance Committee of the NPL recommended that, in order to help create a feeling of institutional camaraderie, Glazebrook and his family take the second floor of the north wing of Bushy House as a private residence. Glazebrook's residence had a wine cellar, a coal cellar, a bicycle shed, and a small enclosed yard—accommodation not unlike that of an Oxbridge college, and Glazebrook had similar status to the head of a college.[8]

The work of the NPL to 1918

In 1901 work began on converting the ground floor and basement of Bushy House into a physics laboratory. Other parts of the building were arranged as temporary laboratories for electrical, magnetic, and thermometric work, in addition to metallurgical and chemical research, all of which were considered to be the most fundamental areas, and had to be accommodated first. By April the contract for an engineering building had been settled and construction had begun. In the early part of 1905 plans were made to erect an electrotechnics building and a building for metrology.[9]

By 1908 the Executive Committee had extended its research programmes to include the study of problems of travel by air and sea. A tank was installed to enable shipping research to be carried out and plans were drawn up for a pioneering programme of aeronautical research. A division was set up to deal with the testing of road materials. With the approach of war, a programme was put in place for research into the production of optical glass—required urgently for telescopes, binoculars, range-finders, prismatic compasses, and periscopes.[10]

In 1915 the Executive Committee of the NPL considered for the first time the idea of employing women. Some members of the Committee thought that this would lead to objections from among the gauge-makers, and the scheme was dropped. Just a few months later, however, increasing demands from the Munitions Department resulted in the enlistment of women. On the insistence of Glazebrook and the Executive Committee, the women who joined the NPL were paid at the same rates as men.

The interwar period

In 1916 Lord Haldane had announced the government's intention of forming a Department of Scientific and Industrial Research (DSIR).[11] Heated debate followed over whether the Royal Society or the new department should control the NPL. Glazebrook wanted the President and Council of the Royal Society to have scientific control of the NPL, and at first the DSIR was out-manoeuvred by the Royal Society.[12] However, with Glazebrook's retirement in 1919 and the impact on the NPL of the unsatisfactory economic climate of the 1920s, the Royal Society lost effective control of the Laboratory.

The interwar period saw a decline of the older established industries like heavy engineering and shipbuilding, and the growth of new science-based industries such as radio, electrical power generation and transmission, non-ferrous metallurgy, synthetic polymers, aeronautics, motor engineering, and motion pictures. Industry became more responsive to the need for scientific expertise. In 1919 the engineer and physicist Sir Joseph Petavel was appointed director of the NPL. Previously Petavel had worked at the Davy–Faraday Laboratory at the Royal Institution, where he had established the primary standard of light and had designed an indicator for measuring pressures set up in exploding gaseous mixtures (later known as the 'Petavel Gauge').[13] With his background in engineering, Petavel tended to value short-term research for direct industrial gain more than longer-term speculative research. This attitude found favour with the DSIR. Under Petavel the Laboratory became oriented toward the pursuit of scientific innovation for industrial application.

The Laboratory also became involved in practical projects affecting the daily lives of the public. Illumination problems were regarded as being of great national importance, since they affected the health and safety of the public in relation to the lighting of homes and other buildings, vehicles, streets, and open spaces. In 1924 the DSIR directed the NPL to undertake illumination research for the entire nation. A Sound Division was added to the Laboratory in 1922, carrying out work on the acoustics of buildings and studying sound transmission problems in connection with the telephone, the gramophone, and radio broadcasting. Other research of the Laboratory included investigation into the wind forces on roofs and structures, the vibrations of buildings, the expansion of concrete, and the acoustical problems of the Royal Albert Hall.

Relations between the NPL and the DSIR deteriorated as Britain entered the depression. The Treasury reduced expenditure in all government departments and basic innovative research received little attention from the government. A rapid change of directors at the NPL during the 1930s did not help its position, preventing the implementation of any plans for long-term development. Following Petavel's death in 1936, Sir Frank Edward Smith (employed at the NPL since 1900) became its acting director. In 1937 Smith was replaced by Sir William Lawrence Bragg, who left after only ten months to take up the Chair of Experimental Physics at the Cavendish Laboratory in Cambridge.

Bragg was succeeded by Sir Charles Darwin, grandson of the famous evolutionary biologist. Darwin remained director until 1949. His administrative talents were demonstrated by his reorganization of the Laboratory both before and after the Second World War. Darwin played a leading role in the decision to involve the NPL in the construction of an electronic digital computer, the Automatic Computing Engine.

The Second World War

By 1941 all departments of the Laboratory were providing assistance to the Armed Services and most of its staff were involved full time in the war effort.

Darwin was seconded to the position of director of the Central Scientific Office British Supply Council in Washington DC, his brief to improve Anglo-American scientific cooperation. He stayed in Washington for six months. Edward Appleton, Secretary of the DSIR, acted as director of the NPL during Darwin's absence. When Darwin returned to Britain he was made scientific advisor to the Army Council, in addition to continuing as director at the NPL. He resumed his full-time duties at the Laboratory in 1943.

During the early 1930s a Radio Research Station had been established by the DSIR and a Wireless (later Radio) Division was created at the NPL to cooperate in this work. By 1933 radio direction finding, later known as 'radar', was being pursued. It was claimed at the end of the war that radar (which offered a method of detecting the position of aircraft by bouncing radio waves off them) was the most important national asset to have emerged from the NPL.[14] Other important war work undertaken at the NPL included the organization of the British research on the atomic bomb, and research in 'electronics'—a term that was used to include investigations into the principles and design of electronic valve circuits and also the study of their very wide applications.

Because of the increasing emphasis on industrial quality control, brought about largely by the demands of the war, industry was growing more receptive to the adoption of various statistical tools. It became evident to Darwin that there was a need to establish a centre for statistical and other mathematical research. The result, with the arrival of peace in 1945, was the creation at the NPL of a Mathematics Division (Chapter 2).

In 1945 both mathematics and electronics stood on the brink of a new, digital, future. Alan Turing, who joined the newly formed Mathematics Division in October of that year, could see this future clearly.

Notes

1. Strange, A. (1869) 'On national institutions for practical scientific research', *Quarterly Journal of Science*, 15, 38–50. Also see Morrell J. B. (1973) 'The patronage of mid-Victorian science in the University of Edinburgh', *Science Studies*, 3, 358–78.
2. Lockyer, N. (1906) *Education and National Progress.* London: Macmillan.
3. *Eighth Report of the Royal Commission on Scientific Instruction and the Advancement of Science. The Devonshire Commission*, Parliamentary Papers, C. 1298, xxviii (1875).

4. Galton, D. (1896) 'On the Reichsanstalt, Charlottenburg, Berlin', *Report of the Sixty-Fifth Meeting of the British Association for the Advancement of Science*. London: John Murray, 606–8. Anon. (1938) 'The Physikalisch-Technische Reichsanstalt: fifty years of progress', *Nature*, 142, 352–4.

5. Anon. (1900) 'A modern scientific industry', *Nature*, 63, 173–4. Glazebrook, R. T. (1902) 'The aims of the National Physical Laboratory of Great Britain', in *Annual Report of the Board of Regents of the Smithsonian Institution 1901*. Washington DC: Government Printing Office, 341–57. Glazebrook, R. T. (1899–1900) *Annual Report of the NPL*.

6. Anon. (1980) 'Richard Tetley Glazebrook', *Dictionary of Scientific Biography*, 5, 423–4.

7. Glazebrook, R. T. (1899 and 1900) *Annual Report of the NPL*.

8. Glazebrook, R. T. (1900) *Annual Report of the NPL*.

9. *Ibid.*

10. Glazebrook, R. T. (1915) *Annual Report of the NPL*.

11. Vascoe, I. (1970) 'Scientists, government and organised research. The early history of the DSIR 1914–1916', *Minerva*, 8, 192–217; MacLeod, R., Andrews, E. K. (1970) 'The origins of the DSIR: reflections on idea and men 1915–16', *Public Administration*, Spring, 23–48.

12. Hutchinson, E. (1969) 'Scientists and civil servants: the struggle over the National Physical Laboratory', *Minerva*, 7, 373–98.

13. Kaye, G. W. C. (1936) 'Joseph Petavel, KBC, FRS', *Nature*, 137, 646–7.

14. Watson-Watt, R. (1946) 'The evolution of radiolocation', *Journal of the Institution of Electrical Engineers*, Part 1, 93, 374–82.

2   The creation of the NPL Mathematics Division

Mary Croarken

Introduction

In April 1945 the journal *Nature* announced that the National Physical Laboratory was to 'extend its activities by the establishment of a Mathematics Division'.[1] This announcement coincided with John Womersley's official appointment to the post of superintendent of the Division and saw the beginning of computer research at the NPL. The new Mathematics Division was intended to act as a 'central mathematics station' and was the first of the three main centres of early electronic computer development in Britain. The Division had two main functions: to undertake research into new computing methods and machines, and to provide computing services and advice to government departments and to industry. It was soon providing a national computing service, and became a leading centre for numerical analysis.

This chapter sets the stage for these developments in computing, focusing on the circumstances surrounding the creation of the NPL Mathematics Division. Four questions are discussed. Why was a central mathematics station needed? Why was it established at the NPL? Why was Womersley chosen as superintendent? And finally, to what extent did the NPL Mathematics Division succeed as a central mathematics station?

A Central Mathematics Station

The Second World War had a huge effect on how computation—i.e. calculation—was perceived and undertaken. The war increased the demand for scientific and statistical computation (in terms both of bulk and complexity). The increase was most pronounced in two areas: ballistics, and applied research into new forms of weapons and defence. Several specialist calculating groups were set up around the country in response to the increasing demand for computation.

The newly created Ministry of Supply was particularly active in promoting such groups, since it supported a great deal of applied research. Just before war was declared, the Ministry of Supply's External Ballistics Department took over the newly created Cambridge Mathematical Laboratory.[2] The Cambridge Laboratory had been created by John Lennard-Jones as a central computing resource for Cambridge scientists. (Lennard-Jones was Plummer Professor of Theoretical Chemistry at Cambridge and worked for the Ministry of Supply during the war, holding the positions of chief superintendent of Armament Research and, in 1945, director general of Scientific Research in the Ministry of Supply.) The Cambridge Mathematical Laboratory housed a type of mechanical analogue computing machine called a differential analyser, and was also equipped with desk calculating machines (see the photograph on the following page), a model differential analyser, and a machine for solving simultaneous equations (known as the 'Mallock machine' after its inventor). Lennard-Jones recruited several mathematicians to work for the External Ballistics Department, including E. T. (Charles) Goodwin, James Wilkinson, and Tom Vickers, all of whom were later closely associated with the Automatic Computing Engine (ACE). During the war some of the staff were transferred to computational work at the Armaments Research Department at Fort Halstead, and others moved into the Admiralty.

In addition the Ministry of Supply took over Professor Douglas Hartree's differential analyser at the University of Manchester and also had a group working on internal ballistics problems at Woolwich, where Womersley was employed. In July 1942, the Ministry of Supply created a statistical service called S. R. 17 headed by Womersley.

The Ministry of Aircraft Production used R. V. Southwell's group at the University of Oxford to help with stress calculations for aircraft structures. One of Southwell's young students was Leslie Fox. The Royal Aircraft Establishment at Farnborough was also trying to cope with overwhelming computational difficulties and had a computing section whose staff included T. B. Boss. Boss and Fox were both to join the NPL Mathematics Division.

The War Office, the Air Ministry, and the Ministry of Supply contracted out some of their calculating work to the Scientific Computing Service Ltd. This was a commercial computing bureau set up in 1937 by L. J. Comrie,

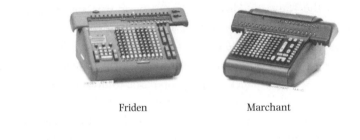

Friden   Marchant

Facit   Monroe   Brunsviga

Types of desk calculating machine. Before the advent of electronic computers such as the Pilot ACE, many large-scale calculations were done by teams of clerks—known as 'human computers'—equipped with desk machines.

Source: National Physical Laboratory. © Crown copyright; reproduced by permission of the Controller of HMSO.

ex-superintendent of the Nautical Almanac Office, and Britain's leading mathematical table maker.[3] Comrie and his staff were in very great demand during the war, working throughout the Blitz to provide calculations for the service ministries. The type of work that they performed varied from producing ballistics tables to calculations for locating German radio transmitters.[4]

The Ministry that took the most decisive steps to resolve its computational problems was the Admiralty. The Admiralty already had an extensive computing facility in the form of the Nautical Almanac Office, which produced astronomical and navigational tables on an annual basis. By the standards of the day it was very well equipped with desk calculators and accounting machines, and it had a well-trained staff. During the 1930s the then superintendent, Comrie, had carried out work for the War Office using Nautical Almanac Office facilities and had been dismissed for doing so.[5] It was ironic that it was the Nautical Almanac Office to which the Admiralty and others turned for computational help during the war. By January 1942,

30 per cent of the Nautical Almanac Office's work was specifically war-related. The Office was under-resourced for the amount of work it was being asked to take on, and as early as 1941 the superintendent, Donald Sadler, suggested to the Hydrographer of the Navy that a National Computing Service be set up.[6] Sadler's suggestion came to nothing at the time, however.

The less ambitious idea to create an Admiralty Computing Service came from elsewhere. John Todd was on the staff of the director of Scientific Research Admiralty. Todd's previous postings had involved him in calculations concerning the design of mines and had convinced him that computations within applied departments were being carried out by inexperienced workers who regarded calculation as a chore.[7] He concluded that it would be both more effective and more efficient to centralize computing efforts within the Admiralty. Todd's superior, J. A. Carroll (an astronomer in peacetime), suggested that the Nautical Almanac Office would be a good place to carry out the actual computations involved. In late 1942 Sadler was asked to report on the suggestion that an Admiralty Computing Service be created. He endorsed the idea and by March 1943 the Admiralty Computing Service had been set up, its brief to advise on and to carry out computational tasks for Admiralty establishments.

The Service was administered in London by Todd and a small mathematical staff, and the computations that were required were carried out at the Nautical Almanac Office (evacuated to Bath by that time). Sadler began to recruit staff for the Service, which at its peak employed some 15–20 people. Those recruited included Goodwin, Fox, and Frank Olver. In addition the Admiralty Computing Service used a variety of mathematical consultants to help guide their work. This is not the place to give a full account of the work of the Service, but of its usefulness there is no doubt.[8] It carried out two kinds of work: large, repetitive calculations and complex mathematics. Its success was one of the main factors in the creation of the NPL Mathematics Division.

Todd and Sadler realized the limitations of the Admiralty Computing Service within a year of its getting started. It did not operate on a large enough scale to run a fully equipped computing service, and was too small to justify the purchase of punched-card tabulating machines, a differential analyser, or a more diverse selection of hand calculating machines. Consequently Todd, Sadler, and Arthur Érdelyi (a mathematical consultant who worked for the Admiralty Computing Service) wrote their *Memorandum on the Centralization of Computation in a National Mathematical Laboratory*[9] and sent it to Sir Edward Appleton, Secretary of the Department of Scientific and Industrial Research (DSIR). In it they presented a case for a National Mathematical Laboratory (emphasizing operating efficiency and economies of scale). They also recognized that the Admiralty Computing Service's role as a computing bureau could be extended to include research, and pointed out that the development of new computing methods and machines would be a valuable additional function of the National Laboratory.

For a number of reasons Appleton and the DSIR took the proposal to create a National Laboratory seriously. First, the proposals were based on the practical experience of running the Admiralty Computing Service. Second, the extensive use that the armed services were making of Comrie's Scientific Computing Service proved that government scientists needed extra computing resources. (Appleton himself had made use of Comrie's Service for calculations concerning the height variations in the E-layer of the ionosphere.) Third, a similar suggestion had been voiced in other influential quarters.

That voice belonged to Sir Charles Galton Darwin, director of the NPL. In March 1943 Darwin had remarked to a meeting of the DSIR Advisory Council that 'He was inclining more and more to the opinion that a Mathematical Department should be established at the National Physical Laboratory'.[10] Appleton had been at that meeting. Darwin followed up these remarks in a paper to the NPL Executive Committee sometime in mid- to late-1943.[11] He identified a need for a statistical department at the NPL, aimed especially at quality control problems for mass production in industry. Darwin also commented, in a small paragraph tucked away in the middle of the paper, 'that there may well be scope in making new inventions of mathematical machines'.

Earlier in the war Darwin had spent a year in Washington as director of a project to improve liaison between Britain and the United States over the scientific war effort (in what later became known as the British Central Scientific Office). Darwin was privy to the work of the MAUD committee on the atomic bomb and would probably have heard about calculating machine projects such as Howard Aiken's Sequence Controlled Calculator at Harvard University (see Chapter 3). He may also have been familiar with the computational work being done with differential analysers at the Moore School of Electrical Engineering, part of the University of Pennsylvania. (Later, in the spring of 1943, the influential ENIAC project started at the Moore School.)

Perhaps Darwin's suggestion that computing machine research be carried out at the NPL was to some extent based on his knowledge of American developments.

Darwin sounded out other senior scientific figures. Darwin, Lennard-Jones (originator of the Cambridge Mathematical Laboratory), and R. H. Fowler (Plummer Professor of Mathematical Physics at Cambridge and a member of the NPL Executive Committee) discussed the possibility of a National Mathematical Laboratory over lunch on 27 May 1943. From Lennard-Jones's notes of the meeting[12] it is clear that Hartree too had been consulted. Hartree, a leading expert on calculating machines (and soon to become a member of the NPL Executive Committee), was privy to information about the US wartime calculating machine projects.

Overall, then, the time was ripe for the creation of a National Mathematical Laboratory. The DSIR set up an Interdepartmental Technical Committee to report on the issue and to work out the details.

The DSIR Interdepartmental Technical Committee

The Committee consisted of 20 members, drawn from 11 different government departments (see Table 1). Familiar faces included Darwin as chairman, Sadler, Todd, Hartree, and Womersley.

Many members of the Committee had practical computing or statistical experience. The Committee's report reflected a realization that computing machinery research needed to be sponsored by the government—a year ahead of von Neumann's 1945 draft report on the EDVAC.

The Committee reported to the DSIR Advisory Council on 10 May 1944,[13] recommending that a Central Mathematical Station be established which would:

1. Undertake research into new computing methods and machines.
2. Encourage the development of new computing methods and machines by the dissemination of knowledge.
3. Deal with statistical problems arising from industry, the physical sciences and engineering.
4. Advise on and, if necessary, prepare mathematical tables.
5. Provide computing services for government departments, industry, and universities.
6. Act as a consultant on mathematical and statistical techniques.

Table I The DSIR Interdepartmental Technical Committee membership[a]

Member	Representing
Sir Harold Spencer-Jones	Admiralty
Prof. J. A. Carroll	Admiralty
Mr D. H. Sadler	Admiralty
Mr J. Todd	Admiralty
Dr F. Yates	Agricultural Research Council
Prof. W. J. Duncan	Ministry of Aircraft Production
Dr S. H. Hollingdale	Ministry of Aircraft Production
Mr J. R. N. Stone	War Office Central Statistical Office
Mr A.W. Taylor	Customs and Excise
Dr Christopherson	Ministry of Home Security
Dr David	Ministry of Home Security
Prof. D. R. Hartree	Ministry of Supply
Dr J. W. Maccoll	Ministry of Supply
Mr J. R. Womersley	Ministry of Supply
Major Gen. G. Cheetham	Ordnance Survey Department
Mr A. W. Mattocks	Treasury
Mr G. F. Peaker	Treasury
Major E. H. Thompson	War Office, Directorate of Military Survey
Dr S. Goldstein	DSIR

[a] Chairman—Sir Charles Darwin

Source: Report on an Interdepartmental Technical Committee. Public Records Office 1944 DSIR 2/204.

The Committee recommended that the Station be staffed by 25 scientific officers and by 50 ancillary staff. The next question was where such an institution should be based. The Committee considered three main points in the choice of a site (all of which were based on earlier suggestions by Darwin to the NPL Executive Committee). The Laboratory should be attached to an intellectual centre, should be conveniently placed for government departments and industry, and should be close to engineering workshops that could be used for the development of new machines.

Cambridge, while offering the benefits of a thriving intellectual centre, was at that time very inconvenient for industry. It was felt that London, although home to many industries and intellectual centres, was short of easily accessible workshop space. The Committee's recommendation was that a new Division be set up at the NPL. Given Darwin's interest in establishing a mathematical and statistical section at the NPL, and his chairmanship of the Committee, this was more or less a foregone conclusion. By July 1944 the NPL Executive Committee approved the proposal, and the search was on for a suitable superintendent.

The appointment of Womersley

There were only two candidates for the post of superintendent of the proposed new Division, Womersley and Sadler, both of whom had sat on the Interdepartmental Technical Committee. Sadler, on paper much the stronger candidate, had been pressed into applying by Darwin, but did not really want to make the move from the Nautical Almanac Office.[14] Womersley had worked with Hartree on the Manchester differential analyser before the war and had set up S. R. 17. He had also, according to his own recollection,[15] read Turing's 'On Computable Numbers'[16] and had already considered the possibility of using telephone relays to build a machine implementing Turing's ideas (see Chapter 3). Womersley indeed made this suggestion to the Interdepartmental Technical Committee.

Womersley was appointed superintendent of the new NPL Mathematics Division in September 1944, but did not take up his post until the following April. In the light of the subsequent electronic computer developments at the NPL, the Mathematics Division might have benefited from the appointment of a more technically able superintendent. But in many other ways Womersley was a good choice. His forte was the political manoeuvring required to set up a new organization. He had the personality needed to address meetings, and the political acumen to court favour in order to push plans through. His friendly relationships with both Darwin and Hartree ensured Womersley cooperation from the NPL Executive Committee and, crucially, gave him access to information about the then classified computing developments in the United States. It was through Hartree and Darwin that Womersley gained access to the ENIAC during his visit to the United States in spring 1945, and also access to von Neumann's draft report on the EDVAC.[17] Womersley's key role was to show Turing von Neumann's draft report and to recruit Turing for the NPL.

Womersley also did a good job of recruiting other senior staff to the Maths Division. Goodwin, Fox, Olver, Robertson and others came directly from the Admiralty Computing Service, Wilkinson and Vickers came from the

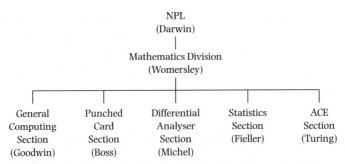

Fig. 1 Organization of the NPL Mathematics Division 1946.

Armaments Research Department at Fort Halstead, and Boss came from the Royal Aircraft Establishment.

Initial structure of the NPL Maths Division

The Mathematics Division was up and running by 1946. It was divided into five sections (see fig. 1). The General Computing Section, headed by Goodwin, was divided into two parts. The first concentrated primarily on numerical analysis and consisted of a strong team of mathematicians, including Fox, Olver, and Wilkinson (part time). The other arm of the General Computing Section was made up of a well-trained junior team of desk calculator operators, led by the experienced Vickers. Vickers and his team carried out many of the computational problems submitted to the Mathematics Division.

The Punched Card Machine Section, initially headed by Boss and his deputy F. Rigg, was staffed predominantly by school leavers trained at the NPL. They worked in groups of up to three, applying Hollerith Punched Card machines to a range of statistical and mathematical jobs for a variety of users. (See the photograph on the following page.) Where necessary the mathematicians of the General Computing Section contributed mathematical methods for use with the machines.

The Differential Analyser Section was headed by Jack Michel and took over the Manchester differential analyser. The staff was quite small and much time was taken up with moving the differential analyser to Teddington (which did not happen until 1948) and also with planning for the installation of a larger machine. The Section took on only a small amount of outside computational work. It did, however, provide an advisory service on all types

The Hollerith Room in the Babbage Building at the NPL. Turing adopted Hollerith punched-card equipment to provide input/output for the ACE.

Source: National Physical Laboratory. © Crown copyright; reproduced by permission of the Controller of HMSO.

of analogue computing (including nomograms, harmonic analysers, and differential analysers).

The Statistics Section staff, under E. C. Fieller, ran a statistical service for government and industry. Its role was predominantly advisory, and the service work it did carry out was performed by staff of the General Computing or Punched Card Sections. Its main work was data analysis, production analysis, and quality control analysis. In 1951 the whole section was transferred to the Ministry of Supply.

The ACE Section, established to design and develop a large-scale electronic digital computer, was headed by Turing. Chapter 3 describes the origins and development of the ACE project.

Did the NPL Mathematics Division succeed as a Central Mathematics Station?

The initial organization of the Mathematics Division reflected the proposals of the 1944 Interdepartmental Technical Committee. Did the Maths Division fulfil the expectations of the Technical Committee? This section considers in turn each of the six tasks set by the Committee.

Undertake research into new computing methods and machines. In this the Maths Division was spectacularly successful. The Pilot ACE went into regular service in 1952 and was the first of a series of computer developments undertaken by the Maths Division. The Division also became a leading centre for the new discipline of numerical analysis. Goodwin, Fox, Olver, and Wilkinson all became well-known numerical analysts.

Encourage the development of new computing methods and machines by the dissemination of knowledge. Maths Division staff, and particularly the numerical analysts, published the results of their work widely in leading academic journals (as well as in internal reports). As early as December 1946 Turing gave a series of lectures about the ACE to an invited audience at the Ministry of Supply in London (see Chapter 22, 'The Turing–Wilkinson Lecture Series (1946–7)'). The Maths Division also hosted a computer conference in 1953 to help disseminate knowledge.

Deal with statistical problems arising from industry, the physical sciences, and engineering. Between 1946 and 1951 (when it was transferred to the Ministry of Supply) the Statistics Section dealt with a range of problems. Much of the Section's work was on an advisory level.

Advise on and, if necessary, prepare mathematical tables. The General Computing Section prepared tables for users. In addition, Goodwin and Olver were important members of the Royal Society Mathematical Tables Committee, which prepared and published high quality mathematical tables in the post-war period. They were also involved in discussions concerning the future of mathematical tables in the light of computer developments.[18]

Provide computing services for Government departments, industry, and universities. From the beginning the Punched Card Machine Section and the desk machine arm of the General Computing Section carried out computing work for a very wide variety of clients. In the period 1946–51 the Mathematics Division had approximately 50 users (see Table 2). The scale of the service increased tremendously when the Pilot ACE became available.

Act as a consultant on mathematical and statistical techniques. The Maths Division acted as consultant to many organizations on digital and analogue computing, on statistics, on numerical analysis, on the features of different makes of desk calculators, and on many other topics.

Table 2   NPL Mathematics Division users 1946–51

Admiralty	Medical Research Council
Armament Research Establishment	Ministry of Agriculture
AERE, Harwell	Ministry of Civil Aviation
Australian National Standards Lab	Ministry of Education
Bank of England	Ministry of Health
Birmingham University	Ministry of Supply
Board of Trade	Ministry of Works
British Cotton Industry Research	NPL Aerodynamics Division
Association	NPL Metallurgy Division
British Electricity Authority	NPL Metrology Division
British Iron and Steel Association	NPL Physics Division
British Railways	NPL Ship Division
British Standards Institute	Ordnance Survey
Building Research Station	Oxford University
Civil Service Commission	Road Research Laboratory
Colonial Survey Department	Royal Aircraft Establishment
DSIR HQ and 'F' Division	Royal Society Mathematical Tables
Electrical Research Association	Committee
Fuel Research Station	Sir Edward Appleton
Home Office	Swedish Government Computer
Committee on Servo Mechanisms	Laboratory
London Passenger Transport	Sperry Gyroscope Company
London University	Treasury Training Division
Manchester University	United Steel Companies Ltd.
Mechanical Engineering Research	UCL Statistics Deptartment
Organization	War Office

Source: NPL Annual Reports.

Notes

The bulk of the research for this chapter was carried out as part of a 1986 Ph.D. thesis while I was in receipt of a grant from the Science and Engineering Research Council. I would like to thank the many ex-employees of the Admiralty Computing Service and the NPL Mathematics Division, including E. T. Goodwin, D. H. Sadler, J. Todd, T. Vickers, M. Woodger, and many more, who have patiently answered my questions and put me straight on a number of things. Thanks also to Jack Copeland for his editorial work on the chapter.

1. Anon. (1945) 'Mathematics at the National Physical Laboratory', *Nature*, 155 (April 7), 431.

2. Croarken, M. (1992) 'The emergence of computing science research and teaching at Cambridge, 1936–49', *IEEE Annals of the History of Computing*, 14(4), 10–15.

3. Croarken, M. (2000) 'L. J. Comrie: A forgotten figure in numerical computation', *Mathematics Today*, 36(4), 114–18.

4. Croarken, M. (1990) *Early Scientific Computing in Britain*. Oxford: Oxford University Press.

5. Croarken, M. (1999) 'Case 5656: L. J. Comrie and the origins of the Scientific Computing Service Ltd.', *IEEE Annals of the History of Computing*, 21(4), 70–1.

6. Sadler, D. H. 'A personal history of H. M. Nautical Almanac Office 30 October 1930–18 February 1972', Ts. edited by G. A. Wilkins. May 1993. Royal Greenwich Observatory Archives, Cambridge University Library.

7. Letter from John Todd to Mary Croarken, July 28, 1983.

8. See Croarken, M. (1990) *Early Scientific Computing in Britain*, pp. 67–73.

9. Anon. '*Memorandum on the Centralization of Computation in a National Mathematical Laboratory*', Ts. probably written by Todd, Sadler and Érdelyi in 1943/4 and presented to Sir E. V. Appleton, Secretary to the DSIR. Received from John Todd 1983.

10. DSIR Advisory Council Minutes 1942–3, Special Meeting of Council, 10 March 1943, Minute 61. Public Record Office (PRO) DSIR 1/10.

11. Darwin. C. G. 'Establishment of a mathematical department', NPL Executive Committee paper E.832. Undated: not later than October 1943.

12. John Lennard-Jones, Daily Journals, Churchill College Archives, Cambridge. Item LEJO 24.

13. DSIR 'Report of Interdepartmental Technical Committee on a Proposed Central Mathematical Station', DSIR Advisory Council, 10 May 1944. PRO DSIR 2/204.

14. Sadler, D. H. 'A personal history of H. M. Nautical Almanac Office 30 October 1930–18 February 1972'.

15. Womersley, J. 'A.C.E. Project—History and Origins', Ts, 26 November 1946. PRO DSIR 10/385.

16. Turing, A. (1936) 'On computable numbers, with an application to the Entscheidungsproblem', *Proceedings of the London Mathematical Society*, Series 2, 42 (1936–7), 230–67.

17. von Neumann, J. 'First Draft of a Report on the EDVAC', Contract No. W-670-ORD-492. Moore School of Electrical Engineering, University of Pennsylvania, 30 June 1945.

18. Croarken, M. and Campbell-Kelly, M. (2000) 'Beautiful Numbers: The rise and decline of the mathematical tables committee, 1871–1965', *IEEE Annals of the History of Computing*, 22(4), 44–61.

3 The origins and development of the ACE project

B. Jack Copeland

Womersley and Turing join the NPL

The name 'Automatic Computing Engine' was due to Womersley[1] and the story of the ACE begins with his appointment as superintendent of the newly created Mathematics Division of the National Physical Laboratory (see the previous chapter).[2] Womersley's proposed research programme for his new division included the items 'To explore the application of switching methods (mechanical, electrical and electronic) to computations of all kinds', 'Investigation of the possible adaptation of automatic telephone equipment to scientific computing', and 'Development of electronic counting device suitable for rapid computing'.[3]

Womersley had himself been a member of the Interdepartmental Technical Committee that in April 1944 had recommended the creation at the NPL of a Mathematics Division whose primary objective was to 'undertake research into new computing methods and machines'.[4] In its report the Committee emphasized that the new division should be provided with 'facilities for designing new machines and perhaps for constructing pioneer ones', noting 'it is probable that new machines may be called for of patterns that cannot be foreseen now'.[5]

In December 1944 Womersley addressed the Executive Committee of the NPL on the potential of electronic computing. The minutes of the meeting summarize his speech:

```
Electronic counting devices ... can be used
and machines can be constructed which have a
high degree of flexibility and which can be
continually improved and extended. Electronic
```

counting can be done at the rate of one
operation per microsecond, a vast improvement
on anything previously attempted. All the
processes of arithmetic can be performed and
by suitable inter-connections operated by
uniselectors a machine can be made to perform
certain cycles of operations mechanically. ...
[T]here is no reason why the instructions to
the machine should not depend on the result of
previous operations so that various iterative
types of method could become fully automatic.[6]

In November 1946 Womersley wrote a synopsis of the principal events that led to the establishment of the ACE project:

1936-37 Publication of paper by A. M. Turing
'On Computable Numbers, with an Application to
the Entscheidungsproblem'. ...
1937-38 Paper seen by J.R.W. [J. R. Womersley]
and read. J.R.W. met C. L. Norfolk, a telephone
engineer who had specialised in totalisator
design and discussed with him the planning of
a 'Turing machine' using automatic telephone
equipment. Rough schematics prepared, and
possibility of submitting a proposal to N.P.L.
discussed. It was decided that machine would
be too slow to be effective.
June 1938 J.R.W. purchased a uniselector and
some relays on Petty Cash at R. D. Woolwich for
spare-time experiments. Experiments abandoned
owing to pressure of work on ballistics. ...
1942 Aiken's machine [the Sequence-Controlled
Calculator at Harvard University] completed and
working.
1943 Stibitz constructed the Relay Computor at
Bell Telephone Laboratories.[7]
Late 1943 J.R.W. first heard of these American
machines.
1944 Interdepartmental Committee on a Central
Mathematical Station. D. R. Hartree mentioned
at one meeting the possible use of automatic

telephone equipment in the design of large
calculating machines. J.R.W. submitted
suggestions for a research programme to
be included in Committee's Report.

1944 Sept. J.R.W. chosen for Maths. Division.

1944 Oct. J.R.W. prepares research programme
for Maths. Division which includes an item
covering the A.C.E.

1944 Nov.[8] J.R.W. addresses Executive
Committee of N.P.L. Quotation from M/S
(delivered verbatim) ... 'Are we to have a
mixed team developing gadgets of many kinds ...
Or are we, following Comrie ... to rely on sheer
virtuosity in the handling of the ordinary
types of calculating machines? I think either
attitude would be disastrous ... We can gain
the advantages of both methods by adopting
electronic counting and by making the
instructions to the machine automatic ...'

1945 Feb-May J.R.W. sent to the U.S.A. by
Director. Sees Harvard machine and calls it
'Turing in hardware'. (Can be confirmed by
reference to letters to wife during visit).
J.R.W. sees ENIAC and is given information
about EDVAC by Von Neumann and Goldstine.

1945 June J.R.W. meets Professor M. H. A.
Newman. Tells Newman he wishes to meet Turing.
Meets Turing same day and invites him home.
J.R.W. shows Turing the first report on the
EDVAC and persuades him to join N.P.L. staff,
arranges interview and convinces Director and
Secretary.[9]

Persuading Turing to join the embryonic ACE project was a great
coup, testifying to Womersley's vision and initiative (even locating Turing,
who was at that time engaged in secret work, could not have been
straightforward). Turing was even more highly qualified for the job than
Womersley realized. While Womersley clearly understood the importance of
Turing's pre-war article 'On computable numbers, with an application to the
Entscheidungsproblem'—the birthplace of the stored-program concept—he

Colossus at Bletchley Park.

Source: The National Archives. © Crown copyright; reproduced by permission of the National Archives Image Library.

was completely unaware of the highly secret developments in electronic computing that had taken place during the war at Bletchley Park (headquarters of the codebreaking organization known as the Government Code and Cypher School). At Bletchley Park Turing was among the few who knew of Colossus, the first large-scale electronic digital computer (see Chapter 5, 'Turing and the Computer'). Designed by Thomas Flowers, Colossus made its first successful codebreaking runs at Bletchley Park in December 1943 (two years before the American ENIAC was operational). Once Turing had seen Colossus it was, according to Flowers, just a matter of his waiting for an opportunity to put the ideas of his 1936 article into practice.[10] Probably Turing did not need much persuasion to join Mathematics Division.

Proposed electronic calculator

Turing's employment commenced on 1 October 1945, by which time Mathematics Division was 'functioning on a limited scale'.[11] Turing set to work on the design of the Automatic Computing Engine. By the end of 1945 he had completed his technical report 'Proposed Electronic Calculator'.[12] 'Proposed

Electronic Calculator' contained the first relatively complete specification of an electronic stored-program digital computer (see the Introduction to this book).

The next step was to present Turing's design to the director of the NPL, Darwin, in order to secure the support necessary for the project. Womersley wrote to Darwin:

<u>"ACE" Machine Project.</u>

With this minute I present three reports. The first is a short account, by Hartree and myself, of recent developments in the U.S.A. in the field of automatically controlled calculating machines. The second is a report by Dr. A. M. Turing which shows how such a machine could be constructed (by combining electrical apparatus already well-developed and having known properties) which would be capable of solving a wide variety of problems at speeds hitherto unattainable.

It is very important to mention that this device is not a calculating machine in the ordinary sense of the word. One does not need to limit its functions to arithmetic. It is just as much at home in algebra, i.e. it can work out matrix multiplications in which the elements are algebraic polynomials, or problems in Boolean Algebra, or the enumeration of group characters. Methods of successive approximation, i.e. the Southwell 'Relaxation' process, are equally possible, since the machine will contain a device which enables it to choose between two sets of instructions according to the sign of some number in it.

The cost is, naturally, a doubtful point. I put it, after careful consideration, at £60,000--£70,000, though it is difficult to be sure of a "ceiling". It will, I believe, be one of the best bargains the D.S.I.R. [Department of Scientific and Industrial Research] has ever made. To give some idea of the speed of the

machine, it will calculate a gun trajectory,
from muzzle to point of fall, in less than
30 seconds, and it would carry through the
preparation of the whole of the ballistic
bombing tables for the R.A.F. in a few weeks,
apart from printing. By its use we can explore
whole fields of both pure and applied
mathematics at present closed to us by the
formidable magnitude of the computing
programmes involved.

We can attack complicated integral equations,
integro-differential equations and partial
differential equations by replacing them by
large blocks of simultaneous linear equations
in 700--1000 unknowns and solve them with ease
and speed. We can take T. Smith's theory of the
design of optical instruments and use it on
practical design problems at a speed which will
enable answers to be given to the firms by
telephone in a few hours. We can revolutionise
the study of compressible fluid flow, and of
aircraft stability. Problems now slowly
attacked piecemeal will be capable of solution
as a whole. The machine will also grapple
successfully with problems of heat-flow in
non-uniform substances, or substances in which
heat is being continuously generated. It will
enable the study of materials with peculiar
elastic properties (e.g. plastics) to be
advanced in a way that is impossible with
present computing resources. ... [W]e could
alter the whole tempo of the numerical
mathematical work associated with the
scientific research of this country if the
machine were available.

The possibilities inherent in this equipment
are so tremendous that it is difficult to state
a practical case to those who are not au
fait with the American developments without
it sounding completely fantastic. But if anyone
is going to suggest that this equipment

is expensive, may I point out that two machines in the U.S.A., the Harvard Sequence Controlled Calculator, and the Bell Telephone Laboratory Relay Computer, cost as much as this, work at 1/1000th the speed, will have neither the versatility nor the storage capacity, and yet were thought by the Americans to be worth while.

The third document is an attempt to state a practical case for the equipment. In view of the unique nature of the equipment this is difficult, but I believe that in this direction the promised support of Commander Sir Edward Travis, of the Foreign Office, will be invaluable.[13] ...

As regards the manufacture of the machine, I think that the Post Office Engineering Research Station is the right place, if they can see their way to do it. Mr. Flowers, of that Station, has had wartime experience in the right field, and, during his recent visit to the U.S.A., visited the places where these developments have been going on.[14]

Approval for the ACE project rested with the Executive Committee of the NPL and Womersley duly prepared a paper for presentation to the Committee:

Memorandum by Mr. J. R. Womersley, Superintendent, Mathematics Division

The research programme of the Mathematics Division contains an item "To explore the application of switching methods (mechanical, electrical and electronic) to computations of all kinds." ... Dr. A. M. Turing was appointed to the staff of the Division, and began to consider the possibilities of electronic methods ... Dr. Turing has now completed a long report, which makes definite proposals for the construction of a machine, capable of solving a wide variety of problems at speeds hitherto unattainable. ...

Summary of Part I of Dr. Turing's Report

It is intended that the ACE machine shall tackle whole problems, i.e. that instead of repeatedly using human labour for taking material out of the machine and putting it back at the appropriate moment, all this will be done by the machine itself. It will not be limited to carrying out a sequence of prescribed operations. Provision is made for making the behaviour of the machine to depend on the results of its own calculations.

Once the human element is eliminated, the increase in speed is enormous. For example, it is intended that the multiplication of two ten-figure numbers shall be carried out in 500 microseconds, about 20,000 times the speed of a normal calculating machine. This speed is not attained by making the equipment more expensive and more elaborate than it need be. It is the natural result of the unconventional methods used, and once this is granted, there is no economy to be obtained by reducing it.

The basic principle is that numbers contained in the machine are stored dynamically, not statically as in other machines. The internal working of the machine is entirely in the binary system, and a number is represented by a series of 1's and 0's, the 1's being pulses, and the 0's the spaces between them. The digit of least significance comes first in point of time. The problem is to find a way of storing a number in this form, so that it can be kept circulating in the machine until it is needed again for use in a subsequent calculation. Dr. Turing describes a 'delay line,' the 'circulating memory' used in radar, which he shows to be suitable for this purpose. The manufacture of memories capable of accommodating 1000 binary digits is shown to be practicable.

The machine then divides into two main parts,
an arithmetical organ, and a logical control,
with input and output organs for communication
with the outside world. The instructions to the
machine can themselves be expressed in the form
of numbers in the binary scale, and fed into
the logical control part of the machine at the
beginning of a problem. There they circulate in
the appropriate delay lines until the numerical
information ('initial conditions') has been fed
in, the final information being another
instruction, the 'instruction to proceed.'

... The original intention was that each unit
would consist of 32 binary digits, the first
digit being '0' or '1' according to the nature
of the 'number' i.e. whether it is a number or
an instruction, the second being an indication
of sign. Dr. Turing has extended this idea to
make it more flexible. Each 'number' consists
of <u>two</u> units of 32 binary digits. One of these
contains a number in the binary system between
zero and unity, the other gives its
'significance,' i.e. a power of 2 by which it
must be multiplied, its sign, and some spare
digits which are used to identify the number.

The processes of arithmetic are very simple.
To add two numbers they are taken out of
storage and by passing them through a simple
network of radio valves, they are 'glued'
together and the sum deposited in another
storage element.

Multiplication is done by repetitions of this
process, with appropriate time delays. It is
proposed to do division by multiplication by
the reciprocal, the reciprocal being calculated
by successive approximation. This is quite
practicable in view of the high speed of the
unit processes.

The scope of the machine is very wide. It is
envisaged that the construction of a set of

range tables could be treated as a single
problem, and that once experience is gained,
could be run off in a few days. As an
indication of speed, it should be possible to
calculate a gun trajectory by 'small arcs' from
muzzle to point of fall, in about half a
minute. All types of 'relaxation' calculation
can be fully mechanised---indeed all methods
of successive approximation. Matrices of degree
less than 30 whose elements are polynomials of
the tenth degree can be multiplied, giving
another matrix with polynomial coefficients.
Other examples are given in the report. Some
attention is also given to the problem of
checking, since the machine must inspire
confidence in its results. This cannot be dealt
with adequately until actual manufacture
begins. ...

    The cost of the proposed machine will be
about the same as the ENIAC constructed for the
Aberdeen Proving Ground, and somewhat less than
the Bell Telephone Relay Computer. The cost of
the Harvard University machine is not known
with accuracy, but is reputed to be
half-a-million dollars. If it is granted that
this country should possess one of these large
machines, the Mathematics Division of the
N.P.L. is the obvious place for it. The machine
envisaged by Dr. Turing would have an output
equal to the total output of all the large
machines so far constructed in the U.S.A.
In fact we are now in a position to reap
handsome benefits from the pioneer work done in
the United States, and it is undoubtedly
advisable that we should build this type of
machine at once, rather than begin with relay
equipment.[15]

As mentioned previously, Womersley had no knowledge of Colossus.
Consequently his view of recent technological developments in computing
was distorted and he tended to exaggerate the intellectual debt owed by the

ACE to the ENIAC and to the relay (i.e. non-electronic) machines at Harvard and Bell Labs.

Womersley's Memorandum and Turing's 'Proposed Electronic Calculator' were submitted to the Executive Committee of the NPL in February 1946. Discussion was deferred until the March meeting, when the 'Committee resolved unanimously to support with enthusiasm the proposal that Mathematics Division should undertake the development and construction of an automatic computing engine of the type proposed by Dr. A. M. Turing'.[16] The minutes of the meeting recorded the historic discussion:

Large Electronic Calculating Machine ACE

The Committee had before it Paper E.881 (Memorandum by Mr. J. R. Womersley, Superintendent, Mathematics Division, concerning the "ACE" Machine Project) and Paper E.882 (Report by Dr. A. M. Turing On Proposals for the Development of an Automatic Computing Engine (ACE)).

The Chairman invited Mr. Womersley to outline the reasons why a machine of the type proposed by Dr. Turing should be constructed for the Mathematics Division.

Mr. Womersley said that he believed it possible to see these proposals in their true perspective if one had knowledge of what had been done in the U.S.A. during the war. He therefore proposed to begin by sketching in some of the background of recent developments.

Mr. Womersley then gave a brief description of three large calculating devices constructed in the U.S.A. during the war---the I.B.M. Sequence Controlled Calculator at Harvard University, the Relay Computer designed by Dr. G. R. Stibitz of Bell's Telephone Laboratories, and the ENIAC, constructed at the Moore School of Electrical Engineering, University of Pennsylvania for the Ballistics Research Laboratories, Aberdeen Proving Ground. Each one of these machines had cost from £80,000--£100,000. It was interesting to note

that Dr. Turing's machine, when fully
completed, would have a potential output
of work greater than the three of them put
together. In other words, we now had the
opportunity to begin work in this field at
a most favourable moment. The machine proposed
had not only a greater output, but greater
versatility than the machines so far made,
because of the greater elaboration of the
logical controls proposed by Dr. Turing. This
was possible by reason of the different mode
of operation, and the fact that in Dr. Turing
we had available an expert in the field of
mathematical logic.

Mr. Womersley then introduced Dr. A. M. Turing,
who was invited to describe the principles
underlying his proposal.

Dr. Turing explained that if a high overall
computing speed was to be obtained it was
necessary to do all operations automatically.
It was not sufficient to do the arithmetical
operations at electronic speeds: provision must
also be made for the transfer of data (numbers,
etc.) from place to place. This led to two
further requirements---'storage' or 'memory' for
the numbers not immediately in use, and means
for instructing the machine to do the right
operations in the right order. There were then
four problems, two of which were engineering
problems and two mathematical or combinatory.
Problem (1) (Engineering). To provide a
suitable storage system.
Problem (2) (Engineering). To provide high
speed electronic switching units.
Problem (3) (Mathematical). To design circuits
for the ACE, building these circuits up from
the storage and switching units described under
Problems 1 and 2.
Problem (4) (Mathematical). To break down the
computing jobs which are to be done on the ACE

into the elementary processes which the ACE is designed to carry out (as determined in the solution of Problem 3). To devise tables of instructions which translate the jobs into a form which is understood by the machine.

Taking these four problems in order, Dr. Turing said that a storage system must be both economical and accessible. Teleprinter tape provided an example of a highly economical but inaccessible system. It was possible to store about ten million binary digits at a cost of £1, but one might spend minutes in unrolling tape to find a single figure. Trigger circuits ['flip-flops'---Ed.] incorporating radio valves on the other hand provided an example of a highly accessible but highly uneconomical form of storage; the value of any desired figure could be obtained within a microsecond or less, but only one or two digits could be stored for £1. A compromise was required; one suitable system was the 'acoustic delay line' which provided storage for 1000 binary digits at a cost of a few pounds, and any required information could be made available within a millisecond. Dr. Turing explained the principle of the delay line, which involved transmitting compression waves down a tube of liquid, using piezo crystals both as transmitters and receivers of sound. The output of the receiving crystal was amplified, restored to its ideal shape and fed back to the transmitter. On account of the shape restoring process it was possible for a signal to travel down the tube many millions of times and remain recognisable.

Dr. Turing then gave a brief account of the high speed switching problems involved in the ACE (Problems (2) and (3)). Numbers were to be represented in the binary scale and valves were only to be used as 'on-off' devices.

Although switching units had not yet been made there was every hope of success because the limiting factors were electron transit time, and the allied quantity O_{ag}/g_m, and the frequencies at which these became serious were well above those at which the ACE would be operated (viz. about 1 Mc/s). Switching circuit design (Problem (3)) was illustrated by means of the adder circuit.

Time did not permit of an adequate account of Problem (4).

In reply to a question from Professor Tyndall, Dr. Turing explained that an elaborate system of checks would be incorporated, so that the failure of any part would be immediately indicated. It was not possible entirely to guard against the failure of the checking system itself, but the probability of an undetected error could be reduced to very small proportions. The majority of the checks would be introduced through instruction tables and would therefore require no special equipment.

The Director asked what would happen if the machine were instructed to sum a series which actually diverged although thought to converge. Dr. Turing replied that it was left to the discretion of the man who constructed the instruction tables (the controller) to state what the machine should do in these cases. The summation could be specified to a given number of terms regardless of convergence, or until the last term was less than a given amount; preferably the controller should have worked out the theory of the convergence to some extent so as to be able to incorporate a suitable test. The Director asked what would happen in cases where the machine was instructed to solve an equation with several roots. Dr. Turing replied that the controller would have to take all these possibilities into account, so that the construction of instruction tables might be a somewhat 'finicky' business.

Professor Hartree pointed out that the serial operation of the machine makes it very economical in its use of radio valves. It requires only 2000 valves as against 18000 in the ENIAC, and gives a 'memory' capacity of 6000 numbers compared with the 20 numbers of the ENIAC. This greater capacity (and the higher speed) are attained at no greater cost than the ENIAC. The greater storage capacity is facilitated by the high speed, and this is an important factor in gaining the economy in equipment.

Professor Hartree also pointed out that if the ACE is not developed in this country the U.S.A. will sweep the field, and reminded the Committee that this country has shown much greater flexibility than the Americans in the use of mathematical hardware. He urged that the machine should have every priority over the existing proposal for the construction of a large differential analyser.

Director enquired whether the machine could be used for other purposes if it did not fulfil completely Dr. Turing's hopes. Dr. Turing replied that this would depend largely on what part of the machine failed to operate, but that in general he felt that many purposes could be served by it.

There was next some discussion as to the possible cost of the machine and Mr. Womersley said that a pilot set-up could possibly be built for approximately £10,000, and it was generally agreed that no close estimate of the overall cost of the full machine could be made at this stage. As regards financing the project, the Secretary stated that the initiation of the work could be undertaken

within the terms of the existing research
programme of Mathematics Division and that
it was probable that sufficient financial
provision had been made for possible
expenditure during the 1946--7 financial year.
It appeared desirable, however, that the
possible magnitude of the project should be
understood by Headquarters and the Treasury
so that, as far as possible, assurance could be
obtained that work on the machine could be
continued in succeeding years in so far as
financial provisions are concerned.

Professor Hartree said that such a machine
would be a means for tackling completely new
ranges of problems, and Dr. Southwell reported
that Professor G. I. Taylor had expressed the
opinion that the machine would be more useful
than ENIAC. He hoped, however, that when the
machine was constructed, charges made for its
use would not be so high as to discourage its
use by Universities, etc. He felt that there
should be no attempt at amortisation of the
cost of the machine, which might well run into
£100,000, but that this should be regarded
as a contribution of the Laboratory to the
general good of the country.

The Committee resolved unanimously to
support with enthusiasm the proposal that
Mathematics Division should undertake the
development and construction of an automatic
computing engine of the type proposed by
Dr. A. M. Turing and Director agreed to
discuss the financial and other aspects of
the matter with Headquarters.

In a letter, Dr. E. T. Paris expressed
the interest of Ministry of Supply in the
development of such a machine, and stated
that he had consulted Colonel Phillips,
Superintendent of Applied Ballistics, and
Dr. McColl, Superintendent of Theoretical
Research in Armaments, who foresaw that

from their point of view the main use of the
machine would be for---(a) The calculations
of trajectories for all types of projectiles
(shell, bombs, rockets and guided missiles);
(b) Internal ballistic calculations (the motion
of projectiles down the bore); (c) The solution
of partial differential equations by
relaxation and characteristic methods.
Dr. Paris also enquired how soon the machine
was likely to be a working concern and to what
extent it was likely to be available for use by
his Ministry, adding that if the project goes
well, at a later date it would be possible to
consider whether it would be advisable to have
a machine of the same general type made for
armaments research and allied work.

    In reply to a question by Dr. Carroll,
Mr. Womersley said that as a very rough
estimate he thought the cost of duplicating
the machine would be approximately 25 per cent
of the original cost, although the capacity of
the machine would be such that duplicates would
not be needed. He pointed out, however, that
when such a machine is in use in his Division
it will require the employment of a higher
proportion of senior officers in the scientific
class, because the machine itself will do much
of the work of the lower staff classes.[17]

    In April Darwin wrote as promised to 'Headquarters', the Department of
Scientific and Industrial Research (DSIR). Notice his claim that a single electronic computer might suffice for 'the whole country'—a national computer,
housed at the NPL!

              Automatic Computing Engine (ACE)
    This is a proposal to construct a computing
machine of very much greater potentialities
than anything done hitherto, though a similar
project is being worked out at present in
America. The proposal has already been
foreshadowed in the research programme of

Mathematics Division, Item 6502 'To explore the application of switching methods (mechanical, electrical and electronic) to computations of all kinds.' In the past the processes of computation ran in three stages, the mathematician, the computer [i.e. the human computer---Ed.], the machine. The mathematician set the problem and laid down detailed instructions which might be so exact that the computer could do his work completely without any understanding of the real nature of the problem; the computer would then use the arithmetical machine to perform his operations of addition, multiplication, etc. In recent times, especially with use of punched card machines, it has been possible gradually for the machine to encroach on the computer's field, but all these processes have been essentially controlled by the rate at which a man can work.

The possibility of the new machine started from a paper by Dr. A. M. Turing some years ago, when he showed what a wide range of mathematical problems could be solved, in idea at any rate, by laying down the rules and leaving a machine to do the rest. Dr. Turing is now on the staff of N.P.L., and is responsible for the theoretical side of the present project, and also for the design of many of the more practical details. The principles he enunciated have now become practicable since it is possible to use electronics in the machine so that its rate of operation is about a hundred thousand times as fast as a man's. The proposed machine is primarily a system of electronic circuits some of which do the arithmetic, while others give the instructions in a codified form, also as numbers. But there is another feature necessary in computation, which may be called the memory;

this corresponds to the fact that the human computer has at intervals to turn back to the results of some previous calculation and bring them forward again. In many ways the memory is the most serious problem in the machine, but a variety of methods have been proposed and some instruments have been made, and the choice among them is largely a matter of economy, since there will need to be several hundred organs of this type. At present it appears that the best solution may be one developed for use in radar, which consists in sending a stream of ultra-sonic pulses down a tube of liquid. These are known as delay lines and it is proposed that attention shall primarily be concentrated on their use for this purpose. As at present planned the electronics will work at a rate of microseconds, and the memory tubes will store the information for a millisecond, or for any desired multiple of a millisecond.

Dr. Turing's proposals are set out in a paper (E.882) considered at the March, 1946 meeting of the Executive Committee of N.P.L. The Committee after discussing the problem with Mr. Womersley, Superintendent of Mathematics Division, and Dr. Turing, resolved unanimously to give the project its enthusiastic support.

An example of the sort of problem that could be solved is the calculation of ballistic trajectories. It is estimated that a full trajectory from muzzle to strike, worked out by small arcs, should be solved in half a minute. Or again a large number of simultaneous equations, as in a geodetic survey, could be solved in a few minutes: or the distribution of electric field round a charged conductor of specified shape.

The complete machine will naturally be costly; it is estimated that it may call for over £50,000, but probably not twice as much.

A smaller one, containing the essential
characteristics, could be constructed first,
perhaps for a cost of £10,000, but its chief
function would be to reveal some of the details
of design that cannot be planned without trial,
and its scope would be too limited to be worth
constructing for its own sake. This would
involve development work on delay lines and
trigger circuits and this part of the work
would be undertaken by the Post Office where
facilities and specially trained staff exist,
with the collaboration of Dr. Turing and his
assistants. The Post Office are expecting to be
able to profit by the development for their own
purposes.

The small machine would not be a miniature
substitute for the large machine but would
later constitute a part of the full scale
machine in due course. It is hoped that
the complete machine can be constructed in
three years and the financial requirement will
be heaviest in the final year. It is proposed
to proceed immediately, and with high priority,
in the design and construction of this
preliminary machine, but in doing so it is
important to know that if it fulfils its
promise there will be full backing for the
greater sums required for the real operating
machine. In view of its rapidity of action, and
of the ease with which it can be switched over
from one type of problem to another it is very
possible that the one machine would suffice to
solve all the problems that are demanded of it
from the whole country. As far as can be
estimated at the moment, two Scientific and one
Experimental Officer will be required for this
work in addition to Dr. Turing. Part of this
staff would work at the Post Office Research
Station during the development period. No
estimate can as yet be given of the staff

```
required to use the machine when completed.
This staff would, however, in the main need
to be in the Scientific Class since the machine
itself will do the work of a very large number
of Laboratory Assistants and Experimental
Officers.18
```

The Advisory Council of the DSIR agreed in May 1946 that 'in the event of the first-stage machine fulfilling expectations, they would be prepared to recommend further expenditure on a complete machine, bringing the total up to perhaps £100,000 in the next three years'.[19] It only remained to build the computer. However, as a result of ineffective management this would take much longer than anyone expected.

Further development of the ACE design

In May 1946 Wilkinson joined the NPL and was assigned to Turing on a half-time basis (see Chapter 4, 'The Pilot ACE at the National Physical Laboratory' by J. H. Wilkinson).[20] The ACE Section grew to a staff of three with the arrival of Woodger (who also joined the NPL in May but soon after fell ill with glandular fever and was absent from work until September).[21]

During 1946 Turing continually modified the design of the ACE. By the time of Wilkinson's arrival Turing had reached what he called 'Version V' of the design. (Woodger explains: 'My understanding is that the original report ['Proposed Electronic Calculator'] was not a Version as such but a general proposal. There is no trace of Versions I to IV; I assume they were sketches in Turing's possession, probably done between March and May 1946.'[22]) By the end of 1946 Turing had reached Version VII. From December 1946 to February 1947 Turing and Wilkinson gave a series of nine lectures on Versions V, VI, and VII of the design (see the Chapter 22, 'The Turing–Wilkinson Lecture Series').

In 'Proposed Electronic Calculator' Turing had emphasized that work on instruction tables—that is to say, programs—should 'start almost immediately' since the 'earlier stages of the making of instruction tables will have serious repercussions on the design' of the ACE. Moreover, this policy would, he said, 'avoid some of the delay between the delivery of the machine and the production of results'. Turing made this point again in a letter to Darwin: 'A large body of programming must be completed beforehand, if any serious work is to be done on the machine when it is made'.[23] So it was that during 1946 Turing and Wilkinson, joined later by Woodger, pioneered the science of computer programming. An end-of-year report summarized their achievements:

> Unless means can be found for the rapid preparation of problems in a form suitable for the machine the value of its high speed will largely be lost, though many other advantages would still remain. The first step, therefore, in planning such a machine, is to study the way in which programmes of work should be prepared. This has formed the main work of the A.C.E. section during the past year. It is intended to prepare the instructions to the machine on Hollerith cards, and it is proposed to maintain a library of these cards with programmes for standard operations. In setting up the machine to do a particular job all that will be necessary (instead of preparing the instructions in detail) will be to select a number of groups of standard instructions and to link them together with a few special cards. In planning the oganisation of the work on the machine it has become clear that a long and careful study of the many possible ways of setting out these instructions is essential, and even after more than a year of work on this problem, the final form is only now being reached. Apart from these general questions of organisation a number of basic instruction routines have been decided upon and prepared in detail. These are:-
>
> Division.
> Extracting square roots.
> Indication of failures.
> Testing the accuracy of a given table.
> Exponential function.
> Sine and cosine.

```
Logarithm.
Multiplication of complex numbers.
Formation of scalar product of vectors.
Formation of product of two matrices.
Solution of sets of linear equations.
```

```
Some work has also been done on developing
known methods of solving both ordinary and
partial differential equations in a form
suitable for use on the A.C.E.[24]
```

All this work was, of course, directed toward a machine that existed only on paper. The ACE Section, consisting only of three mathematicians, had no facilities to construct the computer.

The first attempt to build the ACE: the Flowers era

Turing knew that Flowers, who had designed and built Colossus, was uniquely qualified to undertake the construction of the ACE. Flowers was asked to organize the building of the ACE at his headquarters, the Post Office Research Station at Dollis Hill in North London (where Colossus was built).[25] He agreed. Early in 1946 W. G. Radley, Controller of Research at Dollis Hill (and knowledgeable about Colossus), wrote to Womersley to confirm the arrangement:

```
    Thank you for your letter received on Saturday
morning. I had heard from Flowers that you were
thinking of constructing an electronic
numerical computing machine at the N.P.L. and
share your hopes that it will far transcend in
both facility and speed anything previously
attempted. We should be very happy to
co-operate, firstly by giving assistance with
regard to the technical design of the machine
and, later on when this has taken shape, by
arranging to have it constructed for you within
the Post Office organisation. ...
    With regard to the other aspect, that of
time, we have very considerable arrears of work
to overtake for our own Department. Some of it
is most urgently required and the manpower
```

```
position is difficult. I fully appreciate what
you say, however, with regard to the
fascinating nature of the task and the prestige
value and it shall have the highest degree of
priority possible in the circumstances.²⁶
```

At first things seemed to be going rather well:

```
       Status of the Delay Line Computing Machine
              at the P.O. Research Station.
                     March 7, 1946
   They have constructed a long and a short
delay line and have more or less successfully
circulated pulses in these.²⁷ ... They plan to
start very soon work on clock and pulse source
circuits and on frequency control (use of
a pilot mercury line to vary the frequency
to compensate for change of temperature in
the delay lines).
   Mr. Flowers states that they can have ready
for N.P.L. a minimal ACE by August or
September. ... Mr. Flowers, in fact, proposed
the following time table:

By May 1st:     Have circulating circuits
                decided upon, and the frequency
                control circuits ready.
By June 1st:    Have a 32 long delay line unit
                built, have made the clock and
                pulse source circuits, and have
                made the short delay lines.
By July 1st:    Have made 32 circulating
                circuits for the delay lines.²⁸
```

Unfortunately it proved impossible to keep to Flowers' time-table. Dollis Hill was occupied with a backlog of urgent work on the national telephone system (at that time managed by the Post Office). Flowers could spare only two people to work on the ACE, William Chandler and Allen Coombs, his right-hand men from the Colossus days. His section was, he said, 'too busy to do other people's work'.²⁹ By the summer of 1946 warning bells had begun to ring at the NPL. In August Darwin observed that the Post Office was 'not in a position to plunge very deep' and by November was expressing concern to Radley and others at the Post Office about the 'slow rate of progress' on the ACE.[30]

Clearly it was time to make new arrangements. Initially the NPL persisted with the idea of placing a contract for the construction of the ACE with an outside organization. But this proved very difficult. Engineers experienced in the new art of digital electronics were scarce. Larger firms were 'likely to be too tied up with television and other consumer goods', and a suitable smaller company could not be found.[31]

The NPL also attemped to enlist the help of other public institutions. In August 1946 Darwin wrote to Sir Edward Appleton at the DSIR concerning the possibility of the Telecommunications Research Establishment at Malvern taking on the construction:

> As I told you Womersley was down at T.R.E.
> [Telecommunications Research Establishment]
> to see whether they could do any work about the
> A.C.E. machine. He tells me that it looks
> a most promising chance, and I think we should
> go ahead on it. Their lay-out for the job looks
> good, and I gather it appealed strongly to
> F. C. Williams as a job he would like to do,
> so that it should get a good chance. I am
> kicking myself for not having thought of it
> months ago as a possibility.[32]

Later that year, however, Williams took up the position of Professor of Electro-Technics at the University of Manchester and the desired assistance never materialized. Williams went on to build his own computer at Manchester (see Chapter 5).

The NPL also approached Wilkes, who was planning his own computer at the Cambridge Mathematical Laboratory (the EDSAC, which first ran in 1949). Hartree (Plummer Professor of Mathematical Physics at Cambridge and a member of the NPL Executive Committee) sounded him out and reported to the Executive Committee that Wilkes was 'prepared to give as much help as he could on the ACE'.[33] Wilkes 'had experience of making up [d]elay lines and would exchange information with Dr. Turing', Hartree said.[34] However, the chances of Turing's cooperating fruitfully with Wilkes may be judged from a memo from Turing to Womersley:

> I have read Wilkes' proposals for a pilot
> machine, and agree with him as regards the
> desirability of the construction of some such
> machine somewhere. I also agree with him as
> regards the suitability of the number of delay
> lines he suggests. The 'code' which he suggests
> is however very contrary to the line of
> development here, and much more in the American
> tradition of solving one's difficulties by
> means of much equipment rather than thought.
> I should imagine that to put his code (which is
> advertised as 'reduced to the simplest possible
> form') into effect would require a very much
> more complex control circuit than is proposed
> in our full-size machine. Furthermore certain
> operations which we regard as more fundamental
> than addition and multiplication have been
> omitted.
>
> It might be argued that if one is to have so
> little memory then it is necessary to have a
> complex control to make up. In so far as this
> is true I would say that it is an argument for
> either having no pilot model, or for not using
> it for serious problems. It is clearly rank
> folly to develop a complex control merely for
> the sake of the pilot model. I favour a model
> with a control of negligible size which can
> later be expanded if desired. Only test
> problems would be worked on the minimal
> machine.[35]

By the beginning of 1947 there were no new initiatives to report. 'With regard to the design of actual equipment ... progress has been slow owing to staff shortage' was the lame summary in the end-of-year report for 1946.[36]

The slow pace of the construction work at Dollis Hill was not entirely the fault of the Post Office. As previously mentioned, during 1946 Turing kept changing the logical design of the machine. Moreover, in November 1946 the NPL considered making a radical change to the way the memory was to be constructed, with cathode ray tubes (CRT) taking the place of mercury delay lines. This change would have meant that most of the work done up to that point by Chandler and Coombs was wasted.[37] (In the end CRT memory was used at Manchester but not for the ACE.)

Coombs described the situation from the engineers' point of view:

> One of the problems was, I remember, that NPL kept on changing its ideas, and every time we went down there and said 'Right now! What do you want us to make?', we'd find that the last idea, that they gave us last week, was old hat and they'd got a quite different one, and we couldn't get a consolidated idea at all until eventually we dug our toes in and said 'Stop! Tell us what to make'.[38]

Eight years later Chandler and Coombs finally completed a computer, named the MOSAIC, based on Version VII of Turing's ACE design (see below).

The second attempt to build the ACE: the Huskey era

The situation improved when Huskey—an engineer—arrived in Maths Division on a fixed-term contract. Huskey had worked on the ENIAC project and in 1946 was offered the Directorship of the EDVAC project, although complications prevented him from accepting (see his Chapter 13, 'The ACE Test Assembly, the Pilot ACE, the Big ACE, and the Bendix G15'). Hartree, the ACE's guardian angel on the NPL Executive Committee, had met Huskey while visiting the ENIAC in the spring of 1946; in July Huskey received a telegram offering him a twelve-month visiting position at the NPL. Huskey began work in Maths Division on 4 January 1947.[39] A number of other new recruits joined the ACE Section during the course of 1947, in an expansion initiated by Turing[40]: Gerry Alway in August, Donald Davies in September, Henry John Norton in October, Betty Curtis in November.

Huskey soon suggested that the ACE Section itself make a start on constructing the computer and he proposed to Womersley that a small test assembly be built. With Womersley's blessing, Huskey, Wilkinson, and Woodger began work. They planned to build a simplified form of Turing's Version V known as Version H (for 'Huskey'). The new machine—soon called the 'Test Assembly'—was to be housed in the Babbage Building, a short distance from Maths Division.[41] Womersley summarized the situation in a report written in the middle of 1947:

```
At the beginning of 1947 Dr. H. D. Huskey,
a member of the team working on the EDVAC
```

and ENIAC machines at the University of
Pennsylvania, Philadelphia, joined the staff
of the Mathematics Division for one year.
In the Spring of 1947 it was decided that the
Laboratory itself should undertake some
experimental work and Dr. Huskey, with two
assistants, began the collection of equipment
with a view to constructing a small pilot
model.[42]

Turing was not in favour of this development. On the one hand the Test Assembly was to be a small computer in its own right, involving much more equipment than was strictly necessary to test the fundamentals of Turing's design, and yet on the other it fell far short of being the ACE. Probably Turing saw Huskey's project as diverting effort from his own. Turing 'tended to ignore the Test Assembly', simply 'standing to one side'.[43] (Woodger described how he 'was writing a program for [Version H] when Turing came in ... looked over my shoulder and said, "What is this? What's Version H?". So I said, "It's Huskey's." "W H A T!" ... [T]here was a pretty good scene about that.'[44])

Huskey and the others pushed ahead with the Test Assembly. By about the middle of 1947 the NPL workshops were fabricating a mercury delay line to Huskey's specifications, valve types had been chosen and circuit block diagrams made, source and destination decisions had been taken, and programs were being written to check these decisions.[45] A main frame was built and construction of the plug-in chassis holding the circuitry was planned.[46] Huskey's first goal was to run a simple stored program using an absolute minimum of equipment and a single full-length delay line.[47] Then the group would develop a substantial computer capable of solving practical problems; this would contain approximately thirteen delay lines, and would involve punched card input and output and a hardware multiplier.[48] In October 1947 Womersley and Fieller expected—very optimistically—that the Test Assembly would 'be ready by the end of November'.[49] (Huskey said: 'I never hoped to have the Test Assembly working before I left in December. I certainly hoped the group would have it working in 1948.'[50]) Then, in what was one of the worst administrative decisions of the whole ACE saga, Darwin summarily stopped the work. 'Morale in the Mathematics Division collapsed', Huskey recalled.[51]

The man behind Darwin's decision was Horace Augustus Thomas, head of the newly formed electronics group in the NPL's Radio Division.

The third attempt to build the ACE: the Thomas era

In January 1947 Turing had gone to the United States, visiting several of the groups that were attempting to build an electronic stored-program computer. In his report on his visit he wrote:

> One point concerning the form of organisation
> struck me very strongly. The engineering
> development work was in every case being done
> in the same building with the more mathematical
> work. I am convinced that this is the right
> approach. It is not possible for the two parts
> of the organisation to keep in sufficiently
> close touch otherwise. They are too deeply
> interdependent. We are frequently finding that
> we are held up due to ignorance of some point
> which could be cleared up by a conversation
> with the engineers, and the Post Office find
> similar difficulty; a telephone conversation
> is seldom effective because we cannot use
> diagrams. Probably more important are the
> points which are misunderstood, but which
> would be cleared up if closer contact were
> maintained, because they would come to light
> in casual discussion. It is clear that we
> must have an engineering section at the ACE
> site eventually, the sooner the better,
> I would say. [52]

Darwin decided that NPL's Radio Division was the best place for the experimental engineering work to be carried out. The minutes of the March 1947 meeting of the Executive Committee outlined the new arrangements:

> A.C.E. Director reported that he had had a
> meeting with members of the staff concerned.
> With a view to helping on progress with this
> machine it had been suggested that Dr. Thomas

```
of Radio Division should be put in charge of
the work of making by a suitable firm
a prototype model. Dr. Thomas was very keen on
electronic technique in industrial problems and
he felt that the best way in which we could get
progress would be to have a pre-prototype model
started in Radio and Metrology workshops before
approaching an outside firm. It was agreed that
Dr. Thomas should prove a very suitable man for
this work. In reply to a question from
Professor Hartree, Director said that Metrology
and Radio workshops could get on with the
hardware part of the job straight away.
Director stated that he had had a letter from
the Post Office giving the position of their
work which appeared to be most encouraging.
Sir Edward Appleton said that he had heard from
Dr. Radley that they were finding it difficult
to keep to their delivery dates, and Dr. Radley
will be seeing Director about this.[53]
```

The wheels of administration turned slowly and the idea of an in-house electronics section took several months to implement. At the end of April 1947 a joint minute to Darwin from Womersley and R. L. Smith-Rose, Superintendent of Radio Division, suggested that individuals be transferred from other Divisions to Radio Division in order to form 'the nucleus of a future electronics section':

```
The present state of the project requires
that the group should work together in one
place as a whole in close contact with the
planning staff in the Mathematics Division.
The various parts are so interwoven that
it is not practicable at present to farm out
portions of the work to isolated groups. Our
experience with the Post Office confirms this.
... The re-allocation of staff within the
Laboratory requires ... careful consideration,
but if you are in agreement with our proposal,
could you advise us as to what action could be
taken in the immediate future.[54]
```

By August the months of 'careful consideration' finally came to an end and notes were sent out by E. S. Hiscocks (the Secretary of the NPL) to the superintendents of various divisions instructing them to transfer staff to Radio Division for a period of six months. Smith-Rose reported to Darwin: 'We are now in a position to commence experimental work in the development of the A.C.E. for Mathematics Division'.[55] Two members of Maths Division (Gill and Wise) were transferred, and Wilkinson, although not formally transferred, was expected by Womersley to work with Thomas's group for the first few months of the project.[56] Edward Newman joined the group at the beginning of September.[57] Recruited from Electric and Musical Industries (EMI) Research Laboratories, Newman would play an important role in the eventual construction of the Pilot ACE, as would David Clayden, who followed Newman from EMI later that month (see Chapter 19, 'Circuit Design of the Pilot ACE and the Big ACE' by D. O. Clayden.)

In August 1947 a formal meeting was held to inaugurate the new state of affairs:

```
                   A.C.E. Project
     A meeting was held in the large Conference
  Room on the 18th August to initiate the A.C.E.
  Project in the Radio Division. The following
  were present:-

  Director
  Secretary
  Superintendent, Radio Division
  Superintendent, Mathematics Division
  Mr. F. M. Colebrook
  Dr. A. M. Turing
  Mr. J. W. Christelow
  Dr. H. A. Thomas
  Dr. F. Aughtie
  Dr. H. Huskey
  Mr. M. A. Wright
  Mr. W. Wilson
  Mr. A. F. Brown
  Mr. R. G. Chalmers
  Mr. B. J. Byrne
  Mr. R. F. Braybrook
  Mr. A. I. Williams
```

> The Director opened the meeting by stating
> that this was the first time that a large team
> had been assembled at the Laboratory to
> initiate a research programme, and stressed the
> importance of the A.C.E. project. He pointed
> out that the Mathematics Division was the
> parent Department, and that they had, as it
> were, issued a contract to the Radio Division
> for the Development and completion of the
> machine.
>
> Mr. Womersley followed by giving a historical
> summary commencing from the days of Babbage,
> and pointed out that the N.P.L. project,
> if successful, would give a machine much in
> advance of, and much quicker than, any of
> the American machines. It would afford a new
> approach to the problems of Mathematical
> [Physics] and would be of immense value to the
> country.[58]

Even before the inaugural meeting, trouble was brewing behind the scenes. On 12 August Hiscocks wrote to Darwin to alert him to what seemed to be empire-building on Thomas's part:

> Thomas has apparently shown some signs of
> behaving as if he is starting up a new
> Division, and so as to allay certain qualms
> which both Smith-Rose and Womersley have,
> I think it would be better for it to be
> explained to the whole team that Mathematics
> Division is the parent Division, and the one
> which is to justify the financial outlay on
> this work; that the work is being put out on
> contract, as it were, to Radio Division, and
> that Thomas's team is a part of Radio Division.
> I think, even if only for our own peace of
> mind, this is desirable, because Thomas has
> already shown some signs of wanting to set up a
> separate office, etc.[59]

An unfortunate rivalry quickly sprang up between Thomas's group and the ACE Section in Maths Division. On 17 September Smith-Rose was sent an indignant letter complaining about a raid by Thomas's group on the ACE Section:

> During Dr. Turing's and Dr. Huskey's absence on
> leave, most of the apparatus in our laboratory
> in Teddington Hall has, I believe, been removed
> to your Division. It would be of assistance in
> maintaining our inventory if we could have a
> list of what was, in fact, taken; we understood
> from Dr. Huskey that only two items were to be
> moved.[60]

Thomas the empire-builder soon petitioned Darwin to curtail the construction work in the ACE Section. Wilkinson said in an interview given in 1976: 'Thomas particularly didn't like ... the idea of this group in Mathematics Division ... working independently .... Thomas persuaded the Director to lay it down that all work should be done in the Electronics Section and Darwin decreed that we should stop work on the Test Assembly.'[61]

The result was that the construction of the ACE drew almost to a standstill. Although Newman and Clayden were skilled in digital techniques, Thomas's group had much to learn. Thomas's own background was not in digital electronics at all but in radio and industrial electronics. The group 'began to develop their knowledge of pulse techniques', said Wilkinson, and 'for a while they just did basic things and became more familiar with the electronics they needed to learn to build a computer.'[62]

Then, in February 1948, Thomas delivered another blow to the ACE project, resigning from the NPL to join Unilever Ltd (the manufacturers of 'Sunlight' soap). As Womersley summed up the situation shortly afterwards, hardware development was 'probably as far advanced 18 months ago'.[63]

It seems probable that, given better management at the NPL, a minimal computer based on Turing's Version V could have run a program during 1948. Turing first proposed an in-house electronics section at the NPL in his report of 3 February 1947. The Radio Division group could have been set up in six weeks rather than six months. Clearly the new electronics group should have joined forces with Huskey and the ACE Section to work on the Test Assembly. In August 1947 Womersley had pressed for this course of action, but Thomas threw a spanner in the works.[64] A rudimentary form of the Test Assembly might easily have run a trial program before the middle of 1948, becoming the world's first functioning electronic stored-program digital computer (an honour that in the event went to the Manchester 'Baby' in June 1948, itself a very limited machine).

In mid-1947 Turing applied for a period of sabbatical leave to be spent at his Cambridge college. He proposed to pursue research on machine intelligence.[65] Darwin approved the request, saying in a letter to the DSIR:

```
As you know Dr. A. Turing ... is the
mathematician who has designed the theoretical
part of our big computing engine. This has now
got to the stage of ironmongery, and so for the
time the chief work on it is passing into other
hands. I have discussed the matter both with
Womersley and with Turing, and we are agreed
that it would be best that Turing should go off
it for a spell.[66]
```

Turing left for Cambridge in the autumn of 1947.[67] He returned briefly to Mathematics Division the following spring (winning the 3-mile run at the NPL Annual Sports Event).[68] Then, in May 1948, no doubt disheartened by the complete absence of progress on the construction of the ACE, he gave up his job at the NPL, accepting Newman's offer of a position at the Manchester Computing Machine Laboratory. At Manchester Turing continued his ground-breaking theoretical work on machine intelligence and pioneered the field now known as 'Artificial Life' (see Chapter 5, 'Turing and the Computer'). In Mathematics Division, the struggle to bring the ACE into existence was now led by Wilkinson.

The fourth attempt to build the ACE: the Wilkinson–Colebrook era

Thomas was replaced by Francis Morley Colebrook, who in March 1948 was made head of the newly designated Electronics Section of Radio Division.[69] Colebrook at last got things moving. In 1976 Wilkinson recollected:

> I think quite soon [Colebrook] had an uncomfortable feeling that he'd inherited something which was in danger of floundering, and I had the same feeling back in Maths Division—I'd inherited this team from Turing and everybody was really a little bit demoralised by that time. And Colebrook rang me up and he came over to see me and had a chat

about it and said he was not very happy about the position. And ...
to my absolute astonishment he said: 'I've got a suggestion to make
... You chaps have learned a bit of electronics now ... What about
coming over and joining us and the two groups working together?'
... I felt the whole thing was in such a mess it needed quite a decisive
break in order to get it going. ... Colebrook ... was a great diplomat
... and his goodwill was so evident to everybody that I do think he
played a major part in making it possible for the two groups to go
together.[70]

The combined group decided that the best course of action was to revive
the Test Assembly, now described as a 'pilot model'.[71] (The similarity between
the pilot model they now worked on and the out-of-favour Test Assembly
was, Wilkinson said, 'more than ... it was diplomatic to say much about'.[72])
The ACE Section and the Electronics Section worked harmoniously together,
and at the end of the year Womersley was able to report that '[v]ery good
progress has been made'.[73] Huskey's approach to circuit design was replaced
by the Blumlein approach which Newman and Clayden had brought from
EMI (see Chapter 19, 'Circuit Design of the Pilot ACE and the Big ACE'). The
group completely redesigned the electronics of the machine.

Soon the mathematicians from the ACE Section found themselves in a
novel milieu. Woodger recalled: 'We set ourselves up in a little assembly
line with ... a stack of components ... in front of us. Each of us had a
soldering iron and we produced these things and passed them down the line.
Oh, it was tremendous fun.'[74] Following a period of difficulty with the
delay-line amplifiers at the end of 1949,[75] the group finally tasted success.
On 10 May 1950 the Pilot Model ACE ran its first program. Later known affec-
tionately as 'Succ. Digs' (successive digits), the program turned on a row of
32 lights on the control panel at a speed determined by size of the number on
the input switches.

The Pilot Model ACE was (in Colebrook's words) 'powerful computing
equipment'.[76] Although the ultimate goal remained a large-scale ACE, Maths
Division had been planning since 1948 to use the Pilot Model as a computing
machine in its own right.[77] However, there was still considerable work to be
done before the Pilot ACE could be handed over to Maths Division for customer
service. Unreliable components were a problem and it was September 1950
before the machine had an error-free run of half an hour.[78] During 1951
the delay lines and the control of input and output were redesigned and the
parallel multiplier was added.[79] By about the middle of 1951 the Pilot ACE

The Pilot ACE in Bushy House, December 1950. On the left are the modified Hollerith punched-card unit and the control table. The tray slung below the main frame contains the short delay lines ('Temporary Stores').

Source: National Physical Laboratory. © Crown copyright; reproduced by permission of the Controller of HMSO.

was doing over 50 per cent of the computing work of Maths Division.[80] Nevertheless, by October 1951 it had still 'not yet been put into regular service', said Colebrook, 'and we have not yet built up an adequate "library" of generally useful sub-routines'.[81]

At the end of October 1951 the new Director of the NPL, Sir Edward Bullard, expressed the opinion that

```
most of the electronic machines, including our
own, do not really work regularly and reliably.
They have been very much over-advertised and
there is a lot of work to be done before we
have anything that is much use.82
```

Despite Bullard's pessimism, by February 1952 the Pilot ACE was reliable enough to be dismantled and transferred to Maths Division.[83]

The Pilot ACE in March 1952 after its move to Mathematics Division, where it provided an industrial computing service. The Hollerith punched-card equipment is at the far end, next to the table containing the control panel. The stand beneath the main frame holds the short delay lines. The long delay lines are in the large temperature-controlled cabinet—known as 'the coffin'—situated behind the main frame. Two experimental long delay lines are on stands to the right of the main frame.

Source: National Physical Laboratory. © Crown copyright; reproduced by permission of the Controller of HMSO.

Colebrook wrote:

```
Full operation was resumed within two weeks
after the removal and has been continuous ever
since, mainly on defence problems. ... An
analysis of the first eight weeks of operation,
involving 370 power-on hours, gives the
following figures:-

Routine testing and maintenance  92 hrs.    25%
User training, programme         175 hrs.   47%
  testing, etc.                                   } 75%
Paid-for work84                  103 hrs.   28%
```

The Pilot ACE remained in continuous service until replaced by the first DEUCE in 1955, by which time 'the amount of maintenance it require[d] preclude[d] it from being used economically as a computer'.[85] The Pilot ACE was a huge success. It was used both to carry out research in numerical analysis and to do paid work through Mathematics Division's scientific computing service (see Chapter 12, 'Applications of the Pilot ACE and the DEUCE'). In 1954 alone the Pilot ACE earned the NPL £24,000 for over 80 jobs[86]— Turing's annual salary when he designed the ACE was £800[87]. In the course of its working life the machine earned approximately £100,000.

The DEUCE

Toward the end of 1948, the NPL's efforts to find an engineering company willing to assist with the ACE at last bore fruit. The minutes of the NPL Executive Committee for September 1948 report:

> In connection with the A.C.E., members of Sir George Nelson's staff have visited the Laboratory to see the progress made and to discuss possible collaboration in the more detailed design and, later, the construction of the A.C.E.[88]

Nelson, chairman of the English Electric Company, had been a member of the NPL Executive Committee since 1946 (and was present when Turing addressed the historic meeting of March 1946). Early in 1949 it was proposed that the NPL place a contract with English Electric:

> A.C.E. Project.
> Present Position, and request for financial provision for a Study Contract to be placed with the English Electric Co. Ltd.
> ... In order to expedite the construction of [the] pilot assembly and to make possible the construction of the final machine it is now proposed that a Study Contract be placed with the English Electric Co. Ltd. This will have two great advantages, first it will add to the labour force working on the construction of the pilot model, and should make possible its completion before the end of this year, and it will familiarise the engineers and the staff

```
with the specialised requirements and the
techniques involved. This educational period
before they embark upon the construction of the
final machine is very necessary in view of the
novelty of the whole enterprise.[89]
```

This contract was approved by the Treasury in May 1949.[90] It 'provided for members of the English Electric Company staff to work at the N.P.L., with N.P.L. staff involved'.[91] Engineers and wiremen from English Electric joined the NPL team to assist in the completion of the Pilot ACE.

In 1951 an NPL memorandum set out a number of reasons for desiring to 'continue the collaboration with the E.E.Co.', including the following: 'The experience gained by the E.E.Co. would enable them to reproduce the A.C.E. or a similar machine for any subsequent home or foreign demand.'[92] By December 1951 arrangements for producing a commercial version of the Pilot ACE had been firmed up:

```
ACE---Arrangements with English Electric Co.
... [I]t has been decided that the first move
shall be the construction by the E.E. Co. of an
'engineered version' of the present Pilot ACE.
Simultaneously with this some of the detailed
planning and design for the full-scale ACE will
be undertaken.[93]
```

A memo from Treasury noted approvingly that

```
These plans for the development of the A.C.E.
represent a very favourable turn. English
Electric (through Sir George Nelson) offered to
take it upon themselves to construct a properly
'engineered' version of the Pilot model at a
cost to N.P.L. of no more than f5,000. ... The
likelihood is that [this] by no means
represents the full economic cost. But Sir
George Nelson is prepared to think in terms of
such a figure because he would like to see
English Electric getting into the field.[94]
```

The engineered version of the Pilot ACE was called—naturally enough— 'the DEUCE' (Digital Electronic Universal Computing Engine). The first DEUCE to be produced was delivered to Maths Division in March 1955.[95] The DEUCE

The NPL DEUCE in 1958.

Source: National Physical Laboratory. © Crown copyright; reproduced by permission of the Controller of HMSO.

became a cornerstone of the British computer industry (see Chapter 6, 'The ACE and the Shaping of British Computing').

The Big ACE

Work began on a full-scale ACE in the autumn of 1954.[96] In 1956 J. R. Illingworth outlined the reasons for proceeding to this final stage of the project:

ACE---Final Model
... Experience had already shown ... that the
DEUCE would not be an efficient proposition for
more than four or five years since the speed
and storage facilities required for this type
of machine were becoming greater year by year.
It was therefore decided about two years ago to
commence work on an entirely new machine and
this started in the Autumn of 1954. ...

> The whole position, I think, may be summed up
> by saying that both the importance and the
> significance of this work has changed. In the
> early days when the project was envisaged it
> was not even known whether the conception of
> the ACE would lead to a research "toy" of any
> value, whereas events have proved it to be such
> a first class computing mechanism that we are
> under pressure from the Treasury and other
> departments to develop its use as rapidly as
> possible for work in office mechanisation.
> Moreover, whereas in the early days the project
> was regarded as an additional facet to the
> Laboratory's normal work, the whole of this
> field of work has become an inherent part of
> the Laboratory's programme.[97]

Built and housed at the NPL, the Big ACE was in operation by late 1958. (Only one was made.) Wilkinson, Clayden, Davies, Newman, and Woodger all contributed to the final design.[98] English Electric played a part in 1954 and 1955 during the developmental stage, but at that point their contract with the NPL came to an end.[99] With a clock speed of 1.5 MHz and containing some 6000 valves, the Big ACE filled a room the size of an auditorium.[100] The computer remained in service until 1967.

At a Press Day held in 1958 to inaugurate the Big ACE, A. M. Uttley (superintendent of the Control Mechanisms and Electronics Division, as the Electronics Section had by then become) announced: 'Today, Turing's dream has come true'.[101] If so, it was a dream whose time had passed. The Big ACE was not the revolutionary machine that it would have been if completed six or seven years earlier. Not only did it employ valves in the era of the transistor, the designers also retained the by then outmoded mercury delay-line memory proposed by Turing in 1945.[102] Nevertheless, the Big ACE was a fast machine with a large memory, and the decision to stick with the principles used in the Pilot ACE and the DEUCE was reasonable in the circumstances. In 1953 Colebrook urged that the proposed full-scale ACE 'be based on well proved components and techniques, even when revolutionary developments seem to be just around the corner'.[103] 'Otherwise the [Mathematics] Division will get nothing but a succession of pilot models', Colebrook argued.[104] As for speed, the machine's designers wrote in 1957: 'The ACE appears in fact to be about

The Big ACE, November 1958. Raised panels expose the racks of chassis.

Source: National Physical Laboratory. © Crown copyright; reproduced by permission of the Controller of HMSO.

as fast as present-day parallel core-store computers' (magnetic core memory was the most advanced high-speed storage medium at that time).[105]

A Simple Guide to ACE

One of the world's largest and fastest electronic computing machines---popularly called 'Electronic Brains'---has begun to operate at the National Physical Laboratory. Its name is ACE---and it can carry out any calculation process for which exact rules are known. ...

ACE incorporates many unique design features. The operating 'mechanism' consists of about 6,000 miniature electronic valves, arranged in an impressive array of 10 large cabinets. Each cabinet is fitted with an electrically operated rising door to give rapid access for fault finding. It has a cooling system of circulating air, with heat exchangers and a water cooler.

The numbers and instructions have 48 binary digits (equivalent to 14 decimal digits). The working store consists of 800 words, and the backing store of four magnetic drums will contain a total of 32,768 words.

Mercury delay lines and magnetic drums are used for the storage of numerical data and instruction sequences, and both punched card and magnetic tape equipment are provided for input and output.

The compact control desk has some 160 keys and 300 signal lamps, in addition to audible and visual displays of the computation. Nevertheless, most computations are carried out entirely by pushing one key---marked 'Initial Input'. This key causes punched cards to be read, which tell ACE what to do.

The most interesting mechanical assemblies in ACE are the four magnetic drums, entirely designed and constructed at N.P.L. The drums rotate at 12,000 r.p.m. so accurately that the arrival at a given point of a magnetic spot one hundredth of an inch in length, travelling at 200 miles per hour, is timed to a millionth of a second. The drums have unique rapidly moving recording heads and are only a part of the number store or 'memory' of ACE. An idea of the speed of ACE can be gained from the fact that these drums are the slowest part of its number store!

Technical Notes

ACE has three forms of storage. The largest part of the store consists of four magnetic drums. ... The access to these numbers is limited by the need to wait for the drum to revolve, and may take up to 7 milliseconds. These drums are used by the computer for the large mass of data that it may not want quickly.

For more rapid access, a mercury delay line store of 768 numbers is employed. The time

required for access to these is up to one
millisecond. This store is used for the
program, but instructions are stored in such
a way that no time is wasted in waiting
for them.

The most rapid access store consists of short
delay lines storing one, two or four numbers
each. These are used for the small group of
numbers that the computer is using for a short
while.

By good programming, the full advantages of
the large store can be enjoyed, with a time of
access not much greater than is given by the
short delay lines.

The ACE instructions specify the
addresses---the locations in the store---of two
numbers to be operated on, the operation, and
the addresses of the desired destination of the
result and of the location of the next
instruction. Thus four addresses are used, and
all four accesses occur nearly simultaneously.
[M]ultiplication and division occur in
separate, autonomous devices, so that they can
occur simultaneously with other
operations.

In many cases, the computer will average
15,000 operations per second, each involving
extracting two operands, storing the result and
extracting the next instruction.

Input and output is by punched cards.
Magnetic tapes will be fitted later, but will
be used as an extension to the store. The input
speed from cards will be 7000 binary digits per
second, with a rate of 450 cards per minute
punched in binary.[106]

The MOSAIC

With the inauguration of Thomas's electronics group, the NPL had taken
the opportunity to bow out of its formal arrangement with the Post Office.

In August 1948 Womersley wrote to Darwin:

> Post Office Account for Work on A.C.E.
> During the period to which this account applies
> either Dr. Turing or myself kept in regular
> touch with the engineers concerned by regular
> monthly visits. I am satisfied that this
> account is a reasonable one and represents the
> amount of time and effort put in by the Post
> Office during that period. In relation to this
> amount of time and effort the progress which
> the Post Office engineers have made is
> satisfactory, but we cannot pretend that the
> scale on which the Post Office found it
> possible to work was in any way comparable with
> the effort we would like to have seen made.
>
> Now that we are establishing a special group
> under Dr. Thomas to do development work here,
> it is becoming necessary to decide whether we
> continue to ask the Post Office to do work on
> our behalf. If we discontinue our arrangement
> with them (except on a basis of friendly
> interchange of information) work will still
> continue there. They have, in fact, a contract
> with Ministry of Supply to produce a computer
> for the reduction of certain data from
> ballistic trials, and we shall be expected to
> play some part in the planning of it. Since the
> original approach was made by you personally to
> the Chief Engineer of the Post Office, will it
> be necessary for you to inform them yourself if
> we bring our contract with them to an end?[107]

The contract between the Post Office and the Ministry of Supply mentioned by Womersley led to the MOSAIC (Ministry of Supply Automatic Integrator and Computer), which first ran a program in 1952 or early 1953.[108] The MOSAIC was based on Version VII of Turing's logical design for the ACE.[109] Working alone, Chandler and Coombs carried out the engineering design for the MOSAIC. With a pulse rate of 570 kilocycles per second, the MOSAIC contained approximately 7000 valves and 2000 semiconductors (germanium diodes).[110] Originally a high-speed memory of 96 mercury delay lines was

Three of the MOSAIC's racks.

Source: Heritage, Royal Mail. © Crown copyright; reproduced by permission of Royal Mail Group plc.

planned;[111] in the final form of the machine there were 64 long delay lines and a handful of short delay lines, holding a total of 1030 40-digit words.[112] Of the various ACE-type computers that were built, the MOSAIC was (apart from its pulse rate) the closest to Turing's conception of the ACE.

The MOSAIC was manufactured and assembled by the All-Power Transformer Company and was installed at the Radar Research and Development

Establishment (RRDE) Malvern, in 1954 or early 1955.[113] It was used to calculate aircraft trajectories from radar data, in connection with anti-aircraft measures. (The details of the computer's use appear still to be classified.) Two mobile automatic data-recorders worked in conjunction with a radar tracking system. Each recorder involved approximately 2000 valves, with 'special cathode-ray tube switches and pneumatic gear to provide a record on punched paper tape'.[114]

Given that two engineers working alone succeeded in completing the large MOSAIC (Coombs emphasized: 'it was just Chandler and I—we designed every scrap of that machine'[115]), there seems little doubt that, given sufficient manpower, a computer reasonably close to Turing's Version VII of the ACE could have been operational in the early 1950s.

Thanks to their wartime involvement with Colossus, Chandler and Coombs possessed unrivalled expertise in large-scale digital electronics and had a substantial lead on everyone else in the field. Turing, of course, was well aware of this, but the Official Secrets Act prevented him from sharing his knowledge of Colossus with Darwin. Had he been able to do so, the NPL might have acted to boost the resources available to Chandler and Coombs, and so made Turing's dream a reality much sooner.

Notes

I am grateful to Diane Proudfoot for comments on a draft of this material.

1. Woodger in interview with Copeland (June 1998).
2. 'Superintendent of the Mathematics Division', NPL, report E. 849, 28 September 1944 (NPL library; a digital facsimile is in The Turing Archive for the History of Computing <www.AlanTuring.net/womersley_appointment>).
3. 'Research Programme for the Year 1945–6', NPL, October 1944, items 6502, 6502.1, 6502.2 (NPL library; a digital facsimile is in The Turing Archive for the History of Computing <www.AlanTuring.net/research_programme_1945-46>).
4. 'Report of Interdepartmental Technical Committee on a Proposed Central Mathematical Station', Department of Scientific and Industrial Research, 3 April 1944 (NPL library; a digital facsimile is in The Turing Archive for the History of Computing <www.AlanTuring.net/proposed_central_math_station>).
5. Ibid.
6. Minutes of the Executive Committee of the NPL for 19 December 1944 (NPL library; a digital facsimile is in The Turing Archive for the History of Computing <www.AlanTuring.net/npl_minutes_dec1944>). NPL documents are

© Crown copyright and extracts in this chapter are reproduced by permission of the Controller of HMSO.

7. Editor's note. The Aiken and Stibitz machines were neither electronic nor stored-program.

8. Editor's note. This is probably a reference to the December meeting mentioned above (perhaps Womersley's manuscript was dated November 1944). The minutes of the November meeting of the Executive Committee do not indicate that Womersley was present; they contain the statement 'The next meeting will be held on Tuesday, 19th December ... when Mr. J. Womersley, Superintendent-elect of the proposed Mathematical Division, will be invited to attend.'

9. Womersley, J. R. 'A.C.E. Project – Origin and Early History', NPL, 26 November 1946 (National Archives: Public Record Office, Kew, Richmond, Surrey (document reference DSIR 10/385); a digital facsimile is in The Turing Archive for the History of Computing <www.AlanTuring.net/ace_early_history>).

10. Flowers in interview with Copeland (July 1996).

11. Minutes of the Executive Committee of the NPL for 23 October 1945 (NPL library; a digital facsimile is in The Turing Archive for the History of Computing <www.AlanTuring.net/npl_minutes_oct1945>).

12. Woodger, M. handwritten note, no date (in the Woodger Papers, National Museum of Science and Industry, Kensington, London (catalogue reference M15/78)); letter from Woodger to Copeland, 27 November 1999. Woodger records the existence of an NPL file giving the date of Turing's completed report as 1945; the file was destroyed in 1952.

13. Editor's note. Edward Travis was head of the British government's codebreaking operations from 1942.

14. Womersley, J. R. ' "ACE" Machine Project', NPL, no date (Woodger Papers; a digital facsimile is in The Turing Archive for the History of Computing <www.AlanTuring.net/womersley_ace_machine>).

15. Womersley, J. R. ' "ACE" Machine Project', NPL, paper E.881, 13 February 1946 (Woodger Papers; a digital facsimile is in The Turing Archive for the History of Computing <www.AlanTuring.net/ace_machine_project>). Additional extracts from Womersley's memorandum are quoted in Chapter 7.

16. Minutes of the Executive Committee of the NPL for 19 March 1946 (NPL library; a digital facsimile is in The Turing Archive for the History of Computing <www.AlanTuring.net/npl_minutes_mar1946>).

17. Ibid.

18. Darwin, C. 'Automatic Computing Engine (ACE)', NPL, 17 April 1946 (Public Record Office (document reference DSIR 10/385); a digital facsimile is in The Turing Archive for the History of Computing <www.AlanTuring.net/darwin_ace>).

19. Minutes of the DSIR Advisory Council, 8 May 1946 (Public Record Office (document reference DSIR 10/275); a digital facsimile is in The Turing Archive for the History of Computing <www.AlanTuring.net/dsir_minutes_may1946>).

20. Minutes of the Executive Committee of the NPL for 21 May 1946 (NPL library; a digital facsimile is in The Turing Archive for the History of Computing <www.AlanTuring.net/npl_minutes_may1946>).

21. Letters from Woodger to Copeland, 25 February 2003, 22 May 2003.

22. Letter from Woodger to Copeland, 25 February 2003.

23. Memorandum from Turing to Darwin, 30 August 1947 (Woodger Papers (catalogue reference M11/99); a digital facsimile is in The Turing Archive for the History of Computing <www.AlanTuring.net/turing_darwin_30aug47>).

24. 'Draft Report of the Executive Committee for the Year 1946', NPL, paper E.910, section Ma. 1, anon., but probably by Womersley (NPL library; a digital facsimile is in The Turing Archive for the History of Computing <www.AlanTuring.net/annual_report_1946>).

25. Flowers in interview with Copeland (July 1996).

26. Letter from Radley to Womersley, 25 February 1946 (Public Record Office (document reference DSIR 10/385); a digital facsimile is in The Turing Archive for the History of Computing <www.AlanTuring.net/radley_womersley_25feb46>).

27. Editor's note. It was not until January 1947 that the Post Office succeeded in keeping a pattern of pulses circulating for half an hour (Minutes of the Executive Committee of the NPL for 21 January 1947 (NPL library; a digital facsimile is in The Turing Archive for the History of Computing <www.AlanTuring.net/npl_minutes_jan1947>)).

28. 'Status of the Delay Line Computing Machine at the P.O. Research Station', NPL, 7 March 1946, anon. (Woodger Papers (M12/105); a digital facsimile is in The Turing Archive for the History of Computing <www.AlanTuring.net/delay_line_status>).

29. Flowers in interview with Copeland (July 1998).

30. Letter from Darwin to Sir Edward Appleton, 13 August 1946 (Public Record Office (document reference DSIR 10/275); a digital facsimile is in The Turing Archive for the History of Computing <www.AlanTuring.net/darwin_appleton_13aug46>); letter from Radley to Darwin, 1 November 1946 (Public Record Office (document reference DSIR 10/385); <www.AlanTuring.net/radley_darwin_1nov46>).

31. Letter from W. B. Lewis to Womersley, 14 August 1946 (Woodger Papers (M11)).

32. Letter from Darwin to Appleton, 13 August 1946.

33. Minutes of the Executive Committee of the NPL for 19 November 1946 (NPL library; a digital facsimile is in The Turing Archive for the History of Computing <www.AlanTuring.net/npl_minutes_nov1946>).

34. Ibid.

35. Memorandum from Turing to Womersley, undated, *c*.December 1946 (Woodger Papers (M15/77); a digital facsimile is in The Turing Archive for the History of Computing <www.AlanTuring.net/turing_womersley>).

36. 'Draft Report of the Executive Committee for the Year 1946', section Ma. 1 (see note 24).

37. Letter from Darwin to Radley, 26 November 1946 (Public Record Office (document reference DSIR 10/385); a digital facsimile is in The Turing Archive for the History of Computing <www.AlanTuring.net/darwin_radley_26nov46>).

38. Coombs in interview with Christopher Evans in 1976 (*The Pioneers of Computing: An Oral History of Computing*. London: Science Museum).

39. Minutes of the Executive Committee of the NPL for 21 January 1947 (see note 27).

40. Wilkinson in interview with Evans in 1976 (*The Pioneers of Computing: An Oral History of Computing*. London: Science Museum); Memorandum from Turing to Darwin, 30 August 1947 (see note 23).

41. Woodger in interview with Copeland (June 1998); Woodger, M. 'ACE Test Assembly, Sept./Oct. 1947', NPL, no date (Woodger Papers (M15/84); a digital facsimile is in The Turing Archive for the History of Computing <www.AlanTuring.net/test_assembly>).

42. Womersley, J. R. 'A.C.E. Project', NPL, no date, attached to a letter from Womersley to the Secretary of the NPL dated 21 August 1947 (Public Record Office (document reference DSIR 10/385); a digital facsimile is in The Turing Archive for the History of Computing <www.AlanTuring.net/womersley_ace_project>).

43. Wilkinson in interview with Evans in 1976 (see note 40).

44. Woodger in interview with Copeland (June 1998).

45. Letter from Huskey to Copeland, 3 June 2003.

46. Wilkinson in interview with Evans (see note 40).

47. Letter from Huskey to Copeland, 18 January 2004.

48. Ibid.; Woodger, M., 'ACE Test Assembly, Sept./Oct. 1947' (see note 41).

49. Fieller, E. C. 'Hollerith Equipment for A.C.E. Work – Immediate Requirements', NPL, 16 October 1947 (Public Record Office (document reference DSIR 10/385); a digital facsimile is in The Turing Archive for the History of Computing <www.AlanTuring.net/hollerith_equipment>).

50. Letter from Huskey to Copeland, 3 June 2003.

51. Huskey, H. D. (1984) 'From ACE to the G-15', *Annals of the History of Computing*, 6, 350–71; the quotation is from p. 361.

52. Turing, A. M. 'Report on visit to U.S.A., January 1st–20th, 1947', NPL, 3 February 1947 (Public Record Office (document reference DSIR 10/385); a digital facsimile is in The Turing Archive for the History of Computing <www.AlanTuring.net/turing_usa_visit>). The rest of the report is quoted in Chapter 7.

53. Minutes of the Executive Committee of the NPL for 18 March 1947 (NPL library; a digital facsimile is in The Turing Archive for the History of Computing <www.AlanTuring.net/npl_minutes_mar1947>).

54. Womersley, J. R. and Smith-Rose, R. L. 'A.C.E. Pilot Test Assembly and later Development', NPL, 30 April 1947 (Public Record Office (document reference DSIR 10/385); a digital facsimile is in The Turing Archive for the History of Computing <www.AlanTuring.net/pilot_test_assembly>).

55. Memorandum from Smith-Rose to Darwin, NPL, 5 August 1947 (Public Record Office (document reference DSIR 10/385); a digital facsimile is in The Turing Archive for the History of Computing <www.AlanTuring.net/smith-rose_darwin_5aug47>).

56. Memorandum from Hiscocks to Womersley, NPL, 6 August 1947 (Public Record Office (document reference DSIR 10/385); a digital facsimile is in The Turing Archive for the History of Computing <www.AlanTuring.net/hiscocks_womersley_6aug47>); Womersley, J. L. 'A.C.E. Project. Transfer of Staff.', NPL, 13 August 1947 (Public Record Office (document reference DSIR 10/385); a digital facsimile is in The Turing Archive for the History of Computing <www.AlanTuring.net/staff_transfer>).

57. Memorandum from Smith-Rose to Darwin, 5 August 1947 (see note 55); Minutes of the Executive Committee of the NPL for 23 September 1947 (NPL library; a digital facsimile is in The Turing Archive for the History of Computing <www.AlanTuring.net/npl_minutes_sept1947>).

58. 'A.C.E. Project', NPL, 21 August 1947, initialled 'JWC/JG' (Public Record Office (document reference DSIR 10/385); a digital facsimile is in The Turing Archive for the History of Computing <www.AlanTuring.net/ace_project_meeting>).

59. Letter from Hiscocks to Darwin, 12 August 1947 (Public Record Office (document reference DSIR 10/385); a digital facsimile is in The Turing Archive for the History of Computing <www.AlanTuring.net/hiscocks_darwin_12aug47>).

60. Letter from E. C. Fieller to Smith-Rose, 17 September 1947 (Woodger Papers (M11)).

61. Wilkinson in interview with Evans (see note 40).

62. Ibid.

63. Minutes of the Executive Committee of the NPL for 20 April 1948 (NPL library; a digital facsimile is in The Turing Archive for the History of Computing <www.AlanTuring.net/npl_minutes_apr1948>).

64. Womersley, J. R. 'A.C.E. Project', August 1947 (see note 42).

65. Letter from Darwin to Appleton, 23 July 1947 (Public Record Office (document reference DSIR 10/385); a digital facsimile is in The Turing Archive for the History of Computing <www.AlanTuring.net/darwin_appleton_23jul47>).

66. Ibid.

67. Probably at the end of September. Turing was still at the NPL when Geoff Hayes arrived in Maths Division on 23 September 1947 (communication from Hayes to Woodger, November 1979).

68. Hayes, G. (2000) 'The Place of Pilot Programming', manuscript.

69. *National Physical Laboratory Report for the Year 1948.* London: HMSO, 1950, p. 53.

70. Wilkinson in interview with Evans (see note 40).

71. Womersley, J. R. 'A.C.E. Pilot Models', NPL, 26 April 1948 (Woodger Papers; a digital facsimile is in The Turing Archive for the History of Computing <www.AlanTuring.net/ace_pilot_models>).

72. Wilkinson in interview with Evans (see note 40).

73. *National Physical Laboratory Report for the Year 1948*, p. 29 (see note 69).

74. Woodger in interview with Copeland (June 1998).

75. Wilkinson in interview with Evans (see note 40).

76. Colebrook, F. M. 'Present Position and Future Prospects of Work by the Mathematics Division and Electronics Section, N.P.L., on High Speed Electronic Digital Computation', NPL, 30 August 1949 (Woodger Papers (M11)).

77. Memo from Womersley to Director, 31 May 1948 (Woodger Papers (M11)); 'A.C.E. Project', NPL Mathematics Division, 1 February 1949, anon. (Woodger Papers; a digital facsimile is in The Turing Archive for the History of Computing <www.AlanTuring.net/ace_project_position>); Womersley, J. R. 'Scientific Computing Service Ltd.', NPL, 3 May 1949 (Woodger Papers (M11)).

78. Woodger, M. (1969) 'In the beginning: Pilot ACE made history for the NPL in the former butler's pantry', *Computer Weekly*, 17 April, 8–9; 'ACE Bulletin No. 1', NPL, Electronics Section, 3 April 1951 (Woodger Papers; a digital facsimile is in The Turing Archive for the History of Computing <www.AlanTuring.net/ace_bulletin_1>); 'Bulletin No. 2', 3 May 1951 (Woodger Papers (N30/23); <www.AlanTuring.net/ace_bulletin_2>).

79. Woodger, M. 'In the beginning: Pilot ACE made history for the NPL in the former butler's pantry' (see note 78).

80. Letter from J. Illingworth to Fryer, 6 November 1956 (Woodger Papers (M15/87); a digital facsimile is in The Turing Archive for the History of Computing <www.AlanTuring.net/illingworth_fryer_6nov56>).

81. Letter from Colebrook to E. C. Cork, 8 October 1951 (Woodger Papers (N23)).

82. Letter from Bullard to J. H. C. Whitehead, 31 October 1951 (Woodger Papers (N23)).

83. Woodger, M. 'In the beginning: Pilot ACE made history for the NPL in the former butler's pantry', p. 8 (see note 78).

84. Colebrook, F. M. 'A Note on the ACE Pilot Model for the "Digital Computor News Letter"', NPL, 14 May 1952 (Woodger Papers (N25)).

85. Memorandum from Hiscocks to the DSIR, 30 January 1956 (Public Record Office (document reference DSIR 10/275); a digital facsimile is in The Turing Archive for the History of Computing <www.AlanTuring.net/hiscocks_dsir_30jan1956>).

86. Woodger, M. 'In the beginning: Pilot ACE made history for the NPL in the former butler's pantry', p. 9 (see note 78).

87. Minutes of the Executive Committee of the NPL for 23 October 1945 (see note 11).

88. Minutes of the Executive Committee of the NPL for 28 September 1948 (NPL library; a digital facsimile is in The Turing Archive for the History of Computing <www.AlanTuring.net/npl_minutes_sept1948>).

89. 'A.C.E. Project', 1 February 1949 (see note 77).

90. Letter from I. G. Evans to Darwin, 28 May 1949 (Public Record Office (document reference DSIR 10/275); a digital facsimile is in The Turing Archive for the History of Computing <www.AlanTuring.net/evans_darwin_28may49>).

91. Letter from B. Lockspeiser to the Lord President of the DSIR, 21 September 1950 (Public Record Office (document reference DSIR 10/275); a digital facsimile is in The Turing Archive for the History of Computing <www.AlanTuring.net/lockspeiser_dsir_21sept50>).

92. 'Memorandum on a Proposal to Construct the A.C.E. at the N.P.L.', NPL, Executive Committee paper E.15/51, 12 September 1951, anon. (Public Record Office (document reference DSIR 10/275); a digital facsimile is in The Turing Archive for the History of Computing <www.AlanTuring.net/proposal_construct_ace>).

93. Letter from Hiscocks to C. Jolliffe, 10 December 1951 (Public Record Office (document reference DSIR 10/275); a digital facsimile is in The Turing Archive for the History of Computing <www.AlanTuring.net/hiscocks_jolliffe_10dec51>).

94. Letter from Evans to Jolliffe, 15 December 1951 (Public Record Office (document reference DSIR 10/275); a digital facsimile is in The Turing Archive for the History of Computing <www.AlanTuring.net/evans_jolliffe_15dec51>).

95. Letter from Illingworth to Fryer, 6 November 1956 (see note 80).

96. Ibid.

97. Ibid.

98. Letter from A. M. Uttley to Sara Turing, 19 December 1958 (Modern Archive Centre, King's College, Cambridge (catalogue reference A 11)).

99. Letter from Illingworth to Fryer, 6 November 1956 (see note 80).

100. Blake, F. M., Clayden, D. O., Davies, D. W., Page, L. J. and Stringer, J. B. 'Some Features of the ACE Computer', NPL, 8 May 1957 (Woodger Papers (N12/102); a digital facsimile is in The Turing Archive for the History of Computing <www.AlanTuring.net/ace_features>).

101. Letter from Uttley to Sara Turing (see note 98).

102. An experimental transistorized machine went into operation at Manchester University in 1953 (see Lavington, S. H. (1980) *Early British Computers.* Manchester: Manchester University Press).

103. Colebrook, 4 May 1953, quoted on p. 67 of Yates, D. M. (1997) *Turing's Legacy: A History of Computing at the National Physical Laboratory 1945–1995.* London: Science Museum.

104. Ibid.

105. 'Some Features of the ACE Computer', p. 3 (see note 100).

106. 'A Simple Guide to ACE', NPL, no date, anon., marked 'based on Mr. Davies's notes' (Woodger Papers (N25)).

107. Womersley, J. L. 'Post Office Account for Work on A.C.E.', NPL, 18 August 1947 (Woodger Papers (M15/84); a digital facsimile is in The Turing Archive for the History of Computing <www.AlanTuring.net/ post_office_account>).

108. 'Engineer-in-Chief's Report on the Work of the Engineering Department for the Year 1 April 1952 to 31 March 1953', Post Office Engineering Department (The Post Office Archive, London).

109. Coombs in interview with Evans (see note 38); Coombs, A. W. M. (1954) 'MOSAIC', in anon. *Automatic Digital Computation: Proceedings of a Symposium Held at the National Physical Laboratory.* London: HMSO.

110. Digital facsimiles of a series of Post Office technical reports concerning MOSAIC by Coombs, Chandler, and others, are in The Turing Archive for the History of Computing <www.AlanTuring.net/mosaic>.

111. 'Engineer-in-Chief's Report on the Work of the Engineering Department for the Year 1 April 1949 to 31 March 1950', Post Office Engineering Department (The Post Office Archive, London).

112. 'Engineer-in-Chief's Report on the Work of the Engineering Department for the Year 1 April 1954 to 31 March 1955', Post Office Engineering Department (The Post Office Archive, London); Coombs, 'MOSAIC' (see note 109).

113. The 'Engineer-in-Chief's Report on the Work of the Engineering Department for the Year 1 April 1954 to 31 March 1955' (see note 112) stated 'The digital

computer "Mosaic" has been completed and handed over to the Ministry of Supply'.

114. 'Engineer-in-Chief's Report on the Work of the Engineering Department for the Year 1 April 1951 to 31 March 1952', Post Office Engineering Department (The Post Office Archive, London).

115. Coombs in interview with Evans (see note 38).

4   *The Pilot ACE at the National Physical Laboratory*[1]

James H. Wilkinson

Introduction

The ACE project at the National Physical Laboratory is of particular historical interest because of its connexion with the work of A. M. Turing. He and J. von Neumann are now universally acknowledged to be the two outstanding men of genius of the computer revolution, and it is a matter of some regret that we know so little of the exchanges which took place between them. ... Turing had been interested in automatic digital computing since the thirties when he had written his now world-famous paper 'On computable numbers, with an application to the Entscheidungsproblem'.[2] His early research had been primarily of a theoretical nature but during the war his work at the Foreign Office had given him a knowledge of pulse techniques and it was this that led to an interest in the construction of an electronic computer.

For the first few months at NPL Turing worked on this project entirely on his own, producing a comparatively detailed proposal ['Proposed Electronic Calculator'] for the Executive Committee of the NPL, and this was duly accepted. Professor Williams has referred to the reckless doubling of staff which took place at Manchester University with the addition of a second man to the team.[3] Turing proceeded with greater caution. I was recruited to NPL in May 1946 but was assigned to Turing for half time only; the other half was to be spent in the desk computing section acquiring a knowledge of numerical analysis.

When I arrived Turing was working on what he called Version V of his Automatic Computing Engine (ACE), the use of the word 'Engine' being in recognition of the pioneering work of Babbage on his Analytic Engine. ACE was to be a very large computer with a delay line storage of some 6400 words of 32 binary digits each, held in 200 long delay lines. Turing was confident that a megacycle pulse repetition rate would be perfectly practical and since each delay line stored 32 words the circulation time was 1024 μs, i.e. about one millisecond. The machine had a highly original code, though since at that time I had no other experience of digital computers this was not evident to me, and I only gradually appreciated how far out of step were projects elsewhere! Turing was obsessed with speed of operation; if consecutive instructions were to be stored in consecutive positions in a long delay line it would be possible to perform only one instruction per millisecond. To avoid this, consecutive instructions were stored in such relative positions that each emerged from a delay line just as it was required in so far as this was possible. This subsequently came to be known as 'optimum coding' though Turing never used the expression; he thought of it just as coding. The use of optimum coding made it necessary to indicate in each instruction the storage position of the next. In order further to increase the speed of operation the computer was not based on the use of a central accumulator. Each instruction represented the transfer of information from a source to a destination. Included in the sources and destinations were, of course, the 200 long delay lines but there were also a number of short delay lines storing one or two words each and these were provided with functional sources and destinations. In addition to the natural arithmetic operations all versions of the ACE included a very full set of logical operations. Possibly because of his wartime experience Turing decided quite early on to use punched card equipment for input and output.

Early programming efforts

Later in 1946 M. Woodger joined the ACE Section, followed in 1947 by D. W. Davies and G. G. Alway and finally by Miss B. Curtis and J. H. Norton. Our task was to develop the logical design of the ACE in the light of experience gained in trying to program the basic procedures in mathematical computation. Version V was quickly abandoned and replaced by Versions VI and VII which were essentially four-address code machines in which each instruction was of the form

A FUNCTION B → C, NEXT INSTRUCTION D

This code was adopted partly in order to give a closer relationship between a mathematical algorithm and its coded version and partly to give greater speed; from Turing's point of view there is little doubt that the latter was the more important consideration. In some situations it virtually replaced three single-address instructions of the form

$$A \rightarrow \text{ACCUMULATOR}; B \rightarrow \text{ACCUMULATOR, ADDING};$$
$$\text{ACCUMULATOR} \rightarrow C$$

though this example much exaggerates its overall efficiency. The extra speed was attained at the cost of extra equipment; each delay line had to be fitted with two independent sources leading into two different highways; these highways led into a 'function box' from which emerged the required function of the two sources; the result was then fed directly to the required destination. An operation of the type $A + B \rightarrow C$ took only one word time; in Version VII this had become 40 μs (a minor cycle) since the word length had been increased to 40 binary digits to accommodate the more comprehensive instruction, though by that time we were more than willing to increase the word length for purely computational reasons. The maximum rate at which instructions could be executed was one every two minor cycles, one minor cycle being required for setting up the instruction and one for executing it. An intriguing feature is that the transfer of information from A and B to C could take place for a prolonged period of up to 32 minor cycles (the period of circulation of a long delay line), the period being determined by the position of the next instruction D. This made coding excessively untidy but in some situations it was a very powerful feature. One could, for example, by means of an instruction

$$A + B \rightarrow C \text{ continued for 32 minor cycles}$$

add all 32 numbers in delay [line] A to the corresponding numbers in delay line B and send the resulting sums to delay line C. By having the carry suppression at the end of every minor cycle, every other minor cycle, or omitting it altogether, one could deal automatically with single-word numbers, double-word numbers or multi-length numbers. In this way the fullest advantage was taken of the fact that ACE was a serial machine.

A great deal of quite detailed coding was done by the ACE team in the period from 1946–1948. It included basic subroutines for such things as multi-length arithmetic (including multiplication, division and square roots), floating-point arithmetic (both single-precision and double-precision) and interval arithmetic. The subroutines for floating-point arithmetic were particularly detailed; they were coded by G. G. Alway and myself in 1947 and were for both Version V and Version VII. They were almost certainly the earliest floating-point subroutines and it is interesting that those for the Version V were essentially the same as the subroutines which were subsequently used on the Pilot ACE itself. At a time when the arithmetic provided on modern computers is often so disappointing it is interesting to recall that the subroutines included provision for accumulating inner-products in double-precision floating-point arithmetic and all rounding was immaculate. During this period Turing introduced some of the earliest automatic coding procedures and he and M. Woodger did a fair amount of work in this area. The rather complicated nature of the code perhaps provided an inducement to develop such techniques though at the same time making it more difficult to do so.

A considerable effort was made on numerical linear algebra and some of this work is included in a Progress Report on the Automatic Computing Engine published in April 1948.[4] Solutions of linear systems by the Gauss-Siedel method and by Gaussian elimination with partial pivoting are included in this report. An interesting feature of these codes is that they make rather intensive use of subroutines; the addition of two vectors, multiplication of a vector by a scalar, inner-products, etc., are all coded in this way.

The hardware effort

The decision was taken quite early on not to set up a hardware section at NPL but to sub-contract this side of the work to some other government department, preferably where there had been previous experience with pulse techniques. Some decisions are seen to have been incorrect only in retrospect; this appeared to me to be a deplorable decision even at the time. Either the chosen department would prove not to be interested in the project in which case it would obviously be a bad decision or it *would* prove to be interested in which case NPL would inevitably have lost control of an exciting project. For the first year or so at NPL Turing continued his association with two of his former colleagues who were working in the Post Office Station at Dollis Hill but obviously this could not provide any sort of basis on which to embark on such an ambitious project.

In 1947 the policy of trying to get the computer built outside was finally abandoned and an Electronics Section was set up at NPL with responsibility to R. L. Smith-Rose, Superintendent of the Radio Division. The head of the Section was H. A. Thomas, an energetic man, but unfortunately his chief interest was in industrial electronics rather than in the construction of an electronic computer. ... Turing and Thomas had virtually nothing in common and it was evident from the start that collaboration between the two men was unlikely to be satisfactory.

In January 1947 H. D. Huskey, on the advice of D. R. Hartree, came to the ACE section of Mathematics Division for a sabbatical year. Huskey had worked on the ENIAC and had considerable experience in the electronic field. Unfortunately his relationship with Turing was not particularly cordial though he cooperated extremely well with other members of the ACE section.[5] He made no secret of his views that the idea of getting the computer built elsewhere was a mistaken policy and almost from the start pressed for a more active policy on the hardware side in Mathematics Division. Turing did not oppose this idea but never fully associated himself with it, possibly because he was becoming increasingly disillusioned with progress at NPL and had decided to take a Sabbatical at King's College, Cambridge, where he was still a Fellow. Huskey finally persuaded the rest of the group to work with him on the construction of a pilot machine which for simplicity was based on Version V. This machine was christened 'The Test Assembly'. The objective was to build the smallest computer which could successfully demonstrate the feasibility of Turing's grand project. I remember taking as our objective in the design the ability to solve a system of some eight simultaneous linear equations by the elimination method. Since it was not thought of as a permanent computer the full weight was thrown on the programmer and in deciding whether or not a feature should be included the question we asked ourselves was 'Could we possibly make do without it?'.

By the time the Test Assembly was under way Huskey was already half-way through his year at NPL. He was eager to have the computer working before he left and in retrospect this was clearly an impractical objective since he was mainly dependent on the members of the ACE Section of Mathematics Division for the design. Of these, only Davies had any real background experience and Alway, Woodger and myself were just learning 'on the job'. It is not surprising that this project was unwelcome to Thomas since its success would obviously have threatened his Electronics Section. Nevertheless, considerable progress was made with the construction of the Test Assembly before it was ruled that work on it should cease and the electronics side should be left entirely to Thomas's team. By this time morale in the Mathematics group was at its lowest ebb, with Huskey disillusioned and Turing away at Cambridge. For the next few months the group worked on the production of the Progress Report on the ACE. It is a pity that we did not decide to make it more comprehensive since many of the programs we had produced were the first in this field. Turing returned briefly to NPL in May 1948 but was so dissatisfied with progress that he decided to join F. C. Williams and T. Kilburn on the Manchester project.

The Pilot ACE

In 1948 Thomas decided to leave NPL for Unilever's, which was in line with his natural preference for industrial electronics, and F. M. Colebrook of Radio Division replaced him as head of the section. It was soon obvious to Colebrook that the ACE project was in a state of complete disarray and he came to the conclusion that the lack of communication between Mathematics Division and the Electronics Section was a severe handicap to progress. He had heard of the abortive Test Assembly project and was therefore aware that we were interested in the hardware side. Shortly after his appointment he came to see me and told me it was his opinion that the enterprise was likely to founder unless something fairly decisive was done. He then made the remarkable proposal that the four of us who had been mainly concerned with the Test Assembly, Alway, Davies, Woodger and myself, should join his team temporarily and that we should all work together on the construction of some pilot machine. I was a little taken aback by the suggestion but on reflection decided that there was a lot to be said for it. It was clearly unsatisfactory to remain in Mathematics Division coding for a series of hypothetical machines and I was too fascinated by the idea of an automatic digital computer to contemplate giving it up and joining the desk-computer section. I was delighted to find that this view was shared by the other three. There remained the problem of squaring this with Womersley but fortunately I persuaded Goodwin (for whom in theory I was still working half-time!) to back the proposal and this virtually ensured its acceptance.

That Colebrook should make the suggestion, the four of us should be enthusiastic in our acceptance and that Mathematics Division should agree to our going was a remarkably improbable combination of events. There can be no doubt that it was an extremely successful experiment. The two groups were soon collaborating extremely well in spite of the previous unpromising experience. The comparative absence of friction owed a great deal to the tactful administration of F. M. Colebrook.

Early in 1949 we started on the detailed design of the Pilot ACE. In concept this owed more to the abortive Test Assembly than it was wise to emphasize. Again it was based essentially on Turing's Version V and the equipment was kept as simple as possible consistent with the objective of being able to carry out significant computations. The original design included no multiplier since optimum coding made it possible to perform a programmed multiplication in about 10 milliseconds. A very compact form of construction was decided upon, with interchangeable plug-in chassis. The logical design could be taken over without substantial modification from Version V and the earlier Test Assembly; the circuit design was undertaken by members of both groups.

By the Autumn of 1949 the completed chassis were being delivered from the NPL workshops and the assembly was under way. Initially the assembly was done by G. G. Alway and myself but quite soon E. A. Newman joined us and the three of us then worked together until the computer was in operation. The first half of the machine was assembled very rapidly but in December 1949 the main control chassis were added and progress became slower. By February 1950 sufficient number of chassis had been added for it to be capable of storing and carrying out a simple program but it was not until May that it actually did so. In 1950 E. C. Bullard had succeeded Sir Charles Darwin as Director of NPL and on his tour of inspection of the Laboratory he paid us a visit. He had obviously heard of the earlier trouble with the ACE project and was a little sceptical when we said that we would have something working 'almost any day'. We promised to let him know when we had and it was fortunate that this occurred about a week later.

Towards the end of the day on 10th May 1950 we had all the basic pulse circuits working, the control unit, one long delay line and a short delay line fitted with an additive and subtractive input. However, our only method of inserting instructions was via a set of 32 switches on which we could set up one instruction at a time in binary; our only method of output was a set of 32 lights to which a binary number could be sent. Unfortunately the amplifier on the delay line was barely adequate and the probability of remembering a pulse pattern for as long as a minute was not very high. We concocted a very elementary program consisting of a few instructions only; this took the number on the input switches, added it into the short delay line once per millisecond and put on the next light on the set of 32 output lights when the accumulator overflowed. The lights therefore came on at a speed which depended on the size of the number on the input switches. We laboriously fed this program again and again into the computer but each time the memory would fail before we could finish. On about the twentieth occasion we finally succeeded in inserting the whole program and all the lights flashed up instantaneously. We reduced the input number and they came on slowly one by one; we doubled it and the lights came on at twice the rate and we switched off for the day knowing the computer was working. It was another two days before we again reached this high peak and could convince the Director that it *was* working.

From that point onward we were under constant pressure to have a demonstration for the Press, but there was a good deal to do before this was possible and eventually we fixed on a December date.[6] Most early computer builders have a 'hard luck story' for their first demonstration day. Ours was a good luck story. For the demonstration we had three main programs, two popular and one serious. Of the popular programs the first took a date in decimal from input keys and gave out the corresponding day of the week on a cathode-ray tube while the second took a six-figure decimal number and gave its highest factor. A bottle of beer was offered to any member of the Press who could give a six-figure prime. Popular programs provide a merciless test since the slightest error is readily apparent. The serious program traced the path of rays through a complex set of lens; it was virtually impossible for anybody not intimately connected with the computer to know for certain whether this was working correctly. Up to within a few days of the demonstration period the computer had never performed the serious program correctly and we were not even certain that the program was free of errors. In the event the Pilot ACE worked virtually perfectly for the whole of the demonstration period, a level of performance which was not achieved again for some considerable time.

The Pilot ACE and DEUCE

The Pilot ACE had been designed purely as an experimental machine with the object of demonstrating the competence of the team as computer engineers. It was originally intended that when it was successfully completed a full-scale computer would be built. In the event it did not work out like this. At the time when it was successfully demonstrated it was the only electronic computer in a government department and indeed the only other working computers were EDSAC at Cambridge, SEAC at the National Bureau of Standards and an early version of the Manchester machine. We naturally came under very heavy pressure to use the Pilot ACE for serious computing. We accordingly embarked on a small set of modifications which included the addition of an automatic multiplier and improvements to the control unit which made programming a little less arduous. The computer was then put into general use and did yeoman service for a number of years.

Initially its only storage consisted of mercury delay lines which altogether held some 300 words of 32 binary digits each. This had to cover both instructions and numbers and the code was, of course, fairly inefficient in terms of storage space required for a given program. However, for a number of reasons the Pilot ACE proved to be a far more powerful computer than we initially expected. Oddly enough much of its effectiveness sprang from what initially appeared to be weaknesses resulting from the economy in equipment which dictated its design.

Optimum coding was a controversial matter at the time but much of what was said about it appears in retrospect to have been irrelevant. Of course nobody would use optimum coding if the same effect could be produced without it. However, faced with the same choice as we had then I would certainly use it. An illustration of its effectiveness is provided by the floating-point routines on Pilot ACE. Programmed floating-point arithmetic involved a considerable number of instructions and on a conventional computer such as EDSAC, which performed only one instruction per circulation time of a long delay line, it was too slow to be used for any extensive computation. On Pilot ACE it was only marginally slower than fixed point arithmetic thanks to optimum coding. The speed was further increased by the elementary nature of the multiplier. This was an entirely autonomous unit which did not even deal with signed numbers. However, both the sign correction and the manipulations of the exponents could be carried out while the multiplication was proceeding and hence this effectively took no extra time. Even double precision and triple precision floating-point routines were reasonably fast and we gained extensive experience with such computation long before it was much used elsewhere. I think it is not unreasonable to claim that the development of floating-point error analysis at NPL, which was well in advance of that elsewhere, was an indirect consequence of our use of optimum coding.

Again the very elementary way in which input and output from punched cards was organized played a decisive role in their use. A great deal of numerical linear algebra with matrices of comparatively high order was done on the Pilot ACE using storage in binary on cards. The multiplication of a vector held in the store by a matrix stored on cards was performed by reading the cards at full speed through the reader, all the computation and red tape being done in the intervals between rows of cards. This was faster than would have been possible on EDSAC or SEAC even if they had had adequate high-speed store. By putting two numbers on each row of a card we could have doubled this speed.

Since this use of the punched card equipment required the use of an operator, it indirectly encouraged user-participation generally and this was a distinctive feature of Pilot ACE operation. Speaking for myself, I gained a great deal of numerical experience from user-participation and it was this that led to my development of backward error analysis. Present day developments [1974] are again moving in this direction but they do not yet provide as satisfactory participation as we took for granted on the Pilot ACE. It is difficult to imagine greater differences in the philosophies of computer usage than between those which developed at Cambridge on EDSAC and NPL on the Pilot ACE. I have often wondered whether it was the machines which determined the development or whether it was the natural characteristics of the personnel involved which led to the development of such different computers.

During the period when the Pilot ACE was being built the English Electric Company became interested in electronic computers and a small group from the company joined us at NPL, though all the construction was done in NPL workshops. After the success of the pilot machine English Electric decided to gain experience by building engineered versions of this computer. I was not in favour of this decision since it perpetuated something that had originally been designed merely for experimental purposes and it removed any sense of urgency from the development of the full scale ACE. However, before these engineered versions were produced a magnetic drum store was added to the Pilot ACE and with this addition it was, in spite of its obvious shortcomings, a very powerful computer. In the event the engineered version, marketed under the name DEUCE, was undoubtedly a success.

A Daily Mirror cartoon (July 1952) portraying the Pilot ACE and (in order of appearance) the reporter Ruggles, Colebrook (drawings 3 and 4), Wilkinson (drawings 5 and 9–16), York (drawing 6), and Goodwin (drawings 7 and 8)—Ed.

Source: By permission of Mirrorpix. (Thanks to Mike Woodger, Tom Vickers, and Heather Wilkinson for information.)

Notes

1. Presented at the third Colloquium on the 25th Anniversary of the Stored Program Computer held at the Royal Society, London, on 12th November 1974. The text of Wilkinson's address was first printed in *The Radio and Electronic Engineer* 45 (1975), 336–340, and is reprinted by permission of the Institution of Electrical Engineers. Words in square brackets have been added by the editor of this volume; excisions (chiefly of material overlapping with other chapters) are indicated by '. . .'.

2. *Proceedings of the London Mathematical Society*, Series 2, 42 (1936–37), 230–65.

3. Williams, F. C. (1975) 'Early Computers at Manchester University', *The Radio and Electronic Engineer* 45, 237–331.

4. Wilkinson, J. H., 'Progress Report on the Automated Computing Engine'. NPL Divisional report MA/17/1024, 1948. (In the Woodger Papers, National Museum of Science and Industry, Kensington, London (catalogue reference N32/13); a digital facsimile is in the Turing Archive for the History of Computing <www.AlanTuring.net/wilkinson_report_1948>).

5. Editor's note. Huskey comments (March 2003):

   > In all my relations with Turing I found him cooperative and helpful. Considering what he was like I think I got on very well with him. Turing had no patience for stupid questions or time for casual conversation—although on the other hand if someone was working hard on a problem and having difficulty he would go to great lengths to help. In the fall of 1947 the NPL Director, Sir Charles Darwin, gave our hardware project to the Radio Division. After the traumatic meeting, Turing and I were walking back to the Mathematics Division and he said that he was sorry about the Director's action and hoped I would not feel too bad.

6. Editor's note. The Press show was held on 29–30 November 1950, with a further demonstration on 1 December for invited VIPs (Woodger's diary).

Part II
Turing and the History of Computing

5   *Turing and the computer*
B. Jack Copeland and Diane Proudfoot

The Turing machine

In his first major publication, 'On computable numbers, with an application to the Entscheidungsproblem' (1936), Turing introduced his abstract Turing machines.[1] (Turing referred to these simply as 'computing machines'—the American logician Alonzo Church dubbed them 'Turing machines'.[2]) 'On Computable Numbers' pioneered the idea essential to the modern computer—the concept of controlling a computing machine's operations by means of a program of coded instructions stored in the machine's memory. This work had a profound influence on the development in the 1940s of the electronic stored-program digital computer—an influence often neglected or denied by historians of the computer.

A Turing machine is an abstract conceptual model. It consists of a scanner and a limitless memory-tape. The tape is divided into squares, each of which may be blank or may bear a single symbol ('0' or '1', for example, or some other symbol taken from a finite alphabet). The scanner moves back and forth through the memory, examining one square at a time (the 'scanned square'). It reads the symbols on the tape and writes further symbols. The tape is both the memory and the vehicle for input and output. The tape may also contain a program of instructions. (Although the tape itself is limitless—Turing's aim was to show that there are tasks that Turing machines cannot perform, even given unlimited working memory and unlimited time—any input inscribed on the tape must consist of a finite number of symbols.)

A Turing machine has a small repertoire of basic operations: *move left one square*, *move right one square*, *print*, and *change state*. Movement is always by one square at a time. The scanner can print a symbol on the scanned square (after erasing any existing symbol). By changing its state the machine can (as Turing put it) 'remember some of the symbols which it has "seen" (scanned) previously'.[3] Turing did not specify a mechanism for changing state (Turing machines are abstractions and proposing a specific mechanism is unnecessary) but one can easily be imagined. Suppose that a device within the scanner consists of a dial with a finite number of positions, labelled 'a', 'b', 'c', and so on, each position counting as a different state. Changing state consists in shifting the dial's pointer from one labelled position to another. This device functions as a simple memory; for example, a dial with three positions can be used to record whether the square that the scanner has just vacated contained 'o' or '1', or was blank.

The operation of the machine is governed by (what Turing called) a table of instructions. He gave the following simple example.[4] A machine—call it M—begins work with an endless blank tape and with the scanner positioned over any square of the tape. M has four states, labelled 'a', 'b', 'c', and 'd', and is in state **a** when it starts work. In the table below, 'R' is an abbreviation of the instruction 'move right one square', 'P[o]' is an abbreviation of 'print o on the scanned square', and analogously 'P[1]'. The top line of the table reads: if you are in state **a** and the square you are scanning is blank, then print o on the scanned square, move right one square, and go into state **b**.

State	Scanned square	Operations	Next state
a	Blank	P[o], R	**b**
b	Blank	R	**c**
c	Blank	P[1], R	**d**
d	Blank	R	**a**

Acting in accordance with this table of instructions—or program—M prints alternating binary digits on the tape, o 1 o 1 o 1 . . ., working endlessly to the right from its starting place, leaving a blank square in between each digit.

The universal Turing machine (UTM)

The UTM is universal in that it can be programmed to carry out any calculation that could in principle be performed by a 'human computer'—a clerk who works by rote and has unlimited time and an endless supply of paper and pencils. (Before the advent of the electronic computer, many thousands of human computers were employed in business, government, and research establishments.)

The universal machine has a single, fixed table of instructions built into it ('hard-wired', so to speak, into the machine). Operating in accordance with this one fixed table, the UTM can read and execute coded instructions inscribed on its tape. This ingenious idea—the concept of controlling the function of the computing machine by storing a program of instructions in the machine's memory—is fundamental to computer science. An instruction table for carrying out a desired task is placed on the machine's tape in a suitably encoded form, the first line of the table occupying the first so many squares of the tape, the second line the next so many squares, and so on. The UTM reads the instructions and carries them out on its tape. Different programs can be inscribed on the tape, enabling the UTM to carry out any task for which a Turing-machine instruction table can be written—thus a single machine of fixed structure is able to carry out every computation that can be carried out by any Turing machine whatsoever.

In 1936 the UTM existed only as an idea. But right from the start Turing was interested in the possibility of actually building such a machine.[5] His wartime acquaintance with electronics was the key link between his earlier theoretical work and his 1945 design for an electronic stored-program digital computer.

The Church–Turing thesis

The Church–Turing thesis (also known simply as 'Church's thesis' and 'Turing's thesis') played a pivotal role in Turing's argument (in 'On Computable Numbers') that there are well-defined mathematical tasks that cannot be carried out by a rote method or algorithm.[6] One of Turing's most accessible formulations of the Church–Turing thesis is found in a report written in 1948, 'Intelligent Machinery':

> LCMs [Turing machines] can do anything that could be described as 'rule of thumb' or 'purely mechanical'.[7]

Turing remarked:

> This is sufficiently well established that it is now agreed amongst logicians that 'calculable by means of an LCM' is the correct accurate rendering of such phrases.[8]

In 'On Computable Numbers', having proved that there are tasks the universal Turing machine cannot carry out (even given unlimited time and tape),

Turing appealed to the Church–Turing thesis in moving to the conclusion that there are tasks that cannot be accomplished by means of any rote method.

The Church–Turing thesis is sometimes said to state the 'limits of machines'.[9] For example, Newell asserted:

> That there exists a most general formulation of machine and that it leads to a unique set of input–output functions has come to be called *Church's thesis.*[10]

However, this goes well beyond anything that Turing and Church said themselves. Turing's concern was not to give 'a most general formulation of machine', but to state the limits of what can be accomplished by a human being working by rote. Turing, like Church, aimed to show that some well-defined tasks cannot be carried out by a human being working in this way; it does not follow from this (and nor did they claim that it does) that there are *no machines* able to carry out these tasks.[11]

Cryptanalytic machines

Turing completed the logical design of the famous Bombe—built to break German Enigma messages—in the last months of 1939.[12] His designs were handed over to Keen at the British Tabulating Machine Company in Letchworth, where the engineering development was carried out.[13] The first Bombe, named 'Victory', was installed at the Government Code and Cypher School (GC & CS) at Bletchley Park early in 1940.[14] (An improved model—'Agnus', short for 'Agnus Dei', but later corrupted to 'Agnes' and 'Aggie'—which contained Welchman's ingenious diagonal board was installed some months later.[15]) The Bombe was a 'computing machine'—a term for any machine able to do work that could be done by a human computer—but one with a very narrow and specialized purpose, namely searching through the wheel-positions of the Enigma machine, at super-human speed, in order to find the positions at which a German message had been encrypted. The Bombe produced likely candidates, which were tested by hand on an Enigma machine (or a replica of one)—if German emerged (even a few words followed by nonsense), the candidate settings were the right ones. The Bombe was based on the electromagnetic relay, although some later versions were electronic (i.e. valve-based) and in consequence faster. Relays are small switches consisting of a moving metal rod—which opens and closes an electrical circuit—and an electrical coil, the magnetic field of which moves the rod. Electronic valves, called 'vacuum tubes' in the United States, operate very many times faster than relays, as the valve's only moving part is a beam of electrons.

During the attack on Enigma, Bletchley Park approached the Post Office Research Station at Dollis Hill in London to build a relay-based machine for use in conjunction with the Bombe. Once the Bombe had uncovered the Enigma settings for a given day, these settings were to be transferred to the proposed machine, which would then decipher messages and print out the original German text.[16] Dollis Hill sent electronic engineer Thomas Flowers to Bletchley Park. In the end, the machine Flowers built was not used, but he was soon to become one of the great figures of Second World War codebreaking. Thanks to his pre-war research, Flowers was (as he himself remarked) possibly the only person in Britain who realized that valves could be used on a large scale for high-speed digital computing.[17]

The world's first large-scale electronic digital computer, Colossus, was designed and built during 1943 by Flowers and his team at Dollis Hill, in consultation with the Cambridge mathematician Max Newman, head of the section at Bletchley Park known simply as the 'Newmanry'. (Turing attended Newman's lectures on mathematical logic at Cambridge before the war; these lectures launched Turing on the research that led to his 'On Computable Numbers'.[18]) Colossus first worked in December 1943[19] (exactly two years before the first comparable US machine, the ENIAC, was operational[20]). Colossus was used against the Lorenz cipher machine, more advanced than Enigma and introduced in 1941.[21] The British government kept Colossus secret: before the 1970s few had any idea that electronic computation had been used successfully during the Second World War, and it was not until 2000 that Britain and the United States finally declassified the complete account of Colossus' wartime role.[22] So it was that, in the decades following the war, John von Neumann and others told the world that the ENIAC was 'the first electronic computing machine'.[23]

Although Colossus possessed a certain amount of flexibility, it was very far from universal. Nor did it store instructions internally. As with the ENIAC, in order to set Colossus up for a new job it was necessary to modify by hand some of the machine's wiring, by means of switches and plugs. (During the construction of Colossus, Newman showed Flowers Turing's 'On Computable Numbers', with its key idea of storing coded instructions in memory, but Flowers, not being a mathematical logician, 'didn't really understand much of it'.[24]) Nevertheless, Flowers established decisively and for the first time that large-scale electronic computing machinery was practicable.

Flowers said that, once Turing saw Colossus in operation, it was just a matter of Turing's waiting to see what opportunity might arise to put the idea of his universal computing machine into practice.[25] There is little doubt that by 1944 Newman too had firmly in mind the possibility of building a universal machine using electronic technology. In February 1946, a few months after his appointment as Professor of Mathematics at the University of Manchester, Newman wrote to von Neumann in the United States:

> I am ... hoping to embark on a computing machine section here, having got very interested in electronic devices of this kind during the last two or three years. By about eighteen months ago I had decided to try my hand at starting up a machine unit when I got out. . . . I am of course in close touch with Turing.[26]

Turing's own opportunity came when Womersley appeared out of the blue to recruit him to the National Physical Laboratory (see Chapter 3). By then Turing had educated himself in electronic engineering (during the later part of the war he gave a series of evening lectures 'on valve theory').[27]

The ACE and the EDVAC

In the years immediately following the Second World War, the Hungarian American logician and mathematician John von Neumann, through writings and charismatic public addresses, made the concept of the stored-program digital computer widely known. Von Neumann wrote 'First Draft of a Report on the EDVAC' (see the Introduction) and subsequently directed the computer project at the Princeton Institute for Advanced Study. The ensuing machine, the IAS computer, although not the first to run in the United States (it began work in the summer of 1951[28]), was the most influential of the early US computers and the precursor to the IBM 701, the company's first mass-produced stored-program electronic computer.

Von Neumann's 'First Draft of a Report on the EDVAC' was widely read and was used as a blueprint by, among others, Wilkes, whose EDSAC at the University of Cambridge was the second stored-program electronic computer to function (in 1949). Turing certainly expected his readers to be familiar with the 'First Draft'. At the end of the first section of 'Proposed Electronic Calculator' he said:

> The present report gives a fairly complete account of the proposed calculator. It is recommended however that it be read in conjunction with J. von Neumann's '[First Draft of a] Report on the EDVAC'.

To what extent was the content of 'Proposed Electronic Calculator' influenced by the 'First Draft'? The former follows von Neumann's terminology and notation to some extent. This decision was a sensible one in the circumstances, making it likely that Turing's report would be more readily understood. In order to depict the EDVAC's logic gates, von Neumann had used a modified version of a diagrammatic notation introduced by McCulloch and Pitts in connection with neural nets.[29] Turing adopted this modified notation and in fact considerably extended it.[30] There is no doubt that Turing simply borrowed some of the more elementary material from the 'First Draft'. (For example, his diagram of an adder—figure 10 of 'Proposed Electronic Calculator'—is essentially the same as von Neumann's figure 3.[31] A newspaper report of 1946 stated that Turing 'gives credit for the donkey work on the A.C.E. to Americans'.[32]) However, Turing's logic diagrams set out detailed designs for the logical control and the arithmetic part of the calculator and go far beyond anything to be found in the 'First Draft'. The similarities between 'Proposed Electronic Calculator' and the 'First Draft' are relatively minor in comparison to the striking differences in the designs that they contain (see the Introduction and Chapter 8 'Computer Architecture and the ACE Computers'). Moreover, von Neumann's minor influence on 'Proposed Electronic Calculator' should not be allowed to mask the extent to which Turing's universal machine of 1936 was itself a fundamental influence upon von Neumann (see below).

Notoriously, the universal machine of 1936 received no explicit mention in 'Proposed Electronic Calculator'. In his chapter 'The ACE and the Shaping of British Computing' Campbell-Kelly raises the question whether the universal machine was a 'direct ancestor' of the ACE at all, emphasizing that the ACE's 'addressable memory of fixed-length binary numbers had no equivalent in the Turing Machine'. However, Turing's previously unpublished notes on memory (in Part V) cast light on this issue. The notes are fragments of a draft of 'Proposed Electronic Calculator'; in them Turing related the ACE to the universal Turing machine, explaining why the memory arrangement described in 'On Computable Numbers' could not 'be taken over as it stood to give a practical form of machine'.

Turing regarded the ACE as a 'practical version' of the UTM:

> Some years ago I was researching on what might now be described as an investigation of the theoretical possibilities and limitations of digital computing machines. I considered a type of machine which had a central mechanism, and an infinite memory which was contained on an infinite tape . . . It can be shown that a single special machine of that type can be made to do the work of all. . . . The special machine may be called the universal machine; it works in the following quite simple manner. When we have decided what machine we wish to imitate we punch a description of it on the tape of the universal machine. . . . The universal machine has only to keep looking at this description in order to find out what it should do at each stage. Thus the complexity of the machine to be imitated is concentrated in the tape and does not appear in the universal machine proper in any way. . . . [D]igital computing machines such as the ACE . . . are in fact practical versions of the universal machine. There is a certain central pool of electronic equipment, and a large memory. When any particular problem has to be handled the appropriate instructions for the computing process involved are stored in the memory of the ACE . . . [33]

A letter from Turing to the cyberneticist W. Ross Ashby again highlights the fundamental point of similarity between the ACE and the UTM:

> The ACE is in fact, analogous to the 'universal machine' described in my paper on conputable [sic] numbers . . . [W]ithout altering the design of the machine itself, it can, in theory at any rate, be used as a model of any other machine, by making it remember a suitable set of instructions. [34]

Turing's influence on von Neumann

In the secondary literature, von Neumann is often said to have invented the stored-program computer, but he repeatedly emphasized that the fundamental conception was Turing's. Von Neumann became familiar with ideas in 'On Computable Numbers' during Turing's time at Princeton (1936–8) and was to become intrigued by Turing's concept of a universal computing machine. [35] It was von Neumann who placed Turing's concept into the hands of American engineers. Stanley Frankel (the Los Alamos physicist responsible, with von Neumann and others, for mechanizing the large-scale calculations involved in the design of the atomic and hydrogen bombs) recorded von Neumann's view of the importance of 'On Computable Numbers':

> I know that in or about 1943 or '44 von Neumann was well aware of the fundamental importance of Turing's paper of 1936 'On computable numbers ...', which describes in principle the 'Universal Computer' of which every modern computer (perhaps not ENIAC as first completed but certainly all later ones) is a realization. Von Neumann introduced me to that paper and at his urging I studied it with care. Many people have acclaimed von Neumann as the 'father of the computer' (in a modern sense of the term) but I am sure that he would never have made that mistake himself. He might well be called the midwife, perhaps, but he firmly emphasized to me, and to others I am sure, that the fundamental conception is owing to Turing—insofar as not anticipated by Babbage, Lovelace, and others. In my view von Neumann's essential role was in making the world aware of these fundamental concepts introduced by Turing and of the development work carried out in the Moore school and elsewhere.[36]

In 1944, von Neumann joined the Eckert–Mauchly ENIAC group at the Moore School of Electrical Engineering at the University of Pennsylvania. (At that time he was involved in the Manhattan Project at Los Alamos, where roomfuls of clerks armed with desk calculating machines were struggling to carry out the massive calculations required by the physicists.) ENIAC—which had been under construction since 1943—was, as mentioned above, not a stored-program computer: programming consisted of re-routing cables and setting switches. Moreover, the ENIAC was far from universal, having been designed with only one very specific task in mind, the calculation of trajectories of artillery shells. Von Neumann brought his knowledge of 'On Computable Numbers' to the practical arena of the Moore School. Thanks to Turing's abstract logical work, von Neumann knew that, by making use of coded instructions stored in memory, a single machine of fixed structure can in principle carry out any task for which a program can be written. When Eckert explained his idea of using the mercury delay line as a high-speed recirculating memory, von Neumann saw that this was the means to make concrete the abstract universal computing machine of 'On Computable Numbers'.[37]

When, in 1946, von Neumann established his own project to build a stored-program computer at the Princeton Institute for Advanced Study, he gave his engineers 'On Computable Numbers' to read.[38] Bigelow, von Neumann's chief engineer and largely responsible for the engineering design of the computer built at the Institute, said:

> The person who really ... pushed the whole field ahead was von Neumann, because he understood logically what [the stored-program concept] meant in a deeper way than anybody else. ... The reason he understood it is because, among other things, he understood a good deal of the mathematical logic which was implied by the idea, due to the work of A. M. Turing ... in 1936–1937 .... Turing's [universal] machine does not sound much like a modern computer today, but nevertheless it was. It was the germinal idea. ... So ... [von Neumann] saw ... that [the ENIAC] was just the first step, and that great improvement would come.[39]

Von Neumann repeatedly emphasized the fundamental importance of 'On Computable Numbers' in lectures and in correspondence. In 1946 he wrote to the mathematician Norbert Wiener of 'the great positive contribution of Turing'—Turing's mathematical demonstration that 'one, definite mechanism can be "universal" '.[40] In 1948, in a lecture entitled 'The General and Logical Theory of Automata', von Neumann said:

> The English logician, Turing, about twelve years ago attacked the following problem. He wanted to give a general definition of what is meant by a computing automaton. ... Turing carried out a careful analysis of what mathematical processes can be effected by automata of this type. ... He ... also introduce[d] and analyse[d] the concept of a 'universal automaton' ... An automaton is 'universal' if any sequence that can be produced by any automaton at all can also be solved by this particular automaton. It will, of course, require in general a different instruction for this purpose. *The Main Result of the Turing Theory.* We might expect a priori that this is impossible. How can there be an automaton which is at least as effective as any conceivable automaton, including, for example, one of twice its size and complexity? Turing, nevertheless, proved that this is possible.[41]

The following year, in a lecture entitled 'Rigorous Theories of Control and Information', von Neumann said:

> The importance of Turing's research is just this: that if you construct an automaton right, then any additional requirements about the automaton can be handled by sufficiently elaborate instructions.

> This is only true if [the automaton] is sufficiently complicated, if it has
> reached a certain minimal level of complexity. In other words . . . there
> is a very definite finite point where an automaton of this complexity
> can, when given suitable instructions, do anything that can be done
> by automata at all.[42]

Many books on the history of computing in the United States make no
mention of Turing. No doubt this is in part explained by the absence of
any explicit reference to Turing's work in the series of technical reports in
which von Neumann, with various co-authors, set out a logical design for
an electronic stored-program digital computer.[43] Nevertheless there is evidence in these documents of von Neumann's knowledge of 'On Computable
Numbers'. For example, in the report entitled 'Preliminary Discussion
of the Logical Design of an Electronic Computing Instrument' (1946),
von Neumann and his co-authors, Burks and Goldstine—both former
members of the ENIAC group, who had joined von Neumann at the Institute
for Advanced Study—wrote the following:

> First Remarks on the Control and Code: It is easy to see by formal-
> logical methods, that there exist codes that are in abstracto adequate
> to control and cause the execution of any sequence of operations
> which are individually available in the machine and which are, in
> their entirety, conceivable by the problem planner. The really decisive
> considerations from the present point of view, in selecting a code, are
> more of a practical nature: Simplicity of the equipment demanded by
> the code, and the clarity of its application to the actually important
> problems together with the speed of its handling of those problems.[44]

Burks has confirmed that the first sentence of this passage is a reference to the
UTM.[45] (The report was not intended for formal publication and no attempt
was made to indicate those places where reference was being made to the
work of others.)

The passage just quoted is an excellent summary of the situation at that
time. In 'On Computable Numbers' Turing had shown in abstracto that, by
means of instructions expressed in the programming code of his 'standard
descriptions', a single machine of fixed structure is able to carry out any
task that a 'problem planner' is able to analyse into effective steps. By 1945,
considerations in abstracto had given way to the practical problem of devising
an equivalent programming code that could be implemented efficiently by
means of electronic circuits. Von Neumann's embryonic code appeared in

the 'First Draft'. 'Proposed Electronic Calculator' set out Turing's own very different and much more fully developed code.

The Manchester computer

The first stored-program electronic computer, the Manchester 'Baby', came to life in June 1948 in Newman's Computing Machine Laboratory at the University of Manchester.[46]

At the time of the Baby and its successor, the Manchester Mark I, the electronic engineers Williams and Kilburn, who had translated the logico-mathematical idea of the stored-program computer into hardware, were given too little credit by the mathematicians at Manchester—Williams and Kilburn were regarded as excellent engineers but not as 'ideas men'.[47] Nowadays the tables have turned too far and the triumph at Manchester is usually credited to Williams and Kilburn alone. Fortunately the words of the late Freddie Williams survive to set the record straight:

> Now let's be clear before we go any further that neither Tom Kilburn nor I knew the first thing about computers when we arrived in Manchester University ... Newman explained the whole business of how a computer works to us.[48]

> Tom Kilburn and I knew nothing about computers ... Professor Newman and Mr A. M. Turing ... knew a lot about computers ... They took us by the hand and explained how numbers could live in houses with addresses ...[49]

In an address to the Royal Society on 4 March 1948, Newman presented this very explanation:

> In modern times the idea of a universal calculating machine was independently [of Babbage] introduced by Turing ... There is provision for storing numbers, say in the scale of 2, so that each number appears as a row of, say, forty 0's and 1's in certain places or 'houses' in the machine. ... Certain of these numbers, or 'words' are read, one after another, as orders. In one possible type of machine an order consists of four numbers, for example 11, 13, 27, 4. The number 4 signifies 'add', and when control shifts to this word the 'houses' $H11$ and $H13$ will be connected to the adder as inputs, and $H27$ as output. The numbers stored in $H11$ and $H13$ pass through the adder, are added, and the sum is passed on to $H27$. The control then shifts to

the next order. In most real machines the process just described would be done by three separate orders, the first bringing $\langle H_{11} \rangle$ (=content of H_{11}) to a central accumulator, the second adding $\langle H_{13} \rangle$ into the accumulator, and the third sending the result to H_{27}; thus only one address would be required in each order. ... A machine with storage, with this automatic-telephone-exchange arrangement and with the necessary adders, subtractors and so on, is, in a sense, already a universal machine.[50]

Following this explanation of Turing's three-address concept (source 1, source 2, destination, function) Newman went on to describe program storage ('the orders shall be in a series of houses X_1, X_2, ...') and conditional branching. He then summed up:

From this highly simplified account it emerges that the essential internal parts of the machine are, first, a storage for numbers (which may also be orders). ... Secondly, adders, multipliers, etc. Thirdly, an 'automatic telephone exchange' for selecting 'houses', connecting them to the arithmetic organ, and writing the answers in other prescribed houses. Finally, means of moving control at any stage to any chosen order, if a certain condition is satisfied, otherwise passing to the next order in the normal sequence. Besides these there must be ways of setting up the machine at the outset, and extracting the final answer in useable form.[51]

In a letter written in 1972 Williams described in some detail what he and Kilburn were told by Newman:

About the middle of the year [1946] the possibility of an appointment at Manchester University arose and I had a talk with Professor Newman who was already interested in the possibility of developing computers and had acquired a grant from the Royal Society of £30,000 for this purpose. Since he understood computers and I understood electronics the possibilities of fruitful collaboration were obvious. I remember Newman giving us a few lectures in which he outlined the organisation of a computer in terms of numbers being identified by the address of the house in which they were placed and in terms of numbers being transferred from this address, one at a time, to an accumulator where each entering number was added to what was already there. At any time the number in the accumulator could be transferred back to an assigned address in the store and the accumulator cleared for further use. The transfers were to be effected by a stored program in which a list of instructions was obeyed sequentially. Ordered progress through the list could be interrupted by a test instruction which examined the sign of the number in the accumulator. Thereafter operation started from a new point in the list of instructions. This was the first information I received about the organisation of computers. . . . Our first computer was the simplest embodiment of these principles, with the sole difference that it used a subtracting rather than an adding accumulator.[52]

Turing's early input to the developments at Manchester, hinted at by Williams in his above-quoted reference to Turing, may have been via the lectures on computer design that Turing and Wilkinson gave in London during the period December 1946 to February 1947 (see Chapter 22, 'The Turing–Wilkinson Lecture Series'). The lectures were attended by representatives of various organizations planning to use or build an electronic computer. Kilburn was in the audience.[53] (Kilburn usually said, when asked from where he obtained his basic knowledge of the computer, that he could not remember;[54] e.g., in a 1992 interview he said: 'Between early 1945 and early 1947, in that period, somehow or other I knew what a digital computer was . . . Where I got this knowledge from I've no idea'.[55])

Whatever role Turing's lectures may have played in informing Kilburn, there is little doubt that credit for the Manchester computer—called the 'Newman–Williams machine' by Huskey in a report written shortly after his visit in 1947 to the Manchester project (see Chapter 23, 'The State of the Art in Electronic Digital Computing in Britain and the United States')—belongs not only to Williams and Kilburn but also to Newman, and that the influence on Newman of Turing's 1936 paper was crucial (as was the influence of Flowers' Colossus).

The Manchester computer and the EDVAC

The Baby and the Manchester Mark I are sometimes said to have descended from the EDVAC (see, e.g., the American *Family Tree of Computer Design*, fig. 1 in Chapter 6). Newman was well aware of von Neumann's 'First Draft of a Report on the EDVAC'. In the summer of 1946 he sent David Rees, a lecturer in his department at Manchester and an ex-member of the Newmanry (Newman's section at Bletchley Park), to a series of lectures at the Moore School, where Eckert, Mauchly, and other members of the ENIAC–EDVAC group publicized their ideas on computer design.[56] In the autumn of 1946 Newman himself went to Princeton for three months.[57]

Newman's advocacy of 'a central accumulator'—a characteristic feature of the EDVAC but not of the ACE—was probably influenced by his knowledge of the American proposals. However, von Neumann's ideas seem to have had little influence on other members of the Manchester project. Kilburn spoke scathingly of the von Neumann 'dictat'.[58] Tootill said:

> Williams, Kilburn and I (the three designers of the first Manchester machine) had all spent the 1939–1945 war at the Telecommunications Research Establishment doing R & D on radiolocation equipments. The main U.S. ideas that we accepted in return for our initiatives on these and later on computers were the terms 'radar' and 'memory' . . . We disliked the latter term, incidentally, as encouraging the anthropomorphic concept of 'machines that think'.[59]

> To the best of my recollection FC [Williams], Tom [Kilburn] and I never discussed . . . von Neumann's . . . ideas during the development of the Small-Scale Experimental Machine [the Baby], nor did I have any knowledge of them when I designed the Ferranti Mk I. I don't think FC was influenced at all by von Neumann, because I think he was in general quite punctilious in acknowledging other people's ideas.[60]

Tootill added:

> As well as our own ideas, we incorporated functions suggested by Turing and Newman in the improvement and extension of the first machine. When I did the logic design of the Ferranti Mark I, I got them to approve the list of functions.[61]

Turing joins the Manchester project

In May 1948 Turing resigned from the NPL. Work on the ACE had drawn almost to a standstill (see Chapter 3, 'The Origins and Development of the ACE Project'). Newman lured a 'very fed up'[62] Turing to Manchester, where in May 1948 he was appointed Deputy Director of the Computing Machine Laboratory (there being no Director). Turing designed the input mechanism and programming system[63] of, and wrote a programming manual[64] for, the full-scale Manchester computer. (The first of the production models, marketed by Ferranti, was completed in February 1951 and was the first commercially available electronic digital computer.[65] The first US commercial machine, the Eckert-Mauchly UNIVAC, appeared a few months later.) At last Turing had his hands on a stored-program computer.

Artificial Intelligence

The myth

Artificial Intelligence (AI) is often said to have been born in the mid-1950s in the United States. For example:

> Artificial Intelligence, conceived at Carnegie Tech in the autumn of 1955, quickened by Christmas, and delivered on Johnniac in the spring, made a stunning debut at the conference from which it later took its name.[66]

The AI program 'delivered on Johnniac' (a Californian copy of the IAS computer) was the Logic Theorist, written by Newell, Simon, and Shaw and demonstrated at a conference, the Dartmouth Summer Research Project on Artificial Intelligence, held at Dartmouth College, New Hampshire.[67] The Logic Theorist was designed to prove theorems from Whitehead and Russell's *Principia Mathematica*.[68] (In one case the proof devised by the Logic Theorist was several lines shorter than the one given by Whitehead and Russell; Newell, Simon, and Shaw wrote up the proof and sent it to the *Journal of Symbolic Logic*. This was almost certainly the first paper to have a computer listed as a co-author, but unfortunately it was rejected.[69])

The reality

In Britain the term 'machine intelligence' pre-dated 'artificial intelligence', and the field of enquiry itself can be traced much further back than 1955. If anywhere has a claim to be the birthplace of AI, it is Bletchley Park. Turing was the first to carry out substantial research in the area. At least as early as 1941 he was thinking about machine intelligence—in particular the possibility of computing machines that solved problems by means of searching through the space of possible solutions, guided by what would now be called 'heuristic' principles—and about the mechanization of chess.[70] At Bletchley Park, in his spare time, Turing discussed these topics and also machine learning. He circulated a typescript concerning machine intelligence among some of his colleagues.[71] Now lost, this was undoubtedly the earliest paper in the field of AI.

The first AI programs ran in Britain in 1951–2, at Manchester and Cambridge. This was due in part to the fact that the first stored-program electronic computers ran in Britain and in part to Turing's influence on the first generation of computer programmers. (Even in the United States, the

Logic Theorist was not the first AI program to run. Arthur Samuel's Checkers (or Draughts) Player first ran at the end of 1952 on the IBM 701, IBM's first stored-program electronic digital computer.[72] In 1955 Samuel added learning to the program.)

The Bombe

The Bombe is the first milestone in the history of machine intelligence.[73] Central to the Bombe was the idea of solving a problem by means of a guided mechanical search through the space of possible solutions. In this instance, the space of possible solutions consisted of configurations of the Enigma machine (in another case it might consist of configurations of a chess board). The Bombe's search could be guided in various ways; one involved what Turing called the 'multiple encipherment condition' associated with a crib (described in Chapter 6 of Turing's recently declassified *Treatise on the Enigma*, written in the second half of 1940).[74] A search guided in this fashion, Turing said, would 'reduce the possible positions to a number which can be tested by hand methods'.[75] (A crib is a word or phrase that the cryptanalyst believes might be part of the German message. For example, it might be conjectured that a certain message contains 'WETTER FUR DIE NACHT' (weather for the night). Many Enigma networks were good sources of cribs, thanks both to the stereotyped nature of German military messages and to lapses of cipher security. One station sent exactly the same message ('beacons lit as ordered') each evening for a period of several months.[76])

Modern AI researchers speak of the method of *generate-and-test*. Potential solutions to a given problem are generated by means of a guided search. These potential solutions are then tested by an auxiliary method to find out if any is actually a solution. Nowadays in AI both processes, generate and test, are typically carried out by the same program. The Bombe mechanized the first process. The testing of the potential solutions (the 'stops') was then carried out manually (by setting up a replica Enigma accordingly, typing in the cipher text, and seeing whether or not German words emerged).

Machine intelligence, 1945–8

In designing the ACE, machine intelligence was not far from Turing's thoughts—he described himself as building 'a brain'[77] and declared 'In working on the ACE I am more interested in the possibility of producing models of the action of the brain than in the practical applications to computing'.[78] In 'Proposed Electronic Calculator' he said:

> 'Can the machine play chess?' It could fairly easily be made to play a rather bad game. It would be bad because chess requires intelligence. We stated at the beginning of this section that the machine should be treated as entirely without intelligence. There are indications however that it is possible to make the machine display intelligence at the risk of its making occasional serious mistakes. By following up this aspect the machine could probably be made to play very good chess.

Turing's point was probably that the use of heuristic search brings with it the risk of the machine's sometimes making mistakes.

In February 1947 (in the rooms of the Royal Astronomical Society in Burlington House, London[79]) Turing gave what is, so far as is known, the earliest public lecture to mention computer intelligence, providing a breathtaking glimpse of a new field.[80] He described the human brain as a 'digital computing machine'[81] and discussed the prospect of machines that act intelligently, learn, and beat human opponents at chess. He stated that '[w]hat we want is a machine that can learn from experience' and that '[t]he possibility of letting the machine alter its own instructions provides the mechanism for this'.[82] (The possibility of a computer's operating on and modifying its own program as it runs, just as it operates on the data in its memory, is implicit in the stored-program concept.) At the end of the 1947 lecture Turing set out what he later called the 'Mathematical Objection' to the hypothesis of machine intelligence. This is now widely known as the Gödel argument, and has been made famous by John Lucas and Roger Penrose. (In fact the objection originated with the mathematical logician Emil Post, as early as 1921.[83]) Turing proposed an interesting and arguably correct solution to the objection.[84]

In mid-1947, with little progress on the physical construction of the ACE, a thoroughly disheartened Turing applied for a twelve-month period of sabbatical leave to be spent in Cambridge. The purpose of the leave, as described by Darwin in July 1947, was to enable Turing

> to extend his work on the [ACE] still further towards the biological side. I can best describe it by saying that hitherto the machine has been planned for work equivalent to that of the lower parts of the brain, and [Turing] wants to see how much a machine can do for the higher ones; for example, could a machine be made that could learn

by experience? This will be theoretical work, and better done away from here.[85]

Turing left the NPL for Cambridge in the autumn of 1947.[86]

In the summer of 1948 Turing completed a report describing the outcomes of this research. It was entitled 'Intelligent Machinery'.[87] Donald Michie recalls that Turing

> was in a state of some agitation about its reception by his superiors at N.P.L.: 'A bit thin for a year's time off!'.[88]

The headmasterly Darwin—who once complained about the 'smudgy' appearance of Turing's work[89]—was, as Turing predicted, displeased with 'Intelligent Machinery', describing it as a 'schoolboy's essay'[90] and 'not suitable for publication'.[91] In reality this far-sighted paper was the first manifesto of Artificial Intelligence; sadly Turing never published it.

'Intelligent Machinery' is a wide-ranging and strikingly original survey of the prospects of AI. In it Turing brilliantly introduced a number of the concepts that were later to become central in AI, in some cases after reinvention by others. These included the logic-based approach to problem-solving, now widely used in expert systems, and, in a brief passage concerning what he called 'genetical or evolutionary search'[92], the concept of a genetic algorithm—important in both AI and Artificial Life. (The term 'genetic algorithm' was introduced in *c.*1975.[93]) In the light of his work with the Bombe, it is not surprising to find Turing hypothesizing in 'Intelligent Machinery' that 'intellectual activity consists mainly of various kinds of search'.[94] Eight years later the same hypothesis was put forward independently by Newell and Simon and through their influential work[95] became one of the principal tenets of AI. 'Intelligent Machinery' also contains the earliest description of (a restricted form of) what Turing was later to call the 'imitation game' and is now known simply as the Turing test.[96]

The first AI programs

Both during and after the war Turing experimented with machine routines for playing chess: in the absence of a computer, the machine's behaviour was simulated by hand, using paper and pencil. In 1948 Turing and David Champernowne, the mathematical economist, constructed the loose system of rules dubbed the 'Turochamp'.[97] (Champernowne reported that his wife, a beginner at chess, took on the Turochamp and lost.) Turing began to program the Turochamp for the Manchester Ferranti Mark I but unfortunately never completed the task.[98] He later published a classic early article on chess programming.[99] Dietrich Prinz, who worked for Ferranti, wrote the first chess program to be implemented.[100] It ran in November 1951 on the Ferranti Mark I.[101] Unlike the Turochamp, Prinz's program could not play a complete game and operated by exhaustive search rather than under the guidance of heuristics. (Prinz 'learned all about programming the Mark I computer at seminars given by Alan Turing and Cecily Popplewell'.[102] Like Turing, he wrote a programming manual for the Mark I.[103] Prinz also used the Mark I to solve logical problems, and in 1949 and 1951 Ferranti built two small experimental special-purpose computers for theorem-proving and other logical work.[104])

Christopher Strachey's Draughts Player was—apart from Turing's 'paper' chess-players—the first AI program to use heuristic search. He coded it for the Pilot Model ACE in May 1951.[105] Strachey's first attempt to get his program running on the Pilot ACE was defeated by coding errors. When he returned to the NPL with a debugged version of the program, he found that a major hardware change had been made, with the result that the program would not run without substantial revision.[106] He finally got his program working on the Ferranti Mark I in mid-1952 (with Turing's encouragement and utilizing the latter's recently completed *Programmer's Handbook*[107]).[108] By the summer of 1952 the program could play a complete game of draughts at a reasonable speed.[109] The essentials of Strachey's program were taken over by Samuel in the United States.[110]

The first AI programs to incorporate learning, written by Anthony Oettinger at the University of Cambridge, ran in 1951.[111] Oettinger wrote his 'response learning programme' and 'shopping programme' for the Cambridge EDSAC computer. Oettinger was considerably influenced by Turing's views on machine learning,[112] and suggested that the shopping program—which simulated the behaviour of 'a small child sent on a shopping tour'[113]—could pass a version of the Turing test in which 'the questions are restricted to . . . the form "In what shop may article *j* be found?"'[114].

Turing's unorganized computing machines

Turing did not only invent the concept of the stored-program digital computer; he also pioneered the idea of computing by neural networks. The major part of 'Intelligent Machinery' is a discussion of (what Turing called) 'unorganised machines'.[115,116] His unorganized computing machines have been ignored by historians of the computer and merit a detailed introduction.

Turing described three types of unorganized machine. A-type and B-type unorganized machines consist of randomly connected two-state 'neurons' whose operation is synchronized by means of a central digital clock. We call these networks 'Turing Nets'. Turing's P-type unorganized machines are not neuron-like but are modified Turing machines: they have 'only two interfering inputs, one for "pleasure" or "reward" ... and the other for "pain" or "punishment" '.[117] Turing studied P-types in the hope of discovering procedures for 'training' a machine to carry out a task. (It is a P-type machine that Turing was speaking of when, in the course of his famous discussion of strategies for building machines to pass the Turing test, he said 'I have done some experiments with one such child-machine, and succeeded in teaching it a few things'.[118])

Turing had no doubts concerning the significance of his unorganized machines. Of Turing Nets, he said

> [M]achines of this character can behave in a very complicated manner
> when the number of units is large ... A-type unorganised machines
> are of interest as being about the simplest model of a nervous system
> with a random arrangement of neurons. It would therefore be of very
> great interest to find out something about their behaviour.[119]

He theorized that 'the cortex of the infant is an unorganised machine, which can be organised by suitable interfering training'.[120] Turing found 'this picture of the cortex as an unorganised machine ... very satisfactory from the point of view of evolution and genetics'.[121]

A-type unorganized machines

Turing introduced the idea of an unorganized machine by means of an example:

> A typical example of an unorganised machine would be as follows.
> The machine is made up from a rather large number N of similar units.
> Each unit has two input terminals, and has an output terminal which
> can be connected to the input terminals of (o or more) other units.
> We may imagine that for each integer r, $1 \leq r \leq N$, two numbers $i(r)$
> and $j(r)$ are chosen at random from $1 \ldots N$ and that we connect the
> inputs of unit r to the outputs of units $i(r)$ and $j(r)$. All of the units are
> connected to a central synchronising unit from which synchronising

pulses are emitted at more or less equal intervals of time. The times
when these pulses arrive will be called 'moments'. Each unit is capable
of having two states at each moment. These states may be called
0 and 1.[122]

Turing then gave (what would now be called) a propagation rule and an
activation rule for the network. A propagation rule calculates the net input
into a unit, and an activation rule calculates the new state of a unit, given its
net input. The propagation rule is:

The net input into unit r at moment m, net(r, m), is the product
of the state of $i(r)$ at m-1 and the state of $j(r)$ at m-1.

The activation rule is:

The state of r at m is 1-net(r, m).

A network of the sort that Turing described in the above quotation and whose
behaviour is determined by these two rules is an A-type unorganized machine.

In modern terminology an A-type machine is a collection of NAND units.
The propagation rule in effect takes the conjunction of the values on the
unit's two input lines, and the activation rule forms the negation of this
value. Alternative choices of propagation rule and/or activation rule will
cause the units to perform other Boolean operations. As is well known, NAND
is a fundamental operation in the sense that any Boolean operation can be
performed by a circuit consisting entirely of NAND units. Thus any such
operation can be performed by an A-type machine.[123]

B-type unorganized machines

The most significant aspect of Turing's discussion of unorganized machines
is undoubtedly his idea that an initially random network can be organized
to perform a specified task by means of what he described as 'interfering
training'.[124]

Many unorganised machines have configurations such that if once
that configuration is reached, and if the interference thereafter is
appropriately restricted, the machine behaves as one organised for
some definite purpose.[125]

Turing illustrated this idea by means of the circuit shown in fig. 1.[126]
(He stressed that this particular circuit is employed 'for illustrative purposes'

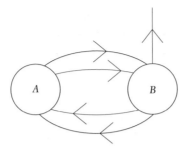

Fig. 1 Introverted pair.

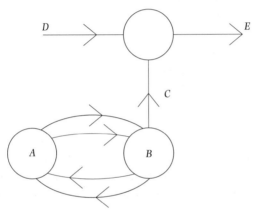

Fig. 2 Connection-modifier.

and not because it is 'of any great intrinsic importance'.[127]) We call a pair of units (*A* and *B*) connected in the way shown an 'introverted pair'. By means of external interference the state of unit *A* may be set to either 0 or 1; this state will be referred to as the 'determining condition' of the pair. The signal produced in unit *B*'s free output connection is constant from moment to moment and the polarity of the signal depends only upon the determining condition of the pair. Thus an introverted pair functions as an elementary memory.[128]

Turing defined B-type machines in terms of a certain process of substitution applied to A-type machines: a B-type results if every unit-to-unit connection within an A-type machine is replaced by the device shown in fig. 2.[129] That is to say, what is in the A-type a simple connection between points *D* and *E* now passes via the depicted device.[130]

Depending on the polarity of the constant signal at *C*, the signal at *E* is either 1 if the signal at *D* is 0 and 0 if the signal at *D* is 1, or always 1 no matter what the signal at D. In the first of these cases the device functions as a negation module. In the second case the device in effect disables the connection to which it is attached. That is to say, a unit with the device attached to one of its input connections delivers an output that is a function only of the signal arriving along its other input connection. (If the devices on both the unit's input connections are placed in disable mode then the unit's output is always 0.) Using these devices an external agent can organize an initially random B-type machine, selectively disabling and enabling connections within it.

Turing claimed that it is a 'general property of B-type machines ... that with suitable initial [i.e. determining] conditions they will do any required job, given sufficient time and provided the number of units is sufficient'.[131] This follows from the more specific claim that given 'a B-type unorganised machine with sufficient units one can find initial conditions which will make it into a universal [Turing] machine with a given storage capacity'.[132] This claim first opened up the possibility, noted by Turing,[133] that the human brain is (in part) a UTM implemented in a neural network.

B-types redefined

Concerning his claim that one can configure a B-type network (with sufficient units) such that it is a UTM with a finite tape, Turing remarked: 'A formal proof to this effect might be of some interest, or even a demonstration of it starting with a particular unorganised B-type machine, but I am not giving it as it lies rather too far outside the main argument'.[134] It is unfortunate that Turing did not give any details of the proof, for this might have cast some light on what appears to be an inconsistency in 'Intelligent Machinery'. It is reasonably obvious that not all Boolean functions can be computed by B-type machines as defined. (A good way to get a feel for the difficulty is to attempt to design a B-type circuit for computing XOR—exclusive disjunction.)

'Intelligent Machinery' contains no clues as to Turing's own procedure for dealing with this problem. The simplest remedy seems to be to modify the substitution in terms of which B-type machines are defined: a B-type results if every unit-to-unit connection within an A-type machine is replaced by *two* of the devices shown in fig. 2, linked in series. That is to say, what is in the A-type a simple connection between two units now passes through two additional units each with its own introverted pair attached. It is trivially the case that if a function can be computed by some A-type machine then it can also be computed by some machine satisfying the modified definition of a B-type.

Turing's anticipation of connectionism

So far as we have been able to discover, Turing was the first person to consider building computing machines out of simple, neuron-like elements connected together into networks in a largely random manner. His account of B-types anticipated the modern approach to AI known as connectionism (i.e. computation by neural networks).[135] Rosenblatt—the inventor of the type of neural net called the 'perceptron'—introduced the term 'connectionist' in the following way:

> [According to] theorists in the empiricist tradition ... the stored information takes the form of new connections, or transmission channels in the nervous system (or the creation of conditions which are functionally equivalent to new connections) ... The theory to be presented here takes the empiricist, or 'connectionist' position.[136]

Turing's arrangement of selectively disabling and enabling connections within a B-type machine is functionally equivalent to one in which the stored information takes the form of new connections within the network.

Turing also envisaged the procedure—nowadays used extensively by connectionists—of programming training algorithms into a computer simulation of the unorganized machine. In modern architectures repeated applications of a training algorithm (e.g. the 'back propagation' algorithm) cause the encoding of the problem solution to develop gradually within the network during the training phase. Turing had no algorithm for training his B-types.[137] He regarded the development of training algorithms for unorganized machines as a central problem. With characteristic farsightedness Turing ended his discussion of unorganized machines by sketching the research programme that connectionists are now pursuing:

> I feel that more should be done on these lines. I would like to invest-igate other types of unorganised machines ... When some electronic machines are in actual operation I hope that they will make this more feasible. It should be easy to make a model of any particular machine that one wishes to work on within such a UPCM [universal practical computing machine] instead of having to work with a paper machine as at present. If also one decided on quite definite 'teaching policies'

these could also be programmed into the machine. One would then allow the whole system to run for an appreciable period, and then break in as a kind of 'inspector of schools' and see what progress had been made.[138]

As a result of his lukewarm interest in publication Turing's work on neuron-like computation remained unknown to others working in the area. Modern connectionists regard the work of Donald Hebb[139] and Frank Rosenblatt[140] as the foundation of their approach. Turing's unorganized machines were not mentioned by the other pioneers of neuron-like computation in Britain—Ross Ashby,[141] Beurle,[142] Taylor,[143] and Uttley.[144] The situation was identical on the other side of the Atlantic. Rosenblatt seemed not to have heard of Turing's unorganized machines.[145] Nor was Turing's work mentioned in Hebb's influential book *The Organization of Behavior*— the source of the so-called Hebbian approach to neural learning studied in connectionism today.[146] Modern discussions of the history of connectionism by Rumelhart, McClelland et al.[147] show no awareness of Turing's early contribution to the field.[148]

Turing himself was unable to pursue his research into unorganized machines very far. At the time, the only electronic stored-program computer in existence was the Manchester Baby. By the time Turing had access to the Ferranti Mark I, in 1951, his interests had shifted and he devoted his time to modelling biological growth. (It was not until 1954, the year of Turing's death, that Farley and Clark, working independently of Turing at MIT, succeeded in running the first computer simulation of a small neural network.[149]) In Mathematics Division Davies and Woodger pursued Turing's ideas on learning (see Chapter 15, 'The ACE Simulator and the Cybernetic Model'). Their Cybernetic Model, constructed in 1949, was a hardware simulation of six Boolean neurons. In a demonstration on BBC TV in 1950, the Cybernetic Model mimicked simple learning in an octopus.

McCulloch and Pitts

It is interesting that Turing made no reference in the 1948 report to the work of McCulloch and Pitts, itself influenced by his 'On Computable Numbers'. Their 1943 article represents the first attempt to apply what they refer to as 'the Turing definition of computability' to the study of neuronal function.[150] McCulloch stressed the extent to which his and Pitts' work was indebted to Turing in the course of some autobiographical remarks (made during the public discussion of a lecture by von Neumann in 1948):

> I started at entirely the wrong angle ... and it was not until I saw
> Turing's paper ['On Computable Numbers'] that I began to get going
> the right way around, and with Pitts' help formulated the required
> logical calculus. What we thought we were doing (and I think we
> succeeded fairly well) was treating the brain as a Turing machine.[151]

Like Turing, McCulloch and Pitts considered Boolean nets of simple two-state 'neurons'. In their 1943 article they showed that such a net augmented by an external tape can compute all (and only) numbers that can be computed by Turing machines; and that, without the external tape, some but not all of these numbers can be computed by nets. Unlike modern connectionists, but like Turing, McCulloch and Pitts made no use of weighted connections or variable thresholds. (Part of the burden of their argument is to show that the behaviour of a net of binary units with variable thresholds can be exactly mimicked by a simple Boolean net without thresholds 'provided the exact time for impulses to pass through the whole net is not crucial'.[152]) McCulloch and Pitts did not discuss universal machines.

Turing had unquestionably heard something of the work of McCulloch and Pitts. Von Neumann mentioned the McCulloch–Pitts article—albeit very briefly—in the 'First Draft of a Report on the EDVAC' and (as noted above) employed a modified version of their diagrammatic notation for neural nets. Wiener would almost certainly have mentioned McCulloch in the course of his 'talk over the fundamental ideas of cybernetics with Mr Turing' at the NPL in the spring of 1947.[153] (Wiener and McCulloch were founding members of the cybernetics movement.) Turing and McCulloch seem not to have met until 1949. After their meeting Turing spoke dismissively of McCulloch, referring to him as 'a charlatan'.[154] It is an open question whether the work of McCulloch and Pitts had any influence whatsoever on the development of the ideas presented in 'Intelligent Machinery'. (Max Newman remarked of Turing 'It was, perhaps, a defect of his qualities that he found it hard to use the work of others, preferring to work things out for himself'.[155])

Whatever the influences were on Turing at that time, there is no doubt that his work on neural nets goes importantly beyond the earlier work of McCulloch and Pitts. The latter gave only a perfunctory discussion of learning, saying no more than that the mechanisms supposedly underlying learning in the brain—they specifically mentioned threshold change and the formation of new synapses—can be mimicked by means of nets whose connections and thresholds remain unaltered.[156] Turing's idea of using supervised interference to train an initially random arrangement of units to compute a specified function was nowhere prefigured.

P-type unorganized machines

Turing's main purpose in studying P-type machines seems to have been to search for general training procedures. A P-type machine is a modified Turing machine. Chief among the modifications is the addition of two input lines, one for reward ('pleasure') and the other for punishment ('pain').[157] Unlike standard Turing machines a P-type has no tape. Initially a P-type machine is unorganized in the sense that its instruction table is 'largely incomplete'.[158] Application of either pleasure or pain by the trainer serves to alter an incomplete table to some successor table. After sufficient training a complete table may emerge.

The P-types that Turing explicitly considered have instruction tables consisting of three columns (unlike the four-column tables of 'On Computable Numbers' described above). An example (a simplified version of Turing's own[159]) is:

State	Control symbol	External action
1	U	A
2	D0	B
3	T1	B
4	U	A
5	D1	B

'U' means 'uncertain', 'T' means 'tentative', and 'D' means 'definite'. (The nature of the external actions A and B is not specified.) This table is incomplete in that no control symbol at all is specified for states 1 and 4 and the control symbol 1 has been entered only tentatively in the line for state 3. Only in the case of states 2 and 5 are definite control symbols listed. The table is complete only when a definite control symbol has been specified for each state.

The control symbol determines the state the machine is to go into once it has performed the specified external action. The rules that Turing gave governing the state transitions are:

1. If the control symbol is 1, either definitely or tentatively, then *next state* is the remainder of $((2 \times \text{present state}) + 1)$ on division by the total number of states (in this case 5).

For example, if the machine is in state 3 then *next state* is 2.

2. If the control symbol is 0, either definitely or tentatively, then *next state* is the remainder of $(2 \times \text{present state})$ on division by the total number of states.

For example, if the machine is in state 2 then *next state* is 4.

Let us suppose that the machine is set in motion in state 2. It performs the external action B, shifts to state 4, and performs the action A. No control symbol is specified in state 4. In this case the machine selects a binary digit at random, say 0, and replaces U by T0. The choice of control symbol determines the next state, in this case 3.

The trainer may apply a pleasure or pain stimulus at any time, with the effect that '[w]hen a pain stimulus occurs all tentative entries are cancelled, and when a pleasure stimulus occurs they are all made permanent'.[160] In other words, pleasure replaces every T in the table by D and pain replaces all occurrences of T0 and T1 by U.

Turing suggested that 'it is probably possible to organise these P-type machines into universal machines' but warned that this 'is not easy'.[161] He continued:

> If, however, we supply the P-type machine with a systematic external memory this organising becomes quite feasible. Such a memory could be provided in the form of a tape, and the [external actions] could include movement to right and left along the tape, and altering the symbol on the tape to 0 or 1 . . . I have succeeded in organising such a (paper) machine into a universal machine . . . This P-type machine with external memory has, it must be admitted, considerably more 'organisation' than say the A-type unorganised machine. Nevertheless the fact that it can be organised into a universal machine still remains interesting.[162]

As a search for 'teaching policies' Turing's experiments with P-types were not a great success. The method he used to train the P-type with external memory required considerable intelligence on the part of the trainer and he described it as 'perhaps a little disappointing', remarking that '[i]t is not sufficiently analogous to the kind of process by which a child would really be taught'.[163]

Artificial Life

In his final years Turing worked on (what since 1987 is called) Artificial Life (A-Life). The central aim of A-Life is a theoretical understanding of naturally occurring biological life—in particular of the most conspicuous feature of living matter, its ability to self-organize (i.e. to develop form and structure spontaneously). A-Life characteristically makes use of computers to simulate living and life-like systems. Langton, who coined the term 'Artificial Life', wrote

> Computers should be thought of as an important laboratory tool for the study of life, substituting for the array of incubators, culture dishes, microscopes, electrophoretic gels, pipettes, centrifuges, and other assorted wet-lab paraphernalia, one simple-to-master piece of experimental equipment.[164]

Turing was the first to use computer simulation to investigate a theory of 'morphogenesis'—the development of organization and pattern in living things.[165] He began this investigation as soon as the first Ferranti Mark I to be produced was installed at Manchester University. He wrote in February 1951:

> Our new machine is to start arriving on Monday. I am hoping as one of the first jobs to do something about 'chemical embryology'. In particular I think one can account for the appearance of Fibonacci numbers in connection with fir-cones.[166]

Shortly before the Ferranti computer arrived, Turing wrote about his work on morphogenesis in a letter to the biologist J. Z. Young.[167] The letter connects Turing's work on morphogenesis with his interest in neural networks, and to some extent explains why he did not follow up his suggestion in 'Intelligent Machinery' and use the Ferranti computer to simulate his unorganized machines.

> I am afraid I am very far from the stage where I feel inclined to start asking any anatomical questions [about the brain]. According to my notions of how to set about it that will not occur until quite a late stage when I have a fairly definite theory about how things are done.

At present I am not working on the problem at all, but on my mathematical theory of embryology . . . This is yielding to treatment, and it will so far as I can see, give satisfactory explanations of -

i)   Gastrulation.

ii) Polyogonally symmetrical structures, e.g., starfish, flowers.

iii) Leaf arrangement, in particular the way the Fibonacci series (0, 1, 1, 2, 3, 5, 8, 13, . . .) comes to be involved.

iv) Colour patterns on animals, e.g., stripes, spots and dappling.

v)   Patterns on nearly spherical structures such as some Radiolaria, but this is more difficult and doubtful.

I am really doing this now because it is yielding more easily to treatment. I think it is not altogether unconnected with the other problem. The brain structure has to be one which can be achieved by the genetical embryological mechanism, and I hope that this theory that I am now working on may make clearer what restrictions this really implies. What you tell me about growth of neurons under stimulation is very interesting in this connection. It suggests means by which the neurons might be made to grow so as to form a particular circuit, rather than to reach a particular place.

In June 1954, while in the midst of this groundbreaking work, Turing died. He left a large pile of handwritten notes concerning morphogenesis, and some programs.[168] This material is still not fully understood.

Notes

1. Turing, A. M. (1936) 'On computable numbers, with an application to the Entscheidungsproblem', *Proceedings of the London Mathematical Society*, Series 2, 42, (1936–7), 230–65. Reprinted in Copeland, B. J. (ed.) (2004) *The Essential Turing*. Oxford & New York: Oxford University Press.

2. Church, A. (1937) 'Review of Turing's "On computable numbers, with an application to the Entscheidungsproblem"', *Journal of Symbolic Logic*, 2, 42–3.

3. 'On computable numbers, with an application to the Entscheidungsproblem', p. 231.

4. Ibid., p. 233.

5. Newman in interview with Christopher Evans (*The Pioneers of Computing: An Oral History of Computing*. London: Science Museum); 'Dr. A. M. Turing', *The Times*, 16 June 1954, p. 10.

6. On the Church–Turing thesis, see Copeland, B. J. (1996) 'The Church–Turing thesis', in E. Zalta (ed.), *The Stanford Encyclopedia of Philosophy* <http://plato.stanford.edu>.

7. Turing, A. M. (1948) 'Intelligent machinery' (National Physical Laboratory Report, 1948), in Copeland (ed.) *The Essential Turing* (see note 1), p. 414. (A copy of the original typescript is in the Woodger Papers, National Museum of Science and Industry, Kensington, London; a digital facsimile is in The Turing Archive for the History of Computing <www.AlanTuring.net/intelligent_machinery>.)

8. Ibid., p. 414.

9. See, for example, White, I. (1988) 'The limits and capabilities of machines—a review', *IEEE Transactions on Systems, Man, and Cybernetics*, 18, 917–38.

10. Newell, A. (1980) 'Physical symbol systems', *Cognitive Science*, 4, 135–83; the quotation is from p. 150.

11. See further Copeland, B. J. (2002) 'Hypercomputation', *Minds and Machines*, 12, 461–502; Copeland, B. J. (2000) 'Narrow versus wide mechanism', *Journal of Philosophy*, 97, 5–32 (reprinted in Scheutz, M. (ed.) (2002) *Computationalism: New Directions*. Cambridge, MA: MIT Press).

12. For a full account of Turing's involvement with the Bletchley Park codebreaking operation, see Copeland (ed.) *The Essential Turing*.

13. A memo, 'Naval Enigma Situation', dated 1 November 1939 and signed by Knox, Twinn, Welchman, and Turing, said: 'A large 30 enigma bomb [sic] machine, adapted to use for cribs, is on order and parts are being made at the British Tabulating Company'. (The memo is in the British National Archives: Public Record Office (PRO), Kew, Richmond, Surrey; document reference HW 14/2.)

14. Mahon, P. 'The History of Hut Eight, 1939–1945' (June 1945), p. 28 (PRO document reference HW 25/2; a digital facsimile of the original typescript is in The Turing Archive for the History of Computing <www.AlanTuring.net/mahon_hut_8>).

15. Welchman, G. (2000) *The Hut Six Story: Breaking the Enigma Codes*, 2nd edn., Cleobury Mortimer: M&M Baldwin; 'Squadron-Leader Jones, Section' (PRO document reference HW 3/164; thanks to Ralph Erskine for sending a copy of this document).

16. Flowers in interview with Copeland (July 1998).

17. Flowers in interview with Copeland (July 1996).

18. Newman in interview with Evans (see note 5).

19. Flowers in interview with Copeland (July 1996).

20. Goldstine, H. (1972) *The Computer from Pascal to von Neumann*. Princeton: Princeton University Press, pp. 225–6.

21. For a history of Colossus see Copeland, B. J. 'Colossus and the Dawning of the Computer Age', in R. Erskine and M. Smith (eds) (2001) *Action This Day*. London: Bantam.

22. 'General Report on Tunny, with Emphasis on Statistical Methods' (PRO document reference HW 25/4, HW 25/5 (2 Vols)). This report was written in 1945 by Good, Michie, and Timms—members of Newman's section at Bletchley Park. A digital facsimile is in The Turing Archive for the History of Computing <www.AlanTuring.net/tunny_report>.

23. Von Neumann, J. (1954) 'The NORC and problems in high speed computing', in A. H. Taub (ed.) (1961) *Collected Works of John von Neumann*, Vol 5. Oxford: Pergamon Press; the quotation is from pp. 238–9.

24. Flowers in interview with Copeland (July 1996).

25. Ibid.

26. Letter from Newman to von Neumann, 8 February 1946 (in the von Neumann Archive at the Library of Congress, Washington, DC; a digital facsimile is in The Turing Archive for the History of Computing <www.AlanTuring.net/newman_vonneumann_8feb46>).

27. Turing, S. (1959) *Alan M. Turing*. Cambridge: W. Heffer, p. 74.

28. Bigelow, J. (1980) 'Computer development at the Institute for Advanced Study', in N. Metropolis, J. Howlett, and G. C. Rota (eds), *A History of Computing in the Twentieth Century*. New York: Academic Press, p. 308.

29. McCulloch, W. S. and Pitts, W. (1943) 'A logical calculus of the ideas immanent in nervous activity', *Bulletin of Mathematical Biophysics*, 5, 115–33.

30. As noted by Hartree (1949) *Calculating Instruments and Machines*. Illinois: University of Illinois Press, pp. 97, 102.

31. Figure 3 is on p. 198 of the reprinting of the 'First Draft' in Stern, N. (1981) *From ENIAC to UNIVAC: An Appraisal of the Eckert-Mauchly Computers*. Bedford, MA: Digital Press.

32. *Evening News*, 23 December 1946. The cutting is among a number placed by Sara Turing in the Modern Archive Centre, King's College, Cambridge (catalogue reference K 5).

33. Turing, A. M. (1947) 'Lecture on the Automatic Computing Engine', in Copeland (ed.), *The Essential Turing* (see note 1); the quotation is from pp. 378, 383.

34. Turing in an undated letter to W. Ross Ashby (in the Woodger Papers (catalogue reference M11/99); a digital facsimile is in The Turing Archive for the History of Computing <www.AlanTuring.net/turing_ashby>).

35. 'I know that von Neumann was influenced by Turing ... during his Princeton stay before the war', said Stanislaw Ulam (in interview with Evans in 1976, *The Pioneers of Computing: An Oral History of Computing*. London: Science Museum). When Ulam and von Neumann were touring in Europe during the summer of 1938, von Neumann devised a mathematical game involving

Turing-machine-like descriptions of numbers (Ulam reported by William Aspray (1990) in his *John von Neumann and the Origins of Modern Computing*. Cambridge, MA: MIT Press, pp. 178, 313). The word 'intrigued' is used in this connection by von Neumann's friend and colleague Herman Goldstine (*The Computer From Pascal to von Neumann*, p. 275 (see note 20)).

36. Letter from Frankel to Brian Randell, 1972 (first published in Randell (1972) 'On Alan Turing and the origins of digital computers' in B. Meltzer and D. Michie (eds) *Machine Intelligence 7*. Edinburgh: Edinburgh University Press. (Copeland is grateful to Randell for giving him a copy of the letter.)

37. Burks (a member of the ENIAC group) summarized matters thus in his 'From ENIAC to the stored-program computer: two revolutions in computers', in Metropolis, Howlett, and Rota (eds), *A History of Computing in the Twentieth Century*.

> Pres [Eckert] and John [Mauchly] invented the circulating mercury delay line store, with enough capacity to store program information as well as data. Von Neumann created the first modern order code and worked out the logical design of an electronic computer to execute it. (p. 312)

Burks also recorded (ibid., p. 341) that von Neumann was the first of the Moore School group to see the possibility, implicit in the stored-program concept, of allowing the computer to modify selected instructions in a program as it runs (e.g. in order to control loops and branching). The same idea lay at the foundation of Turing's theory of machine learning (see below).

38. Letter from Bigelow to Copeland, 12 April 2002. See also Aspray, *John von Neumann and the Origins of Modern Computing*, p. 178.

39. Bigelow in a tape-recorded interview made in 1971 by the Smithsonian Institution and released in 2002. (Copeland is grateful to Bigelow for sending a transcript of excerpts from the interview.)

40. Letter dated 29 November 1946 (in the von Neumann Archive at the Library of Congress, Washington, DC).

41. The text of 'The general and logical theory of automata' is in Taub (ed.), *Collected Works of John von Neumann*, Vol 5; the quotation is from pp. 313–14.

42. The text of 'Rigorous theories of control and information' is printed in von Neumann, J. (1966) *Theory of Self-Reproducing Automata*. Urbana: University of Illinois Press; the quotation is from p. 50.

43. The first papers in the series were the 'First Draft of a Report on the EDVAC', by von Neumann (1945), and 'Preliminary Discussion of the Logical Design of an Electronic Computing Instrument' by Burks, Goldstine, and von Neumann (1946).

44. Section 3.1 of Burks, A. W., Goldstine, H. H., and von Neumann, J. 'Preliminary Discussion of the Logical Design of an Electronic Computing Instrument', 28 June 1946, Institute for Advanced Study; reprinted in Taub (ed.), *Collected Works of John von Neumann*, Vol 5.

45. Letter from Burks to Copeland, 22 April 1998. See also Goldstine, *The Computer from Pascal to von Neumann*, p. 258.

46. Williams (1975) described the Computing Machine Laboratory on p. 328 of his 'Early computers at Manchester University', *The Radio and Electronic Engineer*, 45, 327–31:

> It was one room in a Victorian building whose architectural features are best described as 'late lavatorial'. The walls were of brown glazed brick and the door was labelled 'Magnetism Room'.

47. Peter Hilton in interview with Copeland (June 2001).

48. Williams in interview with Evans in 1976 (*The Pioneers of Computing: An Oral History of Computing*. London: Science Museum).

49. Williams, 'Early Computers at Manchester University', p. 328.

50. Newman, M. H. A. (1948) 'General principles of the design of all-purpose computing machines', *Proceedings of the Royal Society of London*, Series A, 195, 271–4; the quotation is from pp. 271–2.

51. Ibid., pp. 273–4.

52. Letter from Williams to Randell, 1972 (in Randell 'On Alan Turing and the origins of digital computers', p. 9).

53. Bowker, G. and Giordano, R. (1993) 'Interview with Tom Kilburn', *Annals of the History of Computing*, 15, 17–32.

54. Letter from Brian Napper to Copeland, 16 June 2002.

55. Bowker and Giordano, 'Interview with Tom Kilburn', p. 19.

56. Letter from Rees to Copeland, 2 April 2001.

57. Newman, W. 'Max Newman: Mathematician, Codebreaker and Computer Pioneer', to appear.

58. Kilburn in interview with Copeland (July 1997).

59. Letter from Tootill to Copeland, 18 April 2001.

60. Letter from Tootill to Copeland, 16 May 2001.

61. Letter from Tootill to Copeland, 18 April 2001.

62. Gandy in interview with Copeland (November 1995).

63. Letter from Williams to Randell, 1972 (see note 52).

64. 'Programmers Handbook for Manchester Electronic Computer', Computing Machine Laboratory, University of Manchester (no date, *c*.1950); a digital facsimile is available in The Turing Archive for the History of Computing <www.AlanTuring.net/programmers_handbook>).

65. Lavington, S. H. (1975) *A History of Manchester Computers*. Manchester: NCC Publications, p. 20.

66. Haugeland, J. (1985) *Artificial Intelligence: The Very Idea*. Cambridge, MA: MIT Press, p. 176.

67. Newell, A., Shaw, J. C., and Simon, H. A. (1957) 'Empirical explorations with the logic theory machine: a case study in heuristics', *Proceedings of the Western Joint Computer Conference*, 15, 218–39 (reprinted in E. A. Feigenbaum and J. Feldman (eds) (1963) *Computers and Thought*. New York: McGraw-Hill).

68. Whitehead, A. N. and Russell, B. (1910) *Principia Mathematica*, Vol 1. Cambridge: Cambridge University Press.

69. Shaw in interview with Pamela McCorduck (McCorduck, P. (1979) *Machines Who Think*. New York: W. H. Freeman, p. 143).

70. Donald Michie in interview with Copeland (October 1995).

71. Donald Michie in interview with Copeland (February 1998).

72. Letter from Samuel to Copeland, 6 December 1988; Samuel, A. L. (1959) 'Some studies in machine learning using the game of checkers', *IBM Journal of Research and Development*, 3, 211–29 (reprinted in Feigenbaum and Feldman (eds), *Computers and Thought*).

73. We include in this claim the Polish Bomba, a more primitive form of the Bombe which also employed guided search (although cribs were not used). The Polish machine in effect used the heuristic 'Ignore the Stecker'. This heuristic was satisfactory during the period when the Enigma machine's stecker-board affected only 10–16 of the 26 letters of the alphabet (see Rejewski, M. (1980) 'Jak Matematycy polscy rozszyfrowali Enigme' [How the Polish mathematicians broke Enigma], *Annals of the Polish Mathematical Society, Series II: Mathematical News*, 23, 1–28; English translation in Kozaczuk, W. (1984) *Enigma: How the German Machine Cipher Was Broken, and How It Was Read by the Allies in World War Two*. London: Arms and Armour Press, trans. C. Kasparek).

74. The title 'Treatise on the Enigma' was probably added to Turing's document by a third party outside the Government Code and Cypher School, most likely in the United States. The copy of the otherwise untitled document held in the US National Archives and Records Administration (document reference RG 457, Historic Cryptographic Collection, Box 201, NR 964) is prefaced by a page typed some years later than the document itself; it is this page that bears the title 'Turing's Treatise on the Enigma'. Another copy of the document, held in the British Public Record Office (document reference HW 25/3), carries the title 'Mathematical theory of ENIGMA machine by A. M. Turing'; this too was possibly added at a later date. At Bletchley Park the document was referred to as 'Prof's Book'. (A digital facsimile of the PRO copy is in

The Turing Archive for the History of Computing <www.AlanTuring.net/profs_book>. Chapter 6 is printed in Copeland (ed.) *The Essential Turing* (see note 1).)

75. *The Essential Turing*, p. 317.
76. Mahon, 'The History of Hut Eight, 1939–1945', p. 40.
77. Turing, S. *Alan M. Turing*, p. 75; Don Bayley in interview with Copeland (December 1997).
78. Turing in a letter to W. Ross Ashby (see note 34).
79. Entry in Woodger's diary for 20 February 1947. (Copeland is grateful to Woodger for this information.)
80. 'Lecture on the Automatic Computing Engine' (see note 33).
81. Ibid., p. 382.
82. Ibid., p. 393.
83. Post, E. L. 'Absolutely unsolvable problems and relatively undecidable propositions: account of an anticipation', in M. Davis (ed.) (1965) *The Undecidable: Basic Papers On Undecidable Propositions, Unsolvable Problems And Computable Functions*. New York: Raven, pp. 417, 423.
84. See Copeland, B. J. (2000) 'The Turing test', *Minds and Machines*, 10, 519–39; Copeland, B. J. (1998) 'Turing's O-Machines, Penrose, Searle, and the brain', *Analysis*, 58, 128–38; and Piccinini, G. (2003) 'Alan Turing and the mathematical objection', *Minds and Machines*, 13, 23–48.
85. Letter from Darwin to Appleton, 23 July 1947 (PRO document reference DSIR 10/385; a digital facsimile is in The Turing Archive for the History of Computing <www.AlanTuring.net/darwin_appleton_23jul47>).
86. Probably at the end of September (see Chapter 3, note 67). Turing was on half-pay during his sabbatical (Minutes of the Executive Committee of the NPL for 28 September 1948, p. 4 (NPL library; a digital facsimile is in The Turing Archive for the History of Computing <www.AlanTuring.net/npl_minutes_sept1948>)).
87. Turing, 'Intelligent machinery' (see note 7).
88. Michie, unpublished note (in the Woodger Papers).
89. Letter from Darwin to Turing, 11 November 1947 (in the Modern Archive Centre, King's College, Cambridge (catalogue reference D 5); a digital facsimile is in The Turing Archive for the History of Computing <www.AlanTuring.net/darwin_turing_11nov47>).
90. Gandy in interview with Copeland (November 1995).
91. Minutes of the NPL Executive Committee for 28 September 1948, p. 4 (see note 86).
92. Turing, 'Intelligent machinery', p. 431.
93. Holland, J. H. (1992) *Adaptation in Natural and Artificial Systems*. Cambridge, MA: MIT Press, p. x.
94. Turing, 'Intelligent machinery', p. 431.

95. See, for example, Newell, A. and Simon, H. A. (1976) 'Computer science as empirical inquiry: symbols and search', *Communications of the Association for Computing Machinery*, 19, 113–26.

96. Turing, A. M. (1950) 'Computing machinery and intelligence', *Mind*, 59, 433–60; reprinted in Copeland (ed.), *The Essential Turing* (see note 1).

97. Letter from Champernowne (January 1980) *Computer Chess*, 4, 80–1.

98. Michie, D. (1966) 'Game-playing and game-learning automata', in L. Fox (ed.), *Advances in Programming and Non-numerical Computation*. New York: Pergamon, p. 189.

99. Turing, A. M. (1953) 'Chess', part of ch. 25 in B. V. Bowden (ed.), *Faster Than Thought*, London: Sir Isaac Pitman & Sons; reprinted in Copeland (ed.), *The Essential Turing*.

100. Prinz, D. G. (1952) 'Robot chess', *Research*, 5, 261–6.

101. Bowden, *Faster than Thought*, p. 295.

102. Gradwell, C. 'Early Days', reminiscences in a Newsletter 'For those who worked on the Manchester Mk I computers', April 1994. (Copeland is grateful to Prinz's daughter, Daniela Derbyshire, for sending him a copy of Gradwell's article.)

103. Prinz, D. G. 'Introduction to Programming on the Manchester Electronic Digital Computer', no date, Ferranti Ltd. (a digital facsimile is in The Turing Archive for the History of Computing <www.AlanTuring.net/prinz>). Turing, A. M. 'Programmers' Handbook for Manchester Electronic Computer' (see note 64).

104. Mays, W. and Prinz. D. G. (1950) 'A relay machine for the demonstration of symbolic logic', *Nature*, 165, no. 4188, 197–8; Prinz D. G. and Smith, J. B. 'Machines for the Solution of Logical Problems', in Bowden (ed.), *Faster than Thought*.

105. Letter from Strachey to Woodger, 13 May 1951 (in the Woodger Papers).

106. Letters from Woodger to Copeland, 15 July 1999 and 15 September 1999.

107. Turing, A. M. 'Programmers' Handbook for Manchester Electronic Computer' (see note 64).

108. Campbell-Kelly, M. (1985) 'Christopher Strachey, 1916–1975: a biographical note', *Annals of the History of Computing*, 7, 19–42, p. 24.

109. Strachey, C. S. (1952) 'Logical or non-mathematical programmes', *Proceedings of the Association for Computing Machinery* (Toronto, September, 1952), 46–9, p. 47.

110. Samuel 'Some studies in machine learning using the game of checkers', in Feigenbaum and Feldman (eds), p. 104 (see note 72).

111. Letter from Oettinger to Copeland, 19 June 2000; Oettinger, A. (1952) 'Programming a digital computer to learn', *Philosophical Magazine*, 43, 1243–63.

112. Oettinger in interview with Copeland (January 2000).

113. 'Programming a digital computer to learn', p. 1247 (see note 111).

114. Ibid., p. 1250.

115. This section is an edited version of our paper 'On Alan Turing's anticipation of connectionism', *Synthese*, 108 (1996), 361–77, and appears here by kind permission of Springer Science and Business Media.

116. For additional discussion of these machines, see Copeland, B. J. and Proudfoot, D. (1999) 'Alan Turing's forgotten ideas in computer science', *Scientific American*, 280, 76–81.

117. Turing, 'Intelligent machinery', p. 425 (see note 7).

118. Turing, 'Computing machinery and intelligence', p. 457 (see note 96).

119. Turing, 'Intelligent machinery', pp. 417–18.

120. Ibid., p. 424.

121. Ibid., p. 424.

122. Ibid., p. 417.

123. Each unit of an A-type circuit introduces a delay of one moment into the circuit. Suppose, for example, that the job of some particular unit U in the circuit is to compute XNANDY for some pair of specific truth-functions X and Y. The subcircuits that compute X and Y may deliver their outputs at different moments yet obviously the values of X and Y must reach U at the same moment. Turing did not say how this is to be achieved. A nowadays familiar solution to this problem—which also arises in connection with CPU design—involves the concept of a 'cycle of operation' of n moments duration. Input to the machine is held constant, or 'clamped', throughout each cycle and output is not read until the end of a cycle. Provided n is large enough, by the end of a cycle the output signal will have the desired value.

124. Turing, 'Intelligent machinery', p. 424.

125. Ibid., p. 422.

126. Ibid., p. 418.

127. Ibid., p. 418.

128. Concerning specific interfering mechanisms for changing the state of A, Turing remarked '[i]t is … not difficult to think of appropriate methods by which this could be done' and he gave one simple example of such a mechanism (ibid., pp. 422–3).

129. Turing, 'Intelligent machinery', p. 418.

130. All B-types are A-types but not vice versa.

131. Turing, 'Intelligent machinery', p. 422.

132. Ibid., p. 422.

133. Ibid., p. 424.

134. Ibid., p. 422. Such proofs have been given for a number of modern connectionist architectures, for example, by Pollack, J. B. (1987) *On Connectionist Models of Natural Language Processing* (Ph.D. Dissertation, University of Illinois, Urbana) and by Siegelmann, H. T. and Sontag, E. D. (1992) 'On the computational power

of neural nets', *Proceedings of the 5th Annual ACM Workshop on Computational Learning Theory*, 440–9. Siegelmann and Sontag establish the existence of a network capable of simulating a finite-tape universal Turing machine in linear time. They are able to give an upper bound on the size of the network: at most 1058 units are required.

135. That Turing anticipated connectionism was first suggested to us by Justin Leiber in correspondence. Leiber gives a brief discussion on pp. 117–18 and 158 of his *An Invitation to Cognitive Science*, Oxford: Basil Blackwell (1991), and on p. 59 of his 'On Turing's Turing test and why the matter matters', *Synthese*, 104 (1995), 59–69. (We cannot endorse Leiber's claim (*An Invitation to Cognitive Science*, p. 118) that Turing made use of weighted connections.)

136. Rosenblatt, F. (1958) 'The perceptron: a probabilistic model for information storage and organization in the brain', *Psychological Review*, 65, 386–408; the quotation is from p. 387.

137. The key to success in the search for training algorithms was the use of weighted connections or some equivalent device such as variable thresholds. During training the algorithm increments or decrements the values of the weights by some small fixed amount. The relatively small magnitude of the increment or decrement at each step makes possible a smooth convergence towards the desired configuration. In contrast there is nothing smooth about the atomic steps involved in training a B-type. Switching the determining condition of an introverted pair from 0 to 1 or vice versa is a savage all-or-nothing shift. Turing seems not to have considered employing weighted connections or variable thresholds.

138. Turing, 'Intelligent machinery', p. 428.

139. Hebb, D. O. (1949) *The Organization of Behavior: A Neuropsychological Theory.* New York: John Wiley.

140. Rosenblatt, F. (1957) 'The Perceptron, a Perceiving and Recognizing Automaton', Cornell Aeronautical Laboratory Report No. 85-460-1; (1958) 'The Perceptron: a Theory of Statistical Separability in Cognitive Systems', Cornell Aeronautical Laboratory Report No. VG-1196-G-1; (1958) 'The perceptron: a probabilistic model for information storage and organization in the brain' (see note 136); (1959) 'Two theorems of statistical separability in the perceptron', in (anon.) *Mechanisation of Thought Processes*, Vol 1, London: HMSO; (1962) *Principles of Neurodynamics*, Washington, DC: Spartan.

141. Ross Ashby, W. (1952) *Design for a Brain*. London: Chapman and Hall.

142. Beurle, R. L. (1957) 'Properties of a mass of cells capable of regenerating pulses', *Philosophical Transactions of the Royal Society of London*, Series B, 240, 55–94.

143. Taylor, W. K. (1956) 'Electrical simulation of some nervous system functional activities', in C. Cherry (ed.) *Information Theory*. London: Butterworths.

144. Uttley, A. M. 'Conditional probability machines and conditioned reflexes' and 'Temporal and spatial patterns in a conditional probability machine', both in C. E. Shannon and J. McCarthy (eds) (1956) *Automata Studies*, Princeton: Princeton University Press; and 'Conditional probability computing in a nervous system', in *Mechanisation of Thought Processes*, Vol 1 (see note 140).

145. Rosenblatt, *Principles of Neurodynamics*, especially pp. 5 and 12 ff.

146. Hebb, *The Organization of Behavior*.

147. Rumelhart, D. E., McClelland, J. L., and the PDP Research Group (1986) *Parallel Distributed Processing: Explorations in the Microstructure of Cognition*, Vol 1: *Foundations*, Cambridge, MA: MIT Press; see, for example, pp. 41 ff., 152 ff., 424.

148. The pioneering work of Beurle, Taylor, and Uttley has been neglected almost to the same extent as Turing's. According to connectionist folklore the field of neuron-like computation originated with Rosenblatt, influenced by McCulloch, Pitts, and Hebb. However this is incorrect. Rosenblatt recorded that the 'groundwork of perceptron theory was laid in 1957' (p. 27 of his *Principles of Neurodynamics*). A series of memoranda by Uttley concerning his probabilistic approach to neuron-like computation survives from as early as 1954 (Uttley, A. M. (1954) 'Conditional Probability Machines and Conditioned Reflexes' (RRE Memorandum No. 1045); (1954) 'The Classification of Signals in the Nervous System' (RRE Memorandum No. 1047); (1954) 'The Probability of Neural Connections' (RRE Memorandum No. 1048); (1954) 'The Stability of a Uniform Population of Neurons' (RRE Memorandum No. 1049). Published accounts of the work of Beurle, Taylor, and Uttley appeared prior to 1957 (see the references given above). Rosenblatt's work was also prefigured in the United States by that of Clark and Farley (Farley, B. G. and Clark, W. A. (1954) 'Simulation of self-organizing systems by digital computer', *Institute of Radio Engineers Transactions on Information Theory*, 4, 76–84; Clark, W. A. and Farley, B. G. (1955) 'Generalization of pattern recognition in a self-organizing system', *Proceedings of the Western Joint Computer Conference*, 86–91). In 1954 Clark and Farley simulated a network of threshold units with variable connection weights. The training algorithm (or 'modifier') that they employed to adjust the weights during learning is similar to the algorithms subsequently investigated by Rosenblatt. Rosenblatt acknowledged that 'the mechanism for pattern generalization proposed by Clark and Farley is essentially identical to that found in simple perceptrons' (*Principles of Neurodynamics*, p. 24).

149. Farley and Clark, 'Simulation of self-organizing systems by digital computer'.

150. McCulloch and Pitts, 'A logical calculus of the ideas immanent in nervous activity', p. 129 (see note 29).

151. Taub (ed.), *Collected Works of John von Neumann*, Vol 5, p. 319.

152. McCulloch and Pitts, 'A logical calculus of the ideas immanent in nervous activity', pp. 119, 123–4.

153. Wiener, N. (1948) *Cybernetics*. New York: John Wiley, p. 32.

154. Gandy in interview with Copeland (November 1995).

155. *Manchester Guardian*, 11 June 1954.

156. McCulloch and Pitts, 'A logical calculus of the ideas immanent in nervous activity', pp. 117, 124.

157. Turing considered other modifications, in particular sensory input lines and internal memory units.

158. Turing, 'Intelligent machinery', p. 425.

159. Ibid., pp. 426–7.

160. Ibid., p. 425.

161. Ibid., p. 427.

162. Ibid., pp. 427–8.

163. Ibid., p. 428.

164. Langton, C. G. (1989) 'Artificial life', in C. G. Langton (ed.), *Artificial Life: The Proceedings of an Interdisciplinary Workshop on the Synthesis and Simulation of Living Systems*. Redwood City, CA: Addison-Wesley, p. 32.

165. Turing, A. M. (1952) 'The chemical basis of morphogenesis', *Philosophical Transactions of the Royal Society of London*, Series B, 237, 37–72; reprinted in Copeland (ed.), *The Essential Turing*. Turing employed nonlinear differential equations to describe the chemical interactions hypothesized by his theory and used the Manchester computer to explore instances of such equations. He was probably the first researcher to engage in the computer-assisted exploration of nonlinear systems. It was not until Benoit Mandelbrot's discovery of the 'Mandelbrot set' in 1979 that the computer-assisted investigation of nonlinear systems gained widespread attention.

166. Letter from Turing to Woodger, undated, marked as received on 12 February 1951 (in the Woodger Papers; a digital facsimile is in The Turing Archive for the History of Computing <www.AlanTuring.net/turing_woodger_feb51>).

167. 8 February 1951. A copy of Turing's letter (typed by his mother, Sara Turing) is in the Modern Archive Centre, King's College, Cambridge (catalogue reference K1.78).

168. These are in the Modern Archive Centre, King's College, Cambridge (catalogue reference C 24–C 27).

6 The ACE and the shaping of British computing

Martin Campbell-Kelly

Origins

In 1957 an anonymous researcher at the National Science Foundation created a 'family tree' of the origins of electronic digital computers, reproduced in fig. 1.[1] This is perhaps not the place to dispute the researcher's contention that the Harvard Mark I was the root-stock out of which the modern computer grew, but all computer historians would agree that the ENIAC and the EDVAC have been afforded their proper places at the very centre of things. The reason this diagram is worth re-examining is that it presents a late 1950s view of a world of computing in which the United States had the dominant role, but Britain was a strong second player. No other nation featured. Had this diagram been drawn twenty years later, America would have remained the dominant player, but the secondary players would have included France, Germany, and Japan in addition to Britain.

Britain's early computer activity was remarkably vigorous. In the NSF tree, we can discern no less than three strong branches, representing the three major centres of UK activity—Manchester University, Cambridge University, and the National Physical Laboratory (NPL).[2] Each of these computing activities is shown as descending from the ENIAC–EDVAC line. In the case of Manchester and Cambridge, this was unequivocally true. But for the NPL's ACE, the history is not so straightforward, and it is this history that this book explores and documents.

Before proceeding, some historical background on the US scene is needed.[3] The ENIAC computer, which first operated in November 1945, was built at the Moore School of Electrical Engineering at the University of Pennsylvania.

Fig. 1 The NSF 'family tree' of computer design.

The project was begun in April 1943, and the two principal inventors were J. Presper Eckert and John Mauchly. The ENIAC was an engineering feat for which Eckert as chief engineer has been justly lauded. However, the architectural design of the ENIAC had several shortcomings. One failing was that it was not a general-purpose machine, being designed solely for the integration of ordinary differential equations (ODEs) for its military sponsor, the US Army Ballistics Research Laboratory at the Aberdeen Proving Ground, Maryland. Another shortcoming of the ENIAC was that despite its vast number of tubes (valves)—about 18,000—it could store only 20 numbers. Further, the machine was 'programmed' by inserting hundreds of plug cords into the machine, so that photographs are reminiscent of a manual telephone exchange without the people. It was said that it could take up to three weeks to set up and debug an ENIAC program.

These limitations of the ENIAC were already apparent when, in August 1944, the great American mathematician John von Neumann began to take an interest in the Moore School's computing activity. Von Neumann's interest was due, not least, to the fact that he was a consultant to the Manhattan Project for the creation of the atomic bomb, and his work on shaped explosive charges involved the integration of very large systems of partial differential equations (PDEs).[4] Because of its feeble storage capacity, the ENIAC was quite unsuitable for solving PDEs except by the most convoluted processes.[5] When John von Neumann arrived on the scene, the construction of the ENIAC had passed the point of no return, despite its technical shortcomings, and discussions had already moved on to what kind of a machine should succeed it. Thus von Neumann was one of a group of five involved in these discussions—the others were Eckert and Mauchly, Lt Herman H. Goldstine, a mathematician who was the Ballistics Research Laboratory's liaison officer, and Arthur Burks, a logician also involved with the ENIAC. The discussions went on for the best part of year, and they were finally summarized in the *First Draft of a Report on the EDVAC* dated June 1945.[6] This report was the foundation on which today's world of computers rests. The stored-program computer has been as central to twentieth-century information processing as the internal combustion engine has been to transport. Because von Neumann was the sole author of the *EDVAC Report*, the contribution of his four colleagues has been slighted and to speak of the 'von Neumann architecture' does a considerable historical injustice.

The EDVAC design consisted of five functional units: a control unit, an arithmetic/logic unit, a memory, and input and output devices. The memory was to be very large—of the order of 8K words—and it would store both programs and data. The fact that a program would be able to process its own instructions was central to the design, and this arrangement later became known as the 'stored-program concept'. Diagrams showing the logical inter-connection of the functional units of a computer appear in every basic text book on computer science. What is less well appreciated is that von Neumann drew heavily on a brain metaphor, and the *EDVAC Report* refers to neur-ones, memory, and input/output 'organs'. Only the term 'memory' survives to remind us of the original metaphor. The idea of computer-as-brain was also one that appealed to Alan Turing.

The ENIAC was inaugurated in February 1946, and the event was of wide-spread scientific interest. As a result of this publicity the Moore School was deluged with requests for information. Aware of the need to diffuse the new ideas, but unwilling to have day-to-day operations disrupted by a stream of visitors, the School organized a summer school in July and August 1946.[7] The course was attended by a representative from all the computer groups that were interested—almost exclusively in the United States and Britain. From Britain, Maurice Wilkes attended on behalf of Cambridge University and David Rees for Manchester. The Moore School Lectures had the effect of establishing the stored-program computer as the big idea in computing, and after the lectures computing groups in America and Britain set to work to turn the design into a reality.

As noted earlier, the NSF family tree of fig. 1 was a late 1950s view of computer development. As we move forward to the present day, to under-stand the genesis of the ACE, especially, we have to introduce two factors of which the NSF researcher was unaware: the Colossus codebreaking computer constructed at Bletchley Park during the war, and the universal Turing machine conceived by Turing before the war (see Chapter 5, 'Turing and the Computer').

We know from the many histories that have been written about Bletchley Park that Turing was not directly involved in the design of the Colossus computer, although he had been central to the earlier electromechanical codebreaking machines, the 'Bombes.'[8] It can be argued that although Tur-ing had no direct involvement in the Colossus, his sphere of influence must have encompassed it. However, this is perhaps not an issue we need to address here, because the Colossus was a special purpose computer for codebreak-ing, just as the ENIAC was a special purpose machine for solving ODEs. The modern computer is based on the EDVAC, which owed very little to the design of ENIAC and absolutely nothing to the Colossus. The EDVAC was *sui generis*. My own view is that the Colossus's legacy was not architectural or intellectual, but cultural and only in Britain. People who had worked on the Colossus knew that it was possible to construct very large assemblies of electronic tubes for information processing, and there was a cohort of people in the British Post Office who had actually done it. While the people who worked on the Colossus were bound in utmost secrecy, there was no proscription to their benign optimism regarding the feasibility of constructing computing machines.

The place of the universal Turing machine in the history of computing is difficult to assess at the present time. This is particularly so in Britain where Turing has become lionized, and many people feel that the Turing machine just *has* to be one of the foundations of computing. The story is very complex, and we will probably not resolve it for many years. In all my researches on the early history of computing, I cannot recall a direct reference to the Turing machine in the context of computer design. My own view is that the Turing machine was really a fairly abstruse branch of mathematics that was only appropriated by computing about a decade after the first machines were built when computer scientists first began to study computability. Jack Copeland and Diane Proudfoot present a different view in Chapter 5, 'Turing and the Computer'. None of this is to take away anything from Turing's conception, which remains one of the towering intellectual triumphs of the twentieth century. And of course, the Turing machine had a direct influence on Turing's *other* machine, the ACE.[9]

Three centres, three computers

One can identify approximately a dozen British research organizations that attempted to build stored-program computers in the decade 1945–55 (Table 1). Of these, three were the first movers that started computer projects in the immediate postwar years, 1945–7: Manchester University, Cambridge University, and the NPL. These three centres operated completely independently of one another, and as a result came up with three quite different computer architectures, programming styles, and operating regimes. The other British computer groups were, to a greater or lesser extent, influenced by one of the original three.

Table 1 Early British computer groups, 1945–55

Research group	Computer	Date	Notes
Cambridge University	EDSAC	1949	First 'practical' stored-program computer
Manchester University	Baby Machine	1948	First operational stored-program computer
	Mark I	1949	Prototype of Ferranti Mark I
	Meg	1954	Prototype of Ferranti Mercury
National Physical Laboratory	Pilot ACE	1950	Prototype of English Electric DEUCE
Birkbeck College, University of London	ARC	1948	Based on IAS Princeton computer
	SEC	1952	Re-engineered ARC
	APE( )C	1953	Prototype of BTM HEC
Telecommunications Research Establishment	TREAC	1953	First 'parallel' British computer
Post Office Research Station, and Radar Research and Development Establishment	MOSAIC	1952	Based on NPL ACE computer
British Tabulating Machine Co.	HEC	1953	Prototype of 1200 series computers
Elliott Brothers	Nicholas	1952	Experimental nickel-delay-line computer
	401	1953	Prototype of 402 computer
English Electric	DEUCE	1954	Based on NPL Pilot ACE
Ferranti	Mark I	1951	Based on Manchester Mark I
	Mark I-Star	1953	Based on Manchester Mark I
J. Lyons and Co.	LEO	1951	Based on Cambridge EDSAC

Source: Campbell-Kelly, M. (1989) *ICL: A Business and Technical History*. Oxford University Press, p. 164.

All three centres were led by one or more outstanding scientist–engineers, their contributions subsequently being recognized by their election to the Royal Society. All three centres made lasting contributions to the development of electronic stored-program computing—Manchester University primarily in hardware innovations, Cambridge University in software, and the NPL in mathematical computation. All three centres produced commercial spin-off machines. And, finally, all three centres became foci of national computer activity and the venue for one of the three early British computer conferences, held in 1949, 1951, and 1953.[10]

The first problem facing all the early computer groups was constructing a reliable electronic memory. Williams and Kilburn devised a system based on a cathode-ray tube, around which they built a small test computer, the Manchester 'Baby' Machine.[11] That machine ran its first program on 21 June 1948. It was the first program for the world's first stored-program computer. The storage system—known as the Williams Tube memory—was patented and used in several commercial computers, including the IBM 701. After this first success, Williams withdrew from computer research leaving the field open to Tom Kilburn. Kilburn quickly turned the Baby Machine into a full-scale prototype, which was subsequently manufactured by the local electrical engineering firm of Ferranti. The Ferranti Mark I—first delivered in February 1951—was the world's first commercially manufactured computer. The Mark I introduced several hardware innovations, the most important of which was the B-line, subsequently known as the index register, an invention used in virtually every computer built to the present day. The Mark I also introduced memory 'pages' (a term probably coined by Turing) which was one of the ideas underlying virtual memory a few years later.[12]

Computing activity at Cambridge University had begun before the war with the formation of a Mathematical Laboratory by Sir John Lennard-Jones, professor of chemistry, in 1937; Maurice Wilkes was appointed assistant director.[13] During the war the Mathematical Laboratory was put on hold, and Wilkes was conscripted for work on operations research and radar electronics. For Wilkes, this turned out to be the perfect background for building computers after the war. The Mathematical Laboratory was re-opened in 1946, and during the summer of that year Wilkes attended the Moore School Lectures. Returning home on the *Queen Mary* he began the design of the machine that was to become the EDSAC. The name EDSAC, a slightly contrived acronym standing for Electronic Delay Storage Automatic Calculator, was chosen as conscious echo and tribute to the EDVAC on which it was directly based.

Wilkes was more interested in using, rather than building, computers and he opted for a conservative design that he hoped would quickly result in a working computer on which his group could solve real problems. For example the EDSAC used a pulse repetition rate of 0.5 Mc/s when 'any electronic engineer worth his salt' would have tried for 1 Mc/s.[14] As a result of this conservative design policy, EDSAC ran its first program on 6 May 1949. Unlike the earlier Manchester Baby, the EDSAC was a full-scale machine capable of running realistic problems from the first day of operation. A programming system was quickly devised for the EDSAC, and a regular computing service was offered from early 1950. The programming system, largely the work of Wilkes, his research student David Wheeler, and research officer Stanley Gill, was subsequently described in the classic textbook *The Preparation of Programs for an Electronic Digital Computer*, more commonly known as 'Wilkes, Wheeler, and Gill'—or WWG for short.[15]

Turing began to work on the design of the ACE computer about June 1945. The name ACE, an acronym for Automatic Computing Engine, was chosen as a conscious tribute to Babbage's calculating engines—Babbage's name, it should be noted, was honoured by the British computing community long before his worldwide renaissance in the 1950s. While we can be quite certain that Babbage's calculating engines had no influence on the design of the ACE, one of the questions to be explored here is how much of the ACE design was derived from the *EDVAC Report* and how much from the universal Turing machine. In his well-known lecture to the London Mathematical Society (LMS) in February 1947 Turing stated:

> Some years ago I was researching on what might now be described as an investigation of the theoretical possibilities and limitations of digital computing machines. I considered a type of machine which had a central mechanism, and an infinite memory which was contained on an infinite tape. ... It was essential in these theoretical arguments that the memory should be infinite. It can easily be shown that otherwise the machine could only execute periodic operations. Machines such as the ACE may be regarded as practical versions of this same type of machine.[16]

This suggests that the Turing machine was a rather direct ancestor of the ACE. However, when one studies the text of Turing's lecture closely, it is clear that there is much more EDVAC in it than Turing machine. Further, in the *ACE Report* Turing advises the reader that:

> The present report gives a fairly complete account of the proposed calculator. It is recommended that it be read in conjunction with J. von Neumann's 'Report on EDVAC'.

If one examines the ACE and EDVAC reports side by side, it is plain that much of Turing's notation as well as some aspects of the design of the ACE are derived from the EDVAC. For example, Turing adopted von Neumann's computer–brain metaphor, and his anthropomorphic terms such as 'organ' and 'memory'. The use of 'logical elements' is derivative, and some of Turing's diagrams are almost identical to von Neumann's. Likewise, the addressable memory of fixed-length binary numbers had no equivalent in the Turing machine. Although Turing did not explicitly acknowledge the fact, like everyone else he accepted the *EDVAC Report* as the definitive blueprint for practical electronic computers.

Most computer designers made some enhancement to the EDVAC design. As noted above, the Manchester group added the index register to the EDVAC design. In the United States, Eckert and Mauchly's UNIVAC used binary coded decimal numbers rather than straight binary. There are many other examples of these incremental enhancements of the EDVAC design. Even the EDVAC, when it was finally completed in 1953, had improved on the original design. This phenomenon, of steadily improving a foundation invention, occurs in most technological fields and historians of technology sometimes speak of radical versus incremental innovations.

However, if one compares the ACE with the EDVAC—and one searches for the right way to phrase this eloquently—they were much *more* different than were the Manchester Mark I and the EDVAC, or the UNIVAC and the EDVAC. Perhaps the ACE's most significant departure from the EDVAC design was the way that programs were arranged in the memory. Instructions were laid out sequentially in the EDVAC, whereas in the ACE they could be laid out in any order, because each instruction nominated its successor. This is something rather close to the way in which 'tables' were specified for the Turing machine. Other fundamental differences between the ACE and the EDVAC are described in Chapter 7 and Chapter 8.

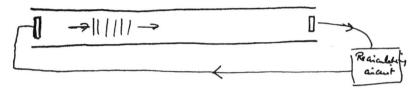

Fig. 2 A delay line as sketched by A. M. Turing.

At the heart of the ACE—optimum coding

The most important single idea in the ACE, and what differentiated it from its contemporaries, was the idea of optimum coding. The term 'optimum coding', it should be noted, was not used until the mid-1950s or later, and Turing had no vocabulary for the idea.

The ACE design of 1946 was based around a mercury delay-line memory. In his LMS lecture, Turing attributed the application of the mercury delay line for computer storage to Eckert (and this was one of the few occasions where Eckert's contribution was explicitly recognized).[17] Eckert had developed the mercury-filled delay line in connection with a moving target indicator (MTI) device for radar, a research project originally quite unconnected with computer development.[18] However, while the ENIAC was under construction, it became clear that the delay line would be far superior to the electronic-tube-based store of the ENIAC, and it was fundamental to the EDVAC design. At the time of the LMS lecture, February 1947, Turing had recently returned from the Harvard *Symposium on Large-Scale Computing Machinery*, where the recorded discussions indicate that he had a strong and perceptive technical interest in memory technology.[19]

As proposed by Eckert, and as subsequently implemented on most serial computers, the delay line consisted of a 5 ft tube of mercury. Figure 2 shows Turing's sketch of a delay line, taken from the manuscript of his LMS lecture. The acoustic delay of a sound pulse transmitted through the tube of mercury was approximately 1 millisecond, and given a pulse repetition rate of 1 microsecond, it was possible to store approximately a thousand (or more conveniently 1024) sonic pulses in the tube before they emerged at the other end. Pulses emerging from the delay line could be regenerated and re-injected into the system. In this way it was possible to store 1024 bits indefinitely. The contents of a delay line were divided up into equal length 'words'. In the case of the ACE it was planned to have 32 words of 32 bits in each delay line. The 1 millisecond recirculation period was known in the Moore School terminology (which Turing adopted) as a major cycle, while the period occupied by a single word was known as a minor cycle.

More memory could be incorporated in the computer by having more delay lines, all operating in parallel. Von Neumann had proposed having 256 delay lines for the EDVAC giving about 8K words of storage, while Turing wanted 200 for the ACE. Both were far too ambitious and the Pilot ACE actually had less than a dozen full-length delay lines, each containing thirty-two 32-bit words.

The problem with the delay-line memory was its 'latency'—the time it took for a word chosen at random to emerge from the end. On average this took half a major cycle. In a classic single-address machine, an instruction such as 'add the number in location n into the accumulator' took an average time of one major cycle—half a major cycle to fetch the instruction, and half a major cycle to fetch the operand. This gave a theoretical maximum instruction execution rate of one per major cycle. The EDSAC, for example, which had a 1 millisecond major cycle, had an instruction rate of about 650 instructions per second. Although the Manchester Mark I had a CRT-based memory which was random access in principle, in practice because of the need to regenerate the memory each instruction cycle, its speed was much the same as a delay-line machine—about 1000 instructions per second.

Turing came up with the brilliant idea of eliminating the effect of latency by having each instruction nominate the address of its successor. In this way, instead of placing instructions in successive memory locations, as in a conventional design, instructions could be placed in optimal locations, so that as one instruction finished, the next instruction would *just* be emerging from a delay line. The same could be done for operands, so that the time to access them could in principle be eliminated as well. This gave the Pilot ACE, with its 1 millisecond major cycle, a theoretical maximum speed of 16,000 instructions per second. Another innovation in the Pilot ACE was to have several short delay lines that could be used to hold the most frequently used operands in a program. This arrangement reduced the need to access the main memory for operands, and hence the need for their optimal placement; while this did not make the machine any faster theoretically, it greatly simplified programming. Even with this innovation, however, it was very difficult to get perfect optimization, and programs typically achieved about 5000 instructions per second. Nevertheless, this was very much faster than either the EDSAC or the Manchester Mark I, and the Pilot ACE used far less equipment than either of them. (Turing was in fact thoroughly dismissive of what he regarded as the unoriginal and pedestrian approaches of Wilkes at Cambridge and Williams and Kilburn at Manchester.)

The issue of optimum coding was hotly debated, and Turing's view was by no means unchallenged. While optimum coding enabled more computing to the pound-sterling, as it were, this came at the cost of greatly complicating the process of programming. In his 1967 Turing Award lecture, Wilkes noted:

> I felt that this kind of human ingenuity was misplaced as a long-term investment, since sooner or later we would have true random access memories. We therefore did not have anything to do with optimum coding in Cambridge.[20]

It is rather difficult to separate the reality from the rhetoric in these early exchanges on optimum programming. First, the Cambridge and Manchester University groups were led by formidable personalities and one cannot discount their natural human tendency to assert that their own way of doing things was the best for their particular circumstances. Second, the evidence for the superior performance on the Pilot ACE was largely anecdotal. No benchmark program had ever been written to run on all three machines. And while it was true that all three machines had similar subroutine libraries and application programs (such as integrating differential equations or inverting matrices) their performance had never been systematically compared.

This was one of the questions I set out to explore in my first foray into computer history in the 1970s. I developed a simulator for each of the three British machines—the EDSAC, the Manchester Mark I, and the Pilot ACE—with the aim of systematically comparing their performance. The benchmark program was the TPK algorithm devised by Donald Knuth and his research student L. Trabb Pardo.[21] The TPK algorithm had originally been devised for comparing a group of early programming languages, in exactly the way I was now proposing to compare three early machines. We do not need to go into the details of the TPK algorithm here, other than to say that it was designed to exercise all the capabilities needed in a program—loops, array accessing, subroutines (both library and user written), and input–output procedures. I like to think of the TPK algorithm as being the programming equivalent of the basket of groceries one uses to compare supermarkets.

The results of this experiment are shown in Table 2. The significant line of the table is the processor time. We see that the Pilot ACE took just 5 seconds to perform a calculation that took 24 seconds on the EDSAC and 37 seconds on the Manchester Mark I—giving the Pilot ACE a speed advantage of a

Table 2 Computing speeds of the Pilot ACE, the EDSAC and the Manchester Mark I with the TPK algorithm

Time (sec)	Pilot ACE	EDSAC	Mark I
Processor	5	24	37
Input–output	17	70	44
Total	21	94	82

Table 3 Computing speeds of the Pilot ACE, the EDSAC and the Manchester Mark I with floating point operations

Time (ms)	Pilot ACE	EDSAC	Mark I
Add/subtract	8	90	60
Multiply	6	105	80
Divide	34	140	150

factor of 5 and 7, respectively. Even when the slow input–output operations are included, the Pilot ACE was still faster by a factor of 4. Table 3 makes the same point in a different way. None of the three machines had hardware floating-point arithmetic and so had to use library subroutines for floating-point operations. The table shows the execution times for the four basic floating-point arithmetic subroutines for each machine. The results are quite astounding, the Pilot ACE being more than 10 times as fast as its rivals (except for the less-used operation of division). It is important to note that these subroutines were written to be as efficient as possible and were honed to perfection by the best programmers the groups had. In all cases the subroutines were as close to optimal as any human could reasonably make them.

Finally, one should note that that the Pilot ACE used 800 electronic tubes compared with the 3000 and 3700 tubes of the Cambridge and Manchester machines, respectively. If we can allow the tube count to stand as a proxy for the cost of a processor, it would be reasonable to argue that in terms of raw computing power, the Pilot ACE was at least 10 times more cost effective than the EDSAC or the Mark I. Most of this extra performance was due to optimum programming. Perhaps we will never know the precise genesis of optimum programming, but the parallel with the Turing machine, in which each instruction also nominated its successor, must surely be more than a coincidence. I would venture to suggest that if Turing had not devised the Turing machine before the war, he might not have designed the ACE in the way he did after the war.

The ACE legacy

Turing's *ACE Report* spawned several derivative machines. A partial genealogy is shown in the NSF family tree of fig. 1, although David Yates's excellent institutional history of computing at the NPL, *Turing's Legacy*, gives a more exhaustive picture (fig. 3).[22]

Two full-scale ACE-type machines were built. The first to be completed was the MOSAIC (Ministry of Supply Automatic Integrator and Computer) which

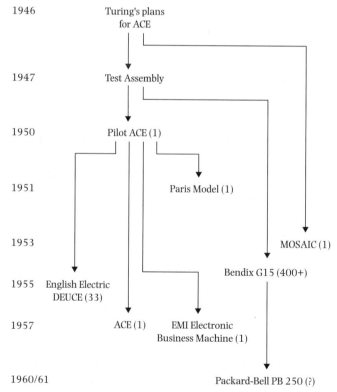

Fig. 3 The ACE family of computers.

was built by the Post Office for the Radar Research and Development Establishment over the period 1947–53. The MOSAIC was one of the largest early British computers, with over 6000 tubes, and was heavily used for data reduction applications. The second machine, the NPL ACE itself, was completed in 1959 (see the Chapter 3, 'The Origins and Development of the ACE Project'). Design of the ACE had begun in earnest in 1953, when the pilot machine was complete, although physical construction did not get started until 1956. By that date it was already apparent to some within the NPL (including Donald Davies) that cyclic memories were a dying technology and the project should have been cancelled. Unfortunately, such is the nature of government bureaucracies, that there was no way of diverting the resources to some more promising development and the ACE stumbled along under its own inertia. When finally completed it was a machine of awesome proportions that remained in use until 1967.

Commercially, the most important machines were those descending from the Test Assembly and the Pilot ACE. In the United States two machines derived from the Test Assembly—the Bendix G15 and the Packard-Bell 250. The G15 was designed by Huskey (see Chapter 13, 'The ACE Test Assembly, the Pilot ACE, the Big ACE, and the Bendix G15'). A comparatively small machine, some 400 G15s were sold from 1955 onwards, many to universities and medium-sized engineering corporations, making it one of the workhorses of the first generation of American stored-program computers. Although the G15 used a drum store rather than a mercury delay line memory, it employed the technique of optimum coding and had a conceptually similar instruction format to the Pilot ACE. In the early 1960s the G15 inspired another American computer, the Packard-Bell 250, of which historians have so far traced few concrete details.

The Pilot ACE had two significant descendants besides the full-scale ACE.[23] EMI produced a 'business machine' under the leadership of Ronald Clayden, who had previously been one of the English Electric team that worked on the Pilot ACE and DEUCE. Like the G15, the EMI Business Machine was drum-based. Only one machine was built, however, the EMI line quickly evolving into transistor-based, core-memory machines in the early 1960s.

Far and away the most important descendant of the Pilot ACE line was the English Electric DEUCE, which can be fairly described as one of the cornerstones of the British computer industry. The DEUCE owed its conception to the presence of Sir George Nelson, chairman of English Electric, on the NPL

Executive Committee. In 1951, after the Pilot ACE had been completed, English Electric decided it would like to manufacture a fully engineered version of the machine, and seconded two of the brightest young engineers at its Nelson Research Laboratories in Stafford to the project—George Davis and Ronald Clayden. The DEUCE was somewhat larger than the Pilot ACE since it included a hardware divider, resulting in a tube count of 1450, 50 per cent more than the Pilot ACE. As it happened the management of the Nelson Research Laboratories was quite bureaucratic, requiring its engineers to produce frequent written reports. Thanks to those reports written in the 1950s, today we have a unique insight into how the technology was transferred during this formative period in the development of the British computer industry. The first DEUCEs were completed in 1955, with machines being produced for English Electric, NPL, and the Royal Aircraft Establishment.

The DEUCE was a considerable commercial success, some 33 machines being produced between 1955 and 1962 (see Chapter 14, 'The DEUCE—a User's View'). This success was primarily due to the superb numerical software produced by the NPL. Perhaps the best known and most widely used program was the GIP (General Interpretive Programme) matrix suite produced by Brian Munday, first for the Pilot ACE and subsequently transferred to the DEUCE.[24] The DEUCE arrived at an opportune moment for aircraft design. The Comet, the first commercial jet aircraft, had experienced widely reported crashes in 1952 and 1953. An official inquiry established that wing 'flutter' during take-off and flight was a potential cause of instability, and from 1954 flutter calculations became a statutory requirement in aircraft design. This led to a number of DEUCE sales to aircraft organizations—the Royal Aircraft Establishment, Short Brothers and Harland, Bristol Aircraft, and of course to English Electric, itself a major aircraft producer. DEUCEs were also popular with engineering and scientific organizations that had heavy computational requirements, such as the National Engineering Laboratory, the Atomic Weapons Research Establishment, and British Petroleum. A number of machines were sold to universities, including Glasgow, Liverpool, and Queen's University Belfast. All of these were major academic centres for numerical computation, and this was the primary reason they bought DEUCEs rather than its more user-friendly competitors.

The DEUCE succeeded in the market place because of its excellent performance in numerical work, itself the consequence of the optimum coding. It was widely anticipated that the arrival of random access core memory would cause the demise of optimum programming. While this may have been true in the United States, in Britain the demise was caused less by technological advance than by a brilliant computer design due to Christopher Strachey—the Pegasus.

Strachey was a school teacher at Harrow School, London, when he made his entrée into computing in 1951. Strachey was a personal friend of Mike Woodger of the NPL Mathematics Division, and when the Pilot ACE began to operate Strachey started to use the machine at weekends. Strachey was, characteristically, more interested in programming logical problems than mathematical ones, and he worked on a draughts-playing program.[25] The Pilot ACE's memory was too small to make much progress, so he transferred his allegiance to the much larger Manchester Mark I. However, in his time with the Pilot ACE Strachey became intrigued with the possibility of writing a program so that the Pilot ACE could do its own optimum coding. He wrote to Woodger in 1951:

> I am sufficiently hopeful about the possibility of making a practical program-coding routine to continue looking into it a bit more, and even if it finally turns out to be of no practical use, it will be an interesting example of making the machine do a logical rather than a mathematical operation.[26]

Strachey worked on the problem on and off for two years, without success. However, there is no question that the experience had a major bearing on the the design of the Pegasus.

In 1952, Strachey was rescued from obscurity as a schoolmaster to become a research officer for the National Research Development Corporation (NRDC). There in 1954 he became principal designer of a medium-sized general purpose computer, to be manufactured by Ferranti. Within the contemporary technological and design constraints, the obvious course was for Strachey to produce a drum-based, optimum-coded machine of a similar type to the Bendix G15, the IBM 650, or several others. However, Strachey came up with a design *tour de force*. It would be out of place to go into the architectural details of the Pegasus here, but suffice it to say that he eliminated optimum coding entirely by a number of ingenious design innovations, producing an easy to use machine with a very respectable performance, and costing approximately the same as the DEUCE.[27]

While the Pegasus was only half as fast as the DEUCE, this was achieved without any use of optimum coding. Indeed, the Pegasus was something of a dream machine to program. The Pegasus' programming systems, whose design was not encumbered by the constraints of optimum programming, marked a high point in the early development of computer programming in Britain. As a result the Pegasus was a much more popular machine for pedagogical purposes, and outsold the DEUCE in British universities.

Of course, whether the Pegasus had come along or not, machines based on optimum programming were doomed to extinction. But it is fitting that in Britain this came about through design rather than the mere advance of technology. Although Turing never lived to see the Pegasus, I think it is fair to speculate that he would have admired Strachey's work, as being—to paraphrase Turing's famous memo to Womersley[28]—in the British tradition of solving one's difficulties by thought rather than equipment.

End of the ACE line

The period 1950 to the early 1960s was an astonishingly fertile time for the British computer industry. In 1963 there were over two dozen computer models available from eight manufacturers (Table 4). Whereas in 1950 a computer had been simply a computer, by the early 1960s the market had fragmented by machine size, application domain, and technological generation. Computers were classified into several categories of size, typically small, medium, large, very large, and giant. In Table 4, small computers were those costing less than about £35,000; medium machines were those costing from £35,000 up to about £150,000; very large computers cost in excess of £250,000; and giant machines, of which the Atlas was the only British example, cost upwards of £1 million. The application domain could be scientific, commercial, or process control. Finally a machine could be classified according to the generation of electronics technology employed: tube-based first generation, or discrete transistor-based second generation. Within this classification scheme, the DEUCE was a first generation, medium-sized scientific computer, and its main competitor was the Pegasus.

Not all manufacturers could compete in all the possible classes of computer, and in the early 1960s English Electric decided to specialize in the development of medium- and large-size scientific machines—the KDF6 and KDF9—and a small process control computer, the KDN2. (The very large KDP10, costing £400,000, was in fact an RCA model 501 manufactured under licence from the American company. This allowed English

Table 4 UK manufactured computers, 1951–63

Manufacturer	First delivery	Model	Average price (£000s)	Generation	Type	Number sold
AEI	1960	1010	250	2	C	B
Elliott-Automation	1955	402	25	1	S	B
	1956	405	125	1	C	C
	1958	802	17	2	P	B
	1959	803	35	2	P/S	E
	1961	503	80	2	C	C
EMI	1959	1100	180	2	C	C
	1961	2400	600	2	C	A
English Electric	1955	DEUCE	50	1	S	C
	1961	KDP10	400	2	C	B
	1962	KDN2	20	2	P	B
	1963	KDF6	60	2	S	B
	1963	KDF9	120	2	S	C
Ferranti	1951	Mark I/I*	45	1	S	B
	1956	Pegasus	50	1	S	C
	1957	Mercury	120	1	S	C
	1959	Pegasus II	120	1	C	B
	1959	Perseus	150	1	C	A
	1961	Sirius	17	1	S	B
	1961	Atlas	2000	2	S	A
	1963	Orion	300	2	C	A
Leo	1957	LEO II	95	1	C	B
	1962	LEO III	200	2	C	C
ICT	1955	HEC 2M	25	1	S	B
	1956	1201	33	1	C	D
	1959	1202	45	1	C	B
	1961	1301	100	2	C	E
STC	1958	Zebra	28	1	S	C

Notes: The table excludes one-of-a-kind machines and prototypes, and imported machines.

Type:	C = commercial	Sales:	A = 5 or less	D = 51–100
	S = scientific		B = 6–15	E = over 100.
	P = process control.		C = 16–50	

Source: Campbell-Kelly, M. (1989) *ICL: A Business and Technical History*. Oxford University Press, p. 216.

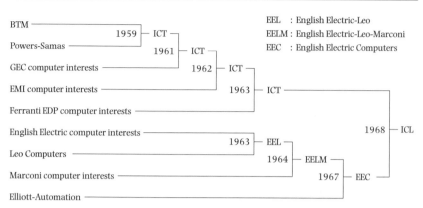

Fig. 4 Evolution of ICL.

Electric to offer a high-end commercial data processing machine with relatively little R&D effort.)

The KDF9 was based on transistor technology and core memory, and can be regarded as the second-generation successor to the first-generation DEUCE. However, the design of the KDF9 owed almost nothing tangible to the DEUCE or the NPL line. This was because the core memory it employed was truly random access and so optimum programming was no longer needed. The KDF9 was, nonetheless, a truly radical machine, as radical in its way as the ACE. It had a zero-address, stack-based architecture, which made it very much a computer scientist's machine. So, although the KDF9 cannot be said to have evolved technologically from the ACE, one can perhaps say that in its architectural radicalism it captured something of its spirit.

The period 1959 to 1963 saw a wave of consolidation in the British computer industry (fig. 4). The very fertility of the industry had resulted in a situation where there were far too many manufacturers and computer models for the relatively small British market. Most manufacturers were losing money on computers, and the escalation of development costs for second-generation machines caused them all to re-evaluate their forward plans. They had a stark choice: either to invest heavily in new machine designs, with inevitable short term losses, or to sell out to one of the other manufacturers. The process of consolidation began in 1959 with the merger of the two British punched-card machine manufacturers BTM and Powers-Samas to form ICT. In the next three years GEC, EMI, and Ferranti all threw in their hands, and ICT—which very much needed to build up its electronics expertise—was a keen buyer in every case.

While English Electric did not need to enhance its electronics capability, it lacked expertise in business data processing. Hence, when J. Lyons decided to sell Leo Computers, a merger with the English Electric computer division made good strategic sense. The merger took place in April 1963, and the enlarged company was named English Electric Leo. Work quickly began on the design of a new range of machines, code-named Project KLX, which would harmonize the KDF9 and LEO III computer lines.[29]

However, on 7 April 1964 IBM astounded the computer world with the announcement of its System/360 range of computers. The System/360 development—heralded as IBM's $5 billion dollar gamble—raised the stakes in the computer industry, and it was clear within English Electric Leo that Project KLX could not hope to compete with IBM's initiative. The company therefore decided to license RCA technology, as it had earlier done with the KDP10. In September 1964, RCA announced its Spectra 70 range of computers, a computer family that was architecturally compatible with the IBM System/360. Shortly afterwards, English Electric announced its System 4 computer series. With System 4, entirely derived from RCA, any vestige of a technical legacy from the ACE line was finally lost.

The arrival of System/360 was not the only trauma for the British computer industry in 1964. In October of that year, Harold Wilson's Labour Government came into power. Wilson wrote in his memoirs:

> My frequent meetings with leading scientists, technologists and indus-
> trialists in the last two or three years of Opposition had convinced me
> that, if action was not taken quickly, the British computer industry
> would rapidly cease to exist, facing, as was the case in other European
> countries the most formidable competition from the American giant.
> When on the evening we took office, I asked Frank Cousins to become
> the first Minister of Technology, I told him that he had, in my view,
> about a month to save the computer industry and that this must be
> his first priority.[30]

The Labour Government completed the consolidation that had already been underway since 1959. First, English Electric absorbed the computer interests of Marconi in 1964. Then, in 1967, English Electric and Elliott-Automation rationalized their computer interests: English Electric took over all mainframe-based computing, while Elliott-Automation took over process control. Finally, in 1968, in the grandest merger of all, ICT and English

Electric Computers merged to form ICL, Britain's only mainframe computer manufacturer.

ICL was the largest non-American computer firm in the world, with a headcount of 34,000. Of these workers, 23,500 came from ICT and 9500 from English Electric Computers. In effect, the ACE legacy represented more than a quarter of the British computer industry. One could characterize the English Electric faction within ICL, still based in Kidsgrove, Staffordshire, as its radical arm: free thinking and lateral thinking. Compared with the stuffy punched-card machine majority, who were doubtless more business like and sales minded, the English Electric newcomers were a breath of fresh air. The English Electric contingent within ICL retained a distinct cultural identity until well into the 1980s.

To conclude on the ACE's legacy to British computing, let me turn to the counter-factual. What if the ACE had never existed? How would the British computer industry have evolved? We can say with some certainty that, first, a major part of the British computer industry of the 1950s— the English Electric computer division—would simply never have existed. Second, Britain's 'national champion' computer firm ICL would have been a very different animal, less colourful and less idiosyncratic. Very possibly, within a risk-averse ICL unleavened by English Electric, ICL's 2900 series of the 1970s—a truly radical computer design—would have been some feeble imitation of an American design. In November 1990, Fujitsu acquired ICL largely on account of its mainframe design, which had the potential to be more cost-effective by far than the IBM-compatible mainframes designed in the 1960s. And had Fujitsu not acquired ICL—well, who knows?

Notes

1. A detailed history of the NSF Family Tree is given in Cohen, I. B. (1999) *Howard Aiken: Portrait of a Computer Pioneer*. Cambridge, MA: MIT Press, pp. 294–6.
2. A careful study of the diagram will also reveal the APEXC computer built at Birkbeck College by A. D. Booth, and which subsequently became the prototype for the very successful BTM 1200 series of computers. However, Birkbeck College was not nearly so influential as Manchester, Cambridge, or the NPL.
3. Goldstine, H. H. (1972) *The Computer from Pascal to von Neumann*. Princeton, NJ: Princeton University Press. Stern, N. (1981) *From ENIAC to UNIVAC: An Appraisal of the Eckert-Mauchly Computers*. Bedford, MA: Digital Press.

4. Aspray, W. F. (1990) *John von Neumann and the Origins of Modern Computing.* Cambridge, MA: MIT Press.

5. The ENIAC was in fact used for solving PDEs after the war by making use of punched cards for backing storage, but it was not an elegant procedure.

6. Von Neumann, J. *First Draft of a Report on the EDVAC*, Moore School of Electrical Engineering, University of Pennsylvania, 30 June 1945. Reprinted in *Annals of the History of Computing*, 5 (1993), 27–75.

7. Campbell-Kelly, M. and Williams, M. R. (eds) (1985) *The Moore School Lectures*, Charles Babbage Reprint Series for the History of Computing, Vol. 9. Cambridge, MA: MIT Press and Los Angeles, CA: Tomash Publishers.

8. Randell, B. (1980) 'The Colossus,' in N. Metropolis, J. Howlett, and G. C. Rota (eds), *A History of Computing in the Twentieth Century.* New York: Academic Press.

9. Doran, R. W. and Carpenter, B. E. (1977) 'The other Turing machine', *Computer Journal*, 20(3), 269–79.

10. Campbell-Kelly, M. and Williams, M. R. (eds) (1989) *The Early British Computer Conferences*, Charles Babbage Reprint Series for the History of Computing, Vol. 14. Cambridge, MA: MIT Press and Los Angeles, CA: Tomash Publishers.

11. Lavington, S. H. (1975) *History of Manchester Computers.* Manchester: NCC Publications.

12. Turing, A. M. 'Programmers' Handbook for the Manchester Electronic Computer Mark II', Computing Machine Laboratory, Manchester University, *c.*March 1951. A digital facsimile is in The Turing Archive for the History of Computing <www.AlanTuring.net/ programmers_handbook>.

13. Wilkes, M. V. (1985) *Memoirs of a Computer Pioneer.* Cambridge, MA: MIT Press.

14. Ibid., p. 129.

15. Wilkes, M. V., Wheeler, D. J., and Gill, S. (1951) *The Preparation of Programs for an Electronic Digital Computer*, Cambridge, MA: Addison-Wesley. Reprinted with an Introduction by Martin Campbell-Kelly as Vol. 1 (1982) of the Charles Babbage Institute Reprint Series for the History of Computing. Cambridge, MA: MIT Press and Los Angeles, CA: Tomash Publishers.

16. Turing, A. M. (1947) 'Lecture to the London Mathematical Society on 20 February 1947', in B. E. Carpenter and R. W. Doran (eds) (1986) *A.M. Turing's ACE Report of 1946 and Other Papers*, Charles Babbage Institute Reprint Series for the History of Computing, Vol. 10. Cambridge, MA: MIT Press and Los Angeles, CA: Tomash Publishers, pp. 106–7.

17. Turing, 'Lecture to the London Mathematical Society on 20 February 1947', p. 108.

18. Goldstine, *The Computer from Pascal to von Neumann*, pp. 188–9.

19. Proceedings of a Symposium of Large-Scale Digital Calculating Machinery, Harvard University Computation Laboratory, 7–10 January 1947. Reprinted with

an Introduction by William Aspray as Vol. 7 (1985) of the Charles Babbage Institute Reprint Series for the History of Computing. Cambridge, MA: MIT Press and Los Angeles, CA: Tomash Publishers.

20. Wilkes, M. V. (1968) 'Computers then and now: 1967 Turing Lecture', *Journal of the ACM*, 15, 1–7.

21. Knuth, D. E. and Pardo, L. T. (1980) 'The early development of programming languages', in Metropolis, Howlett, and Rota (eds) *A History of Computing in the Twentieth Century*.

22. Yates, D. M. (1997) *Turing's Legacy: A History of Computing at the National Physical Laboratory 1945–1995*. London: Science Museum.

23. The 'Paris Model' was an NPL-built pilot machine of little historical significance. See Yates, *Turing's Legacy*, p. 38.

24. Campbell-Kelly, M. (1981) 'Programming the Pilot ACE: early programming activity at the National Physical Laboratory', *Annals of the History of Computing*, 3, 133–68, p. 156.

25. Campbell-Kelly, M. (1985) 'Christopher Strachey 1916–1975: a biographical note', *Annals of the History of Computing*, 7, 19–42.

26. Campbell-Kelly, 'Programming the Pilot ACE', p. 154.

27. For a comprehensive and entertaining historical account of the Pegasus see: Lavington, S. H. (2000) *The Pegasus Story: A History of a Vintage Computer*. London: Science Museum.

28. See Chapter 3, 'The Origins and Development of the ACE Project'.

29. Campbell-Kelly, M. (1989) *ICL: A Business and Technical History*. Oxford: Oxford University Press, p. 240.

30. Wilson, H. (1971) *The Labour Government 1964–1970*. London: Weidenfeld and Nicolson and Michael Joseph, p. 8.

7 From Turing machine to 'electronic brain'

Teresa Numerico

Introduction

This chapter analyses the relationships and differences between the various machines that Alan M. Turing invented, projected, created, and programmed, in his work in fields ranging from logic and the theory of computability to cryptanalysis, computer science and artificial intelligence. Turing's apparent position in his article 'On Computable Numbers' was one of complete trust in the ability of a machine's 'table of instructions' to achieve whatever (computable) result the 'programmer' desired. Nine years later, working on the design of the ACE, he put forward a rather different view, concerning the ability of the stored-program electronic computer to perform 'intelligent' tasks. The first surviving document to show Turing's new attitude towards machines was his 'Proposed Electronic Calculator'.[1] The machine that Turing proposed there was different from both the universal Turing machine and the machine outlined in von Neumann's 'First Draft'.[2] I compare Turing's project with von Neumann's in order to underline similarities and—above all—the differences.

The logical background

Logic, and especially the Hilbertian formalist school, was Turing's principal research subject. The undecidability result of 'On Computable Numbers'[3] contributed to the end of the 'Hilbert program'. According to Hilbert, it was possible to create a formal system that proved all mathematical theorems. Gödel's[4] and Turing's results awoke mathematicians from this dream. In addition to this negative result, 'On Computable Numbers' gave birth to a new field, the theory of computability:

> It was stated above that 'a function is effectively calculable if its values can be found by some purely mechanical process'. We may take this statement literally, understanding by a purely mechanical process one which could be carried out by a machine. It is possible to give a mathematical description, in a certain normal form, of the structures of these machines.[5]

In 1937–8 Turing tried to repair, in a sense, the problems created by Gödel's incompleteness theorems for the formalistic approach to mathematics. Following Gödel's procedure and results, Turing proposed the construction of a succession of 'ordinal logics'. Each of the ordinal logics in the series—one for each ordinal number, up to some transfinite ordinal—was supposed to include the Gödel statement ('I am unprovable in *l*') for the previous logic *l* in the series. This entire succession was supposed to be complete. Reflecting on the meaning of his ordinal logics and the role of intuition and heuristics in mathematical discovery, Turing said:

> In pre-Gödel time it was thought by some that it would probably be possible to carry this programme to such a point that all the intuitive judgements of mathematics could be replaced by a finite number of these rules. The necessity for intuition would then be entirely eliminated. . . .
>
> In consequence of the impossibility of finding a formal logic which wholly eliminates the necessity of using intuition, we naturally turn to 'non-constructive' systems of logic with which not all the steps in a proof are mechanical, some being intuitive.[6]

According to Turing, it was unlikely that interesting theorems would be demonstrated in mathematics without the use of intuition and heuristics.

In a letter to M. H. A. Newman Turing explained the conclusions that he drew from his work on ordinal logics and emphasized the importance of his concept of the 'consequences of an assumption'.

> I think one wants to distinguish two ideas a) consequence of an assumption; b) consequence of an assumption relative to a set of rules of procedure.

> The first of these is an 'intuitive' idea which one tries to approximate
> by the second with suitable sets of rules of procedure. . . . The idea of
> consequences of an assumption relative to given rules of procedure,
> I think explains itself. One tries of course to make the rules of proced-
> ure such that the consequences will be consequences in the sense a),
> but also b) gets as many consequences as are consistent with this.
> Of course one cannot get all such a) with one set of rules.[7]

This position was really innovative. Turing realized that no group of rules
could ever be general enough to include all the possible consequences that
could in principle be drawn from an assumption. According to this approach
the intuitive notion of 'consequence of an assumption' is always broader than
any given system of logic. Turing's new perspective transcended the narrow
borders of the formalist school. In the same letter to Newman, he said:

> I think that you take a much more radically Hilbertian attitude about
> mathematics than I do. . . . If you take this attitude (and it is this one
> that seems to me so extreme . . .) there is little more to be said: we
> simply have to get used to the technique of this machine and resign
> ourselves to the fact that there are some problems to which we can
> never get the answer. . . . However I don't think you really hold quite
> this attitude because you admit that in the case of the Gödel example
> one can decide that the formula is true, i.e. you admit that there is
> a fairly definite idea of a true formula which is quite different from
> the idea of a provable one.

The logic that Turing suggested, and consequently the machine based
on it, had a new structure: it could make mistakes, and it used heuristics,
strategies, and intuition in order to solve problems.

Codebreaking during the Second World War

As is now well known, Turing was a leading figure in the Enigma decryption
project at Bletchley Park. I will not enter too deeply into the details of the
decoding effort but will mention one key element.

Turing's important hand-method known as *Banburismus*, dating from the
end of 1939, eliminated as many wrong possibilities as possible, so that the
manual–mechanical test involving the Bombe had to deal with fewer cases
than would be covered in a brute force exhaustive search. Banburismus was
the system used to break 'Home Waters' Enigma until July 1943.[8] Using an

inferential method it was possible to deduce the 'distance' between the encrypted letters and the real letters, and so to shortlist the possibilities of the Enigma machine's 'wheel order' (to know the wheel order is to know which of the machine's various wheels had been used to encrypt the message and in what order). Banburismus was based on a mathematical system devised by Turing for scoring the probabilities of the distances between letters. By this means it was possible to rule out most of the 336 possible wheel orders. To test all the 336 possible wheel orders would have used more Bombe time than there was available; when Banburismus was successful, it meant that the message could be decrypted before the intelligence it contained was out of date.

This was a new kind of problem-solving procedure, lacking the certainty of mathematical proof. The key benefit lay in the speed of testing and the exclusion of a large number of incorrect solutions. The trade-off was the lack of certainty—the procedure might not work for a given message. Turing accepted the risks because he took the correct attitude: a quick result was better than no result or a certain result too late.

I believe that Turing learnt three lessons from his experience with Enigma:

1. The use of probabilistic techniques. The scoring system underlying Banburismus was later generalized for use against other systems of encryption. The modern term for it is 'sequential analysis'. Probabilistic techniques were essential in Turing's later research in morphogenesis (see Chapter 5, 'Turing and the Computer').

2. The importance of 'just in time' processes. If the Bletchley codebreakers took two weeks to break a message, the intelligence would be useless. In theoretical machines (such as the Turing machine), the process had to be feasible only in principle—it was irrelevant how much time was required. In real world situations—such as codebreaking—a procedure has to be able to solve the problem quickly enough. Mistakes could be acceptable if the trade-off was an increase in execution speed. It was better to have a program that might make some mistakes but which terminated in a reasonable amount of time than to have a precise table of instructions too large to be executed in real time.

3. The importance of sharing information on an interdisciplinary basis. During the war Turing was in contact with groups who were in charge of some of the most advanced technological developments in the United Kingdom, Poland, and the United States.

ACE: the first months

The project of building and programming an electronic general-purpose computer was Turing's inevitable challenge in the years that followed the conflict.

At the beginning of the summer of 1945 Turing had a meeting with John Womersley of the National Physical Laboratory. During this meeting Turing saw von Neumann's 'First Draft' for the first time. Womersley was one of the few in the United Kingdom to receive it from Goldstine. Turing accepted the offer of a job immediately, but could not start until 1 October 1945, when he was released from his war work. It is likely, however, that he started working on the ACE soon after the meeting, because 'Proposed Electronic Calculator' was ready before the end of 1945.

Womersley was a very good manager, according to the documentary evidence and various sources of testimony. He showed considerable diplomacy and was talented in building the right team. Turing's whereabouts at the end of the Second World War were secret but, aware of Turing's work in the theory of computability, Womersley succeeded in recruiting him for Maths Division, having approached Newman for information. Womersley's diplomacy in presenting and supporting projects was well known and considerably appreciated. He was also a good 'marketing manager'; the name 'Automatic Computing Engine' was his. Even Turing, who despised Womersley, had to admit that this name was very well chosen. Donald Davies said about Womersley:

> I think he was a much misunderstood man . . . he was an extraordinarily good manager, adept at fighting for his cause. . . . Womersley, in his inimitable way, sold the project to the Executive Committee. The decision was delayed twice asking for more information, but by May 1946 (not a bad delay) it had been agreed. Womersley proposed to have a Pilot Model Stage which Turing was quite in favour of.[9]

'Proposed Electronic Calculator' was presented to the Executive Committee of the NPL in March 1946 together with a memorandum written by Womersley in February. The following extract from the memorandum gives something of the flavour of those pioneering times.

> The research programme of the Mathematics Division contains an item 'To explore the application of switching methods (mechanical, electrical and electronic) to computations of all kinds.' The first step

in this exploration was the visit of the newly-appointed Superintend-ent to the U.S.A. in February, 1945, to gain acquaintance of recent American developments in this field, and later Dr. A. M. Turing was appointed to the staff of the Division, and began to consider the pos-sibilities of electronic methods, in the light of recent progress in the U.S.A. Dr. Turing has now completed a long report, which makes def-inite proposals for the construction of a machine, capable of solving a wide variety of problems at speeds hitherto unattainable.

The recent development in the United States have been in the direction of automatic operation. . . . It is upon the plans for this last machine [the EDVAC], not as yet constructed, that Dr Turing's proposals are based. Some of the basic ideas are given in Professor von Neumann's 'Report on the EDVAC' a secret report of the Applied Mathematics Panel of the N.R.D.C.,[10] but it contains a number of ideas which are Dr. Turing's own, and which are to be found in a paper published by him in the Proc. Lond. Math. Soc. 1937 . . .

The general case for having such a machine is based upon:

(a) the terms of reference of the Division, which is charged with the duty of machine development,

(b) the fact that with the aid of such a machine problems can be attacked which are at present beyond our powers,

(c) Commander Sir Edward Travis, of the Foreign Office, will give his support,

(d) The Division already has two problems in hand which require from one to three years of work. These could be completed in a few weeks with the aid of such a machine. Others would come forward as soon as the capabilities of the equipment were known.

(e) This country should possess one of these machines, to keep up with world progress. Moreover, we are more resourceful and cunning in our use of machines than the Americans. Past experience has proved this in relation to:

  (i) The National accounting Machine

  (ii) 'Hollerith' Equipment

  (iii) The Differential Analyser.

(f) The machine is not restricted to arithmetic. It is just as much at home in algebra, or the enumeration of group characters. New and fascinating prospects in research in pure mathematics can be foreseen when some experience has been gained in its use.

> This is not a primary reason for building it, but once built, its
> 'spare time' use in this field should not be neglected entirely.[11]

Womersley's report was focused on the economic advantages of the new machine, and on the importance of competing with the United States. Diplomatically, he played down the differences between the ACE and the EDVAC. Turing's style during the same meeting of the Executive Committee was very different. His presentation was hard to understand, technical, and problematic from the diplomatic point of view. He gave cost forecasts for the construction of the machine when this was not in his remit.

It was clear from the beginning that, while the NPL wanted a machine to speed up calculations, Turing wanted to create a genuinely 'general purpose' machine. The ACE, as he said in 'Proposed Electronic Calculator', would be able to solve general problems like:

> *Problem* 8 To count the number of butchers due to be demobilised in
> June 1946 from cards prepared from the army records. The machine
> would be quite capable of doing it, but it would not be a suitable job
> for it. . . .

> *Problem* 9 . . . The calculator could be made to find a solution of the
> jig-saw, and, if they were not too numerous, to list all solutions.
> This particular problem is of no great importance, but it is typical
> of a very large class of non-numerical problems that can be treated
> by the calculator. . . .

> *Problem* 10 Given a position in chess the machine could be made
> to list all the 'winning combinations' to a depth of about three moves
> on either side.

Nowadays it is obvious that all these problems fall within the scope of the computer, but at that time few were aware—as Turing was—of the electronic machine's potential. The members of the Committee would have found it difficult to understand Turing's vision of a truly general-purpose machine. The minutes of the meeting indicate that the Committee was influenced more by Hartree—the only one who could comprehend the meaning of Turing's project—than by Turing himself.

Turing's visit to the United States in January 1947

The relationships between Turing and the US groups building general-purpose electronic computers are elucidated by a report that he wrote of his trip to the United States to attend the *Symposium on Large Scale Digital Calculating Machinery* held at Harvard University from 7–10 January 1947.

Report on visit to US January 1st-20th 1947

My visit to the USA has not brought any very important new technical information to light, largely, I think, because the Americans have kept us so well informed during the last year. I was able, however, to get a useful impression of the values of the various projects, and the scale of their organisation. The number of different computing projects is now so great that it is no longer possible to have a complete list. I think this is a mistake, and that they are dissipating their energies over too wide a range. We ought to be able to do much better if we concentrate all our effort on the one machine, thereby providing a greater drive than they can afford on any single one. At present, however, our effort is puny compared with any one of the larger American projects. To give an idea of the number of people involved in this work in the U.S.A. I may mention that there were between 200 and 300 present at the Symposium at Harvard, and that about 40 technical lectures were given. We are quite unable to match this.

One point concerning the form of organisation struck me very strongly. The engineering development work was in every case being done in the same building with the more mathematical work. I am convinced that this is the right approach. It is not possible for the two parts of the organisation to keep in sufficiently close touch otherwise. They are too deeply interdependent. We are frequently finding that we are held up due to ignorance of some point which could be cleared up by a conversation with the engineers, and the Post Office find similar difficulty; a telephone conversation is seldom effect-ive because we cannot use diagrams. . . . It is clear that we must have an engineering section at the ACE site eventually, the sooner the better, I would say.

Looking on the bright side my visit confirmed that our work so far has been on the right lines. It is probable that the Princeton machine, based on the Selectron, will have some advantages over the ACE in speed, but our proposed machine has some compensating advantages, and I think that, other things being equal, it is better that the two different types should both be tried. The Princeton group seem to me to be much the most clear headed and far sighted of these American organisations, and I shall try to keep in touch with them.

We shall eventually obtain a word-for-word account of the conference. All the information given was 'unclassified'.[12]

A number of conclusions can be drawn from this document:

1. Turing was not impressed by US progress in the field.
2. Turing was reasonably satisfied by the progress of the ACE project at that time, offering a favourable comparison with the results of US projects.
3. Turing understood perfectly—at least in theory—the importance of good organization for the success of the project. In particular, he was well aware that his project required an electronic engineering section at the NPL. He was conscious that the Post Office facilities were not adequate to the production of a pilot machine.

Calculator or electronic brain?

There are very significant differences between Turing's 'Proposed Electronic Calculator' and von Neumann's 'First Draft'. Turing's proposal had the objective of obtaining funding for a real project, while von Neumann's 'First Draft' was written in order to sum up conversations and meetings which had taken place at the Moore School with the aim of improving on the ENIAC. Turing's exposition was in some cases more technical and more detailed than von Neumann's. For example, Turing offered a detailed description of the mercury delay line (the memory device) whereas von Neumann said little about the matter. Von Neumann wrote a generic plan, with no regard to immediate practical output. One of the greatest differences between the two documents, however, lies in the general conception of the purpose and the possibilities of the ACE and the EDVAC. Turing believed that the ACE would have the ability to emulate the human brain, and was well aware that it could solve many problems beyond mathematical ones, the kinds of problems that, if solved by a human being, would be considered 'intelligent' tasks. According to von Neumann:

> An Automatic Computing System is a (usually highly composite) device, which can carry out instructions to perform calculations of a considerable order of complexity—e.g. to solve a non-linear partial differential equation in 2 or 3 independent variables numerically.[13]

This definition implies a very narrow conception of the machine. Turing, on the other hand, said in 'Proposed Electronic Calculator':

> Calculating machinery in the past has been designed to carry out accurately and moderately quickly small parts of calculations which frequently recur. . . . It is intended that the electronic calculator now proposed should be different in that it will tackle whole problems. Instead of repeatedly using human labour for taking material out of the machine and putting it back at the appropriate moment all this will be looked after by the machine itself. This arrangement has very many advantages.
>
> 1) The speed of the machine is no longer limited by the speed of the human operator.
> 2) The human element of fallibility is eliminated, although it may to an extent be replaced by mechanical fallibility.
> 3) Very much more complicated processes can be carried out than could easily be dealt with by human labour.

The machine described by Turing would not only help human beings carry out calculations—it would also be a substitute for human beings, able to carry out some of the tasks that are considered 'intelligent' when accomplished by a human. In Turing's project, but not von Neumann's, we are confronted by a machine different from all previous machines. Machines of the new kind will pursue their own tasks, without being constantly or completely guided by human control. According to Turing, these machines can make mistakes and learn, following different strategies, before finding the right one, in the manner of human researchers who try to demonstrate a theorem or find a solution to a problem.

According to the testimony of Ted Newman (who joined the NPL in September 1947 but was in contact with Turing even before then), Turing was very much focused on his special interest, the emulation of the human brain:

> Turing knew perfectly well what the job was he had to do, which was to manufacture or design a machine that would do the complicated sort of mathematics that had to be done in the Mathematics Division of NPL. But he had all sorts of interesting things that he liked to do: for example, he was really quite obsessed with knowing how the human brain worked and the possible correspondence with what he was doing on computers.

Turing thought that the machine should be made very simple, and at the same time should make everything possible that could be done. His particular purpose was to permit the writing of programs that modify programs, not in the simple way now common but rather in the way that people think.[14]

Programs and programming: Turing's vision

Having contrasted the views of Turing and von Neumann regarding the scope of their machines, I will outline some of the ways in which Turing's thinking about programming was more innovative than von Neumann's.[15]

The stored-program concept

There are various meanings of 'stored program' and these must be clearly distinguished.

1. Instructions can be stored in memory coded as numbers. This idea came directly from the arithmetization technique used in logic by Gödel. In this sense the stored-program concept was present in the universal Turing machine of 1936.
2. Instructions and data can be stored together using the same kinds of symbols. This idea originates with the universal Turing machine.
3. Instructions expressed in the language of numbers can be manipulated like any other numbers, leading to the idea of program *modification*.

Von Neumann certainly knew of Gödel's arithmetization procedure. However, in the 'First Draft' the only instruction element that could be modified by another program was the addressing bit. In subsequent papers, von Neumann did make use of the idea of encoding procedures as if they were data. In contrast, Turing's treatment of conditional branching in 'Proposed Electronic Calculator' was based from the very beginning on the possibility of manipulating instructions as if they were data. Carpenter and Doran say:

> Von Neumann does not take this step [the manipulation of instructions as if they were numbers] but Turing is clear about it, and believes it to be necessary for conditional branching ... As von Neumann gave each word a nonoverrideable tag, he could not manipulate instructions in this way. What we now regard as one of the fundamental characteristics of the von Neumann machine was, as far as we know, suggested independently, if not originally, by Turing.[16]

Machine structure and the centrality of programming

Turing's machine was based on a simple hardware structure. Only elementary operations were provided in the hardware—essentially the operations of moving, erasing, and rewriting, logical operations, and basic arithmetical operations. According to Turing, the most important element of the machine was the instruction table. This approach to designing a computer is strongly influenced by the structure of the universal Turing machine. Turing saw clearly that the power of the machine lay mainly in the memory space and in the ability of the 'programmers'.

Turing's attitude towards computer design was, perhaps, too advanced for his contemporaries. His contribution to machine architecture can be appreciated better now than during the 1940s, when the ACE group was struggling with all kinds of technical problems, obstacles and organizational difficulties. Turing's purist approach was perhaps one of the reasons for the delay in the implementation of the machine at the NPL.

The centrality of programming in Turing's view of the machine enabled him to foresee many key issues. He remarked in 'Proposed Electronic Calculator':

> We also wish to be able to arrange for the splitting up of operations into subsidiary operations. This should be done in such a way that once we have written down how an operation is to be done we can use it as a subsidiary to any other operation.

In order to control the calling of the subroutines during programs, he invented the concept of the stack algorithm ('last in first out' or LIFO). (See his BURY and UNBURY routines in 'Proposed Electronic Calculator'.) This was absolutely new and there was nothing like it in von Neumann's 'First Draft'.

Machine form and popular form of instructions

Turing's vision of the programming process was so clear that even in 'Proposed Electronic Calculator' he saw the necessity of different forms of programming and described different levels of language:

Machine form—the instructions are expressed in full so as to be executed by the machine.

Popular form—the instructions could be easily read and took the form of print on paper rather than punching.

Turing had a clear and accurate view of the programming process as it would be fifty years later. His idea of a popular form of programming language shows that he was aware of the difficulty of dealing with machine code and that he saw the importance of 'high level languages' in order to communicate in a more user-friendly way with the machine. Turing also saw the need for a 'general description of a table that will contain a full description of the process carried out by the machine acting under orders from this table'.

The 'levels of language' picture is necessary linked with another issue: the various simulative layers of the machine and the concept of a virtual machine. The computer that Turing imagined had the capability of emulating different machines, and each virtual machine could be used as a basis for emulating a more complex machine. From this viewpoint, the programmer is permitted to use various different 'high-level' languages and each is translated into the language at the lowest level (the machine code) in order to be executed by the computer.

The forking of the way

When he lectured to the London Mathematical Society on 20 February 1947, Turing still seemed confident regarding the ACE project. A few months later, however, Turing decided to take a sabbatical year, even though the NPL group was working hard on the construction of the 'Test Assembly', under the direction of Huskey. One likely explanation for this sudden lack of confidence was the divergence of interests between Turing and the engineers. The engineers wanted to produce a trial machine quickly to gain experience and then build a small pilot model. This step-by-step strategy was probably better than Turing's own. Turing, however, was confident that the design he had proposed was correct and would work. He did not want to waste time building a reduced prototype, too small for the kind of computing that he envisaged. (He was probably also aware, without waiting for a test machine to be built, that the memory device chosen by Williams—the cathode-ray tube—was better than the mercury delay line.)

Turing declared his interest in 'machine intelligence' at the outset (see 'Proposed Electronic Calculator'). Turing made the following points clear in a letter to the cyberneticist Lord W. Ross Ashby:

1. Turing viewed the ACE as similar to the 'Universal machine' in the sense that the ACE was able to process every instruction without any change to the hardware.

2. Turing planned to program the ACE to try out variations of behaviours and accept or reject them according to some rules.

3. Turing wanted to use the ACE to construct models of the action of the brain.

4. Turing proposed using the ACE to emulate the behaviour of neuron circuits. This appears to be the first time that the simulation method was described in the context of electronic general-purpose devices.[17]

Dear Dr. Ashby,

Sir Charles Darwin has shown me your letter, and I am most interested to find that there is someone working along these lines. In working on the ACE I am more interested in the possibility of producing models of the action of the brain than in the practical applications to computing. I am most anxious to read your paper.

The ACE will be used as you suggest, in the first instance in an entirely disciplined manner, similar to the action of the lower centres, although the reflexes will be extremely complicated. The disciplined action carries with it the disagreeable feature, which you mentioned, that it will be entirely uncritical when anything goes wrong. It will also be necessarily devoid of anything that could be called originality. There is, however, no reason why the machine should always be used in such a manner: there is nothing in its construction which obliges us to do so. It would be quite possible for the machine to try out variations of behaviour and accept or reject them in the manner you describe and I have been hoping to make the machine to do this. This is possible because, without altering the design of the machine itself, it can, in theory at any rate, be used as a model of any other machine, by making it remember a suitable set of instructions.

The ACE is in fact, analogous to the 'universal machine' described in my paper on computable numbers. This theoretical possibility is attainable in practice, in all reasonable cases, at worst at the expense of operating slightly slower than a machine specially designed for the purpose in question. Thus, although the brain may in fact operate by changing its neuron circuits by the growth of axons and dendrites, we could nevertheless make a model, within the ACE, in which this possibility was allowed for, but in which the actual construction of the ACE did not alter, but only the remembered data, describing the model of behaviour applicable at any time. I feel that you would be well advised to take advantage of this principle, and do your experiments on the ACE, instead of building a special machine. I should be very glad to help you over this.

I hope you will find time to visit me here next time you are in town.

Yours sincerely,

A. M. TURING[18]

We can see in this letter some of the reasons for the divorce between Turing and the NPL, given what was said above about the engineers' goals. After February 1947 his role in the team was less and less influential, because the need to achieve results quickly was the main priority. Huskey took control of the team that was supposed to build the Test Assembly and all the effort was concentrated on this project. Turing was not explicitly against the Test Assembly, but was not able to perceive its urgency and strategic importance. He wanted to build the most powerful machine that was feasible and use it to emulate the brain, while the mathematicians, engineers, and managers at the NPL wanted to build a very small machine to prove the technology. There was no room for compromise between these viewpoints and Turing became increasingly marginalized. No doubt there was reciprocal esteem between Turing and those on the other side of the debate but, in the end, the two sides were deaf to one another.

The 1948 report: the start of machine learning studies

Darwin and Womersley agreed when Turing asked to spend a sabbatical year away from the NPL. Darwin planned to have Turing back once the machine was built and envisaged Turing rejoining the team when the machine had to be programmed. The role Darwin imagined for Turing was in fact precisely the one that Turing took up at the University of Manchester at the end of his sabbatical—Darwin's evaluation of the interests and capabilities of this complicated scientist was correct.

Turing's year in Cambridge was a turning point. He wrote a report on the prospects for machine intelligence, a compendium of innovative ideas about computers, machine intelligence, the learning process, and the theory of knowledge.[19] Some of Turing's conclusions were:[20]

1. There is a world of difference between theoretical and practical machines. In the case of practical machines there are relevant space–time

considerations that oblige one to develop appropriate strategies if solutions are to be obtained in a reasonable time. As Turing said:

> Although the operations which can be performed by LCMs [Logical Computing Machines—Turing machines] include every rule-of-thumb process, the number of steps involved tends to be enormous.[21]

2. The universal Turing machine has this similarity to a practical computing machine such as the ACE: in both machines the most important aspect is programming. Each can emulate any other computing machine, if given a correct table of instructions.

> Nearly all of the PCMs [Practical Computing Machines] now under construction have the essential properties of the 'Universal Logical Computing Machines' mentioned earlier. In practice, given any job which could have been done on an LCM one can also do it on one of these digital computers.[22]

3. It is very important to organize knowledge in such a way as to guarantee a flexible, rapid, efficient access to information. The structure and organization of knowledge is vital to practical concerns.

> Two facts which need to be used together may be stored very far apart on the tape. There is also rather little encouragement, when dealing with these machines, to condense the stored expressions at all. ... As the simplified Roman system obeys very much simpler laws one uses it instead of the Arabic system.[23]

4. It is sometimes uninteresting to demonstrate an equivalence between theoretical machines and practical ones:

> It naturally occurs to ask whether, e.g., the ACE would be truly universal if its memory capacity were infinitely extended. I have investigated this question, and the answer appears to be as follows, though I have not proved any formal mathematical theorem about it. ... We should ... have to store n [a number referring to a block of memory], and in theory it would be of indefinite size. This sort of process can be extended in all sorts of ways, but we shall always be left with a positive integer which is of indefinite size and which needs to be stored somewhere, and there seems to be no way out of the difficulty but to introduce a 'tape'. But once this has been done, and since we are only trying to prove a theoretical result, one might as well, whilst proving the theorem, ignore all the other forms

of storage. One will in fact have a ULCM [Universal Logical Computing Machine] with some complications. This in effect means that one will not be able to prove any result of the required kind which gives any intellectual satisfaction.[24]

The central part of the report, however, is a description of experiments that Turing made with what he called 'unorganised machines' (see Chapter 5). Unorganized machines behave according to rules that are not determined in advance. A random element may be involved. Turing's experiments aimed at investigating the possibility of creating machines able to emulate aspects of intelligent behaviour.

Learning and 'interference' were crucial to Turing's method:

> If we are trying to produce an intelligent machine, and are follow-
> ing the human model as closely as we can, we should begin with
> a machine with very little capacity to carry out elaborate opera-
> tions or to react in a disciplined manner to orders (taking the form
> of interference). Then by applying appropriate interference, mimick-
> ing education, we should hope to modify the machine until it could be
> relied on to produce definite reactions to certain commands.[25]

As a result of learning, the machine produces different results for the same input at different times. Turing emphasized that machines can make mistakes during the learning process. It is perfectly possible for the machine to fail to find a solution that does exist or to produce a solution that is incorrect.

External influence is central for the learning process. Man does not develop his intelligence in isolation. Acquiring new knowledge resembles an inter-active social activity. Nor are machines closed systems. Turing compared the training of an unorganized machine with the process of education of a man:

> We might say then that in so far as a man is a machine he is one
> that is subject to very much interference. In fact interference will be
> the rule rather than the exception. He is in frequent communication
> with other men, and is continually receiving visual and other stimuli
> which themselves constitute a form of interference.[26]

Turing's ideas about machine intelligence anticipated the social intelligence model[27] and also the Multi Agent Systems (MAS) approach, now much followed in Artificial Intelligence.[28]

Retrospect

Turing's vision of the computer was many decades ahead of its time. From his perspective, computer science and Artificial Intelligence were almost the same discipline. Despite his consummate familiarity with the technological state of the art, he was impatient of the technological constraints of the time, and he tended to find practical difficulties annoying. Paradoxically, his vision isolated him from the community struggling to create the first general-purpose electronic machines. Turing's contributions to computing were immense. Every programmer should read his *Programmers' Handbook*[29] for some excellent tips! Modern theorists can learn much from his work.

Notes

I am grateful to Jack Copeland for his constant help with the editing of this chapter, which has contributed substantially.

1. Report to the Executive Committee of the National Physical Laboratory; reprinted in this volume.
2. Von Neumann, J. 'First Draft of a Report on the EDVAC', 30 June 1945, Contract W-670-ORD-4926 Moore School of Electrical Engineering, University of Pennsylvania; reprinted in *Annals of the History of Computing*, 15 (1993), 25–75.
3. Turing, A. M. (1936) 'On computable numbers, with an application to the Entscheidungsproblem', *Proceedings of the London Mathematical Society*, Series 2, 42 (1936–7), 230–65; reprinted in Copeland B. J. (ed.) (2004) *The Essential Turing*. Oxford: Oxford University Press.
4. Gödel, K. (1931) 'On formally undecidable propositions of Principia Mathematica and related systems I' and (1934) 'On undecidable propositions of formal mathematical systems', in Davis, M. (ed.) (1965) *The Undecidable*, New York: Raven.
5. Turing A. M. (1939) 'Systems of logic based on ordinals', *Proceedings of the London Mathematical Society*, Series 2, 45, 161–228; reprinted in Copeland (ed.), *The Essential Turing*, the quotation appearing on p. 150.
6. Ibid., pp. 192–3.
7. Letter from A. M. Turing to M. H. A. Newman, undated, but probably written at the beginning of 1940. A copy is in the Modern Archive Centre, King's College, Cambridge (catalogue reference D2); reprinted in Copeland (ed.), *The Essential Turing*.
8. Mahon, A. P. 'The History of Hut Eight, 1939–1945' (June 1945), p. 28 (a copy of the typescript is in the Public Record Office (PRO), Kew, Richmond, Surrey

(document reference HW 25/2); a digital facsimile is in The Turing Archive for the History of Computing <www.AlanTuring.net/mahon_hut_8>).

9. Davies, D. (1993) 'Early computer development at NPL', *Computer Resurrection*, 8, 11–12.

10. National Research Development Corporation.

11. Womersley, J. R. '"ACE" Machine Project', 13 February 1946, National Physical Laboratory Executive Committee paper E.881. A handwritten copy is in a note by M. Woodger dated 8 September 1975 (in the Woodger Papers, National Museum of Science and Industry, Kensington, London (catalogue reference M15)); a digital facsimile of the memorandum is in The Turing Archive for the History of Computing <www.AlanTuring.net/ace_machine_project>.

12. 'Report on Visit to USA January 1st–20th 1947', 3 February 1947 (PRO document reference DSIR 10/385; a digital facsimile of the original is in The Turing Archive for the History of Computing <www.AlanTuring.net/turing_usa_visit>).

13. Von Neumann, 'First Draft of a Report on the EDVAC', p. 33.

14. Newman, E. (1994) 'Memories of the Pilot ACE', *Computer Resurrection*, 9, 11–14, p. 12.

15. I am grateful to Bob Doran for a discussion with him about the differences between Turing's position and von Neumann's. See also Carpenter, B. E. and Doran, R. W. (1977) 'The other Turing machine', *Computer Journal*, 20, 269–79.

16. Carpenter and Doran, 'The other Turing machine', pp. 270–1.

17. Proposals for simulation were more common in the cybernetics and pre-cybernetics field; see Cordeschi, R. (2002) *The Discovery of the Artificial*, Dordrecht: Kluwer. The relationship between Turing and the British cybernetics group is an important issue that for reasons of space will not be tackled here.

18. Undated letter from Turing to W. Ross Ashby (in the Woodger Papers (catalogue reference M11/99); a digital facsimile is in The Turing Archive for the History of Computing <www.AlanTuring.net/turing_ashby>).

19. Turing, A. M. (1948) 'Intelligent machinery', National Physical Laboratory, 1948. A copy of the original typescript is in the Woodger Papers; a digital facsimile is in The Turing Archive for the History of Computing <www.AlanTuring.net/intelligent_machinery>. Reprinted in Copeland (ed.), *The Essential Turing* (page references are to this edition).

20. I am indebted to Jack Copeland for various discussions on this subject. See also Copeland, B. J. (1997) 'The broad conception of computation', *American Behavioral Scientist*, 40, 690–716; Copeland, B. J. and Proudfoot, D. (1996) 'On Alan Turing's anticipation of connectionism', *Synthese*, 108, 361–77.

21. Turing 'Intelligent machinery', p. 414.

22. Ibid., p. 415.

23. Ibid., p. 414.

24. Ibid., pp. 415–16.
25. Ibid., p. 422.
26. Ibid., p. 421.
27. Ibid., p. 431.
28. See, for example, Huhns, M. and Singh, M. (eds) (1997) *Readings in Agents.* San Francisco: Morgan Kaufmann.
29. Turing, A. M. 'Programmers' Handbook for Manchester Electronic Computer', Computing Machine Laboratory, University of Manchester, no date, *c.*1950; a digital facsimile is available in The Turing Archive for the History of Computing <www.AlanTuring.net/programmers_handbook>.

8 Computer architecture and the ACE computers

Robert Doran

Introduction

The style of computer architecture proposed by Turing was very practical, given the circumstances of the time. He anticipated later 'low level' architectures, and higher layers of abstraction for programming. This chapter considers the architecture of the ACE computers in the light of developments in computer architecture over the fifty years that followed. The first part of the chapter reviews the concept of computer architecture and outlines the history of the RISC (Reduced Instruction Set Computer) movement.

A student of computer architecture, if introduced to the Pilot ACE as an historical artefact, might be led to claim that the Pilot ACE was the first RISC architecture, and also the *ultimate* RISC, for it had no operation code and therefore only one instruction! Of course, a claim for such reasons is ill-founded, since there are 'op-codes' in the Pilot ACE, although embedded in the 'register' addresses (see the next chapter). Nevertheless, a closer examination of the Pilot and the earlier proposals for the ACE does show that the ACE's claim to be the first RISC architecture has merit.[1]

Computer architecture

Let's define our terms carefully. The *architecture* of a computer is the interface between its software and its hardware. The architecture includes the instruction set. It is what compiler writers and programmers produce code for and what computer designers implement.

In the beginning, when each individual computer was a new design, the term was not needed—the computer as designed represented the architecture. The concept was required by the late 1950s, when there began to appear sequences of computers which had different designs and yet could run the same software—were 'software compatible', as we would say nowadays. The term 'computer architecture' itself was defined by IBM to describe the hardware–software interface as a construct independent of the design of a particular model of computer.[2]

Although the use of the term 'computer architecture' as just described is standard, there is a formidable confusion of terminology. The high-level aspects of a computer's design (such as caches, pipelines, etc.) are also collectively called the computer's architecture. Sometimes the distinction is marked by using the terms 'instruction set architecture' (ISA) and 'design architecture', but often the word 'architecture' is simply used ambiguously and it is very easy to get the two concepts confused. (The confusion is reinforced by the fact that most of what computer architects discuss, for example at the annual international symposia,[3] is design architecture.) Here we will use the term 'architecture' in the sense of 'computer architecture', the machine-independent interface.

The architect's task

Having defined the term, let us consider in more detail what the computer architect has to achieve. Given that the architecture is the interface between hardware and software, there are two overriding criteria that an architecture should meet. First, the architecture should allow computers to be reasonably easy to design and reasonably cheap to manufacture. Second, the architecture should be suited to programming—it should not be difficult to write, or translate, programs having good performance.[4] There is always a tension between these two goals (and at their extremes they are opposed to each other). For example, a machine without floating-point instructions is simpler to design, but it is then impossible to write floating-point programs with acceptable performance.

The hardware–software trade-off is never simple. If the architecture is to be reasonable to implement it must always be influenced by issues of hardware design, by what is feasible at the time. Sometimes this effect is obvious, as with the restriction of address length in instructions to suit the size of memory that can be provided. Sometimes the effect is more subtle and general, such as allowing instructions to overwrite other instructions because they will be stored in the same kind of memory. In the other direction, the architecture will be influenced by programming needs (e.g. in specifying address modes or supplying some kind of subroutine linkage mechanism). The architect has to make a judgement about what architectural features will ease programming, or make programs run faster. Sometimes the course of action is easy and clear, such as aiding the operating system code with protection-checking instructions (which are otherwise very difficult for software). However, it is important not to include architectural detail that aids one particular program (the operating system excepted), because there are just too many programs to choose from. It is also possible to include features that look useful (such as linked-list lookup instructions) but which end up being unused because they are too restrictive, or because there are better software solutions (e.g. hashing).

Although always keeping technology in mind, the architect is often not a technologist, and may therefore have a mental model of technology and design that is in some respects inaccurate. An architect might well under-estimate the ingenuity of the hardware designer. Some architectural feature that on the surface appears difficult or costly may be avoided by the architect, even though it could in fact be 'designed around'. For example, the extra memory references seemingly implied by virtual memory translation can be bypassed, using a translation look-aside buffer. Or the architect might avoid specifying too many registers because of their cost, yet find that the designer has ways of substituting slower memory so that cost can be reduced without significant loss of performance.

At a given time it might seem appropriate to give more weight to one of the main criteria than to the other. In the 1960s, for example, when programming became recognized as a serious problem—indeed, as the perception changed so that a computer system was viewed as essentially software with hardware support—more emphasis on software seemed reasonable. If an architecture gives more weight to software, it is termed 'high level', whereas if the emphasis is on ease of hardware design, the architecture is termed 'low level'.[5] The question of how high a level the architect should adopt has always been of interest; it was a topical issue in the 1960s and 1970s, and is still with us today.

The choice of architecture and its level is complicated by the fact that neither the hardware design nor the software design are fixtures, but them-selves respond to the architecture. Sometimes this occurs in an obvious way. For example, if a virtual memory mechanism is introduced into the architecture, the hardware designers have to adapt, and software has to use the feature. But the influences are often subtle and long term. Hardware designers might feel constrained in their use of parallelism by the architecture's serial nature. Software designers might view some aspects of the architecture as intrinsic and adapt their programming languages accordingly (as in the case of goto statements, mirroring branches, and variables, mimicking memory locations).

It is clear by this stage of the discussion that specifying an architecture is going to be a delicate task, requiring some wisdom, and the ability to make judgements. But the job is, in fact, made even more difficult by further constraints.

Architecture lines

Very few computer architects have the luxury of starting from scratch. Most computers are software-compatible successors to earlier computers. Because of the great costs involved (for both customer and vendor) in making changes to all the software that is dependent on the architecture, successive computers in a line should be upwards compatible with their predecessors. Even with a new computer, there is a tradition against which the new architecture is to be matched, and in any case the architect must keep in mind that this computer could be the first of a sequence.

Compatibility requires that old architecture be carried forward. For the manufacturer or vendor of a line of computers, it is essential that the architecture be as independent of the quirks of particular implementations as is reasonable. If not, the move from one model to its successor will be accompanied by the cost of modifying software. Also the architecture will grow without bound, as it gradually incorporates the design details of all its implementations. Design detail in the architecture should be avoided, unless it can be argued that the particular detail will clearly continue to be valuable in the future. Nevertheless, it can be tempting to allow the architecture to reflect a feature of a particular design—for example, by specifying the relative execution time of instructions—because this often does allow programs to be written that run faster on that design. This can also ease the design difficulty simply by virtue of describing what has to be implemented.

The importance of design independence was well understood by the large computer manufacturers. At IBM, for example, strict procedures were laid down to maintain architectural purity.[6] However, it seems impossible to keep the architecture completely separate from design, and architectures do gradually grow as they are implemented in successive designs. There are times when there is no other reasonable course than to allow dependence to occur. Address length is the obvious example. Nevertheless, with some thought architects can make sensible provision for future expansion (as in the case of the IBM 360, where the address length was separated physically from the number of bits in the instruction).

Although the architect of a new line of computers is constrained, he or she certainly has more freedom to break free from the past. It is a good time to adapt to current and forthcoming technologies, and to dispose of baggage resulting from architectures' past dependencies. It is a natural time to move the architecture closer to hardware, to a somewhat lower level. The resulting performance boost and design-cost reduction can be very useful in allowing a new enterprise to develop in the face of established competition.

In summary, computer architects need to maintain compatibility with previous architectures, they need to take current technology into account, they must allow for future developments in technology, and yet they must keep the architecture independent of technology!

Evaluating architectures

Given that the architect's task is so difficult, how can one compare proposals for different architectures in order to decide which is superior? Quantitative studies have a role to play (programming instruction frequencies, in particular). The results have to be considered carefully, however, for the studies are generally based on assumptions regarding both the design that is to be used for the hardware and the technology that will be used to generate the software. It is not surprising that the evaluation of computer architectures has to be partly qualitative, and can therefore be very contentious.

It may be that there is no absolute answer as to what architecture is best. Although many details are important and can be judged objectively, perhaps in the end it does not matter what the architecture is, so long as a sensible decision is made and the designers can get on with the job. The fact that such different architectures as Intel, IBM 360, Unisys A-series (Burroughs B6700), IBM AS400, Power PC, and Alpha, among others, are all surviving today[7] and serving useful purposes perhaps tells us something about the importance of arguments about architecture.

The RISC movement

The RISC movement arose in the 1980s with the possibility of high-performance single-chip computers. Most computer architectures at that time were suited to a greater number of logic gates than could be achieved in a single chip (with the circuit densities then possible). It was clear that low-level architectures would fit more easily on a single chip. 'A great deal depends on being able to fit an entire CPU design onto a single chip', announced an article in *Computer Architecture News*.[8]

Although a fast single-chip computer was the overriding goal of the RISC movement, the term 'Reduced Instruction Set Computer' actually means a computer with a small number of simple instructions. The claim was that making computers simple also makes them faster. Some proponents of RISC took this claim to extremes and it is difficult to give credence to some of what was said. For example, a 1985 article stated: 'By leaving out seldom-used instructions, computer designers may improve supermini and mainframe performance by a factor of 2 or 3 while reducing costs by an equal proportion.'[9] Despite the doubtful nature of claims made by some of RISC's supporters, the general thrust of RISC was to simplify the hardware—a pretty sensible design principle.

It was also clear that the architectures of some widely available computers—especially models with which the proponents of RISC were most familiar—were no longer appropriate, given the current technologies. This applied particularly to the computers most commonly used in academia, where the RISC proposals evolved: the DEC PDP11 and its successor, the VAX.[10] (The PDP11 and VAX had many clever address modes; but unfortunately these were obtained with an instruction decode that was intrinsically serial and which, although appropriate for the first PDP11 designs, placed a continuing burden on successors.[11])

So the RISC designers were after single-chip designs, simple designs, and unlike-VAX designs. They were willing to go to the low-level extreme in order to make the architecture appropriate to technology: 'RISC theory ... connotes a willingness to make design tradeoffs freely and consciously across architecture/implementation, hardware/software, and compile-time/run-time boundaries in order to maximise performance.'[12] However, their models of computer design were often misconceived. The following statements by proponents of RISC would have been recognized as questionable at the time by computer designers in industry.

Microcode leads to slower control paths and adds to interpretive overhead.[13]

[T]he peak pipelined execution rate is determined by the longest piece of the pipeline.[14]

Traditional pipelined machines can spend a fair amount of their clock cycle detecting and blocking interlocks.[15]

The early RISC architectures were extremely design-dependent. This lead proponents of RISC to include features in their architectures which proved unwise in terms of product-lines. There were two early proposals, one from the University of California at Berkeley and one from Stanford University. The Berkeley machine[16] had large register files and a complex register-windowing mechanism to use these registers.[17] In both machines the characteristics of their four-stage pipelines appeared in the architecture as delayed branches and register loads. The Stanford machine[18] even brought the pipeline design fully into the architecture and it became the compiler's job to produce sequences of instructions that would work correctly—hence the acronym MIPS: 'Microprocessor with Independent Pipeline Stages'. (The next chapter contains an overview of the MIPS instruction format.)

Was there any understanding among RISC's proponents of the importance of technology independence? A scan of papers from the time reveals no mention of this issue. A major (and sensible) theme underlying the RISC movement was that compiler technology had improved and could be relied upon to bridge a wider gap between hardware and programming. Perhaps it was thought that compilers could be modified for each new model, so that technology independence could be ignored. Certainly this proposal has been made before, but it has never turned out to be successful in practice, because there is just too much software to be adapted to each processor model (and not all the software is controlled by the vendor of the computer).

Nevertheless, there was much that was sensible and good about the early RISC architectures. The Stanford and Berkeley projects led to commercial architectural lines as MIPS and SPARC. Other vendors followed the trend with their own new architecture lines, though the architectures from established companies tended to be less design-dependent. In introducing the RS6000 architecture (later adapted to become the Power PC), IBM even redefined the term RISC as 'reduced instruction set cycles'—which just means 'fast'![19]

For many years it was widely believed that a change of architecture to RISC was necessary in order to gain acceptably high performance. Some companies made a costly across-the-board change to their architectures. However, Intel, the more established vendor, stayed with a compatible line of processors.[20] The MIPS and SPARC lines also developed over time, increasing in performance and gradually becoming design-independent.[21] Yet the long-term outcome seems to be that the complex architecture of Intel has managed to maintain momentum and competitive performance compared to the RISC architectures.

This perhaps means that many of the arguments for the superiority of RISC were ill-founded. There is no doubt that changing to new, simpler architectures made many developments possible which otherwise would have been mired in complexity. But one cannot help wondering whether all the costly moves to RISC architectures were in fact necessary.

The architecture of the ACE

As mentioned in previous chapters, the architecture of the ACE is strikingly different from that of other computers. When Turing wrote the first ACE proposal, he certainly was not constrained by compatibility with the past (except for the input–output devices, perhaps)—and it is exceedingly doubtful that he had in mind a future software-compatible product line. He did have the 'tradition' of the EDVAC proposal to draw on, but he characteristically came up with his own approach to architecture.

Ease of design of a fast computer was the natural immediate goal. (That a machine could be got working quickly, with wartime urgency, permeates Turing's 1945 proposal.) His proposal was at the computer design level. The architecture proper has to be abstracted from the design. Essentially, it is what was appropriate to make the design work. For example, the instructions were of various formats, with the fields of bits corresponding to what was needed to control the hardware directly (or after a simple decode). There were no instructions that actually did a complete and useful piece of work, such as a multiplication; the instructions were rather at the level of the steps that go into performing a multiplication. Nowadays we would classify the ACE as a very low-level architecture, a micromachine or inner computer at the register transfer level.[22]

Turing does not discuss the issue of high versus low levels of architecture,[23] but the modern reader of his proposal gets the impression that he clearly knew what he was doing. Later he had this to say, in comments made at the Inaugural Conference for the Manchester University Computer in July 1951:

> In connection with a remark to the effect that it was difficult to do programming with the A.C.E., Mr Turing (speaking as one who had at one time been connected with that machine) admitted that this was so, and that ease of programming had knowingly been sacrificed to speed.[24]

It seems reasonable to say that Turing was, in giving emphasis to speed and ease of hardware design, proceeding down the same path that would be followed by the RISC movement when the move to a single-chip computer was contemplated.

Turing's approach now seems sensible and appropriate. Strangely, though, it was not the popular path to follow at the time. The fashion was set by the architecture of the Princeton IAS machine, laid out by von Neumann and his collaborators.[25] Although the von Neumann architecture did have design dependencies (two instructions packed per word, for example), it was clearly much more hardware-independent than a computer at Turing's level. It had operation codes that actually did something whole for the programmer. In a sense, the von Neumann machine was a high-level architecture, intended to be easy to program (though the programming technology to which it was directed was hand-coding)—a line of development leading ultimately to the PDP11 and the VAX.

The von Neumann architecture was well publicized and widely followed, while the ACE languished with few successors, and ACE-style architecture for programming died away. I have no understanding of why the low-level approach was not followed at the time, since it seems so sensible in retrospect. Taking a positive view, perhaps it was the hardware-independence of IAS that caused its spread—it was a de facto standard that could be accepted without question, so enabling designers to get on with the job. Less positively, this is perhaps an early example of victory going to the group with the better publicity—certainly not the last time that this would happen in the computer industry!

Layered architectures

There is more to the ACE's architectural legacy than its being low level and RISC-like. One of the most profound developments in our understanding of architecture is that architectures can be *layered*. One architecture can be defined in terms of a lower level which can in turn be defined in terms of a still lower level, and so on. (This is taken to extremes in data communications with the ISO OSI seven-layer model, for example.)

A two-level architecture has become standard for computers. This accounts for some of the confusion in terminology: the two levels are the computer architecture and the design architecture that implements it.

Wilkes introduced microcode,[26] which allowed for two levels of architecture with the mapping between the levels performed by hardware. The inner level or micromachine has an architecture very like Turing's ACE. The level for programming could be as specified by von Neumann, with no significant loss of performance. With microprogramming, the computer may be high level in its architecture but low level in its design architecture, which made the job of satisfying both the main architectural criteria much easier. It also offered the prospect of using a much higher-level architecture than would otherwise be sensible (an opportunity taken to extremes in the 1970s).

Turing recognized the need to simplify programming. He never intended the ACE to be programmed at the machine level, and his 1945 proposal outlined one way in which the programming process could be organized.[27] He proposed that a sequence of subroutines be developed, to be used by the programmer in a 'popular' abbreviated form. The program expressed as the sequence of subroutines was to be expanded into machine language by software before execution (what we would now call assembly, loading and linking).[28] In the examples he gave, Turing showed how the ACE could be programmed as a machine at the level of von Neumann, by using some 'registers' as accumulators, some to handle indices, et cetera (here following the EDVAC architecture). Of course, Turing could also go higher than the von Neumann model had, and he immediately introduced subroutine linkage instructions with a stack and BURY/UNBURY 'instructions' to push and pop data to/from the stack.[29]

This approach to dual levels is a brilliant idea, but has some practical drawbacks. Some of these, such as reporting errors at the programming rather than machine-language level, were faced, and gradually dealt with, by compiler writers over the years. A serious issue arises from the need of a computer system's vendor to control the hardware–software interface. If this interface is itself a software construct, then control is not feasible unless the vendor restricts the ability of the user to access the real hardware. The beauty of Wilkes' approach is that it retains control by the vendor.

(There was a period in the 1970s when there was a flirtation with user-access to microcode, but it was quickly stamped out—although not before a plethora of new instructions was introduced, causing trouble for some architectures for years to follow.)

Only one computer has used an approach similar to that proposed by Turing. In the highly successful IBM AS400, the hardware–software interface is very high level and quite unlike any other (e.g. in its use of instruction numbers, rather than addresses, for branch destinations). However, there is also a lower architecture, not available to the user, to which all programs are translated before execution. No user is allowed to 'see' this lower level, so the vendor maintains control.[30] (The lower level was originally System/370-like, but was later changed to Power PC, with no impact on customers, showing the flexibility of the approach.)

There is a further approach to architecture layering, which is to emulate one computer by another. This, although involving a performance penalty, has many advantages, and is widely used (e.g. for the Java virtual machine today). This was not part of the Turing proposal originally, but he later made the following comments (continuing his remarks at the Inaugural Conference for the Manchester University Computer):

> He suggested that, if this difficulty [of programming the ACE] should prove very burdensome, it might be wise to use the A.C.E. a good deal in connection with interpretive routines.
>
> An interpretive routine is one which enables the computer to be converted into a machine which uses a different instruction code from that originally designed for the computer. The code is applied by the interpretive routine rather than by the computer direct. A good example of such a routine was one to make floating binary (or decimal) point working possible. The chief objection to using such routines lay in the loss of speed which went with them. With the A.C.E. this loss could more easily be afforded than with other machines, and furthermore the programming difficulty would disappear, for the labour of making the interpretive routine itself is small with any (machine) code, whilst the labour of programming with the (interpretive routine) code was independent of the machine code.

Although the kind of interpretation being discussed is different from that in the original ACE proposal, it does appear that Turing could have been aware, even in 1945, that the programming level of use of a computer could be different from the hardware implementation level, and that a simple low-level design could allow a more efficient interpreter to be made than is possible with a higher-level computer.

Conclusion

When reading Turing's work one is always impressed by the broad scope of his thought. He clearly had a very deep understanding of computing and of how it should be organized. Of course, he had been thinking for many years about issues such as that of one machine interpreting another, in connection with his theoretical Turing machines. Naturally enough he could also see the practical applications of these ideas.

It is of great interest that the struggle between high- and low-level architectures existed even before the first computers were designed. We can indeed conclude that the ACE is a RISC machine in the sense of having an architecture heavily influenced by the design of the computer. The concept of programming at a higher level than the raw machine was an integral part of the original ACE proposal. With more understanding of the processes involved, we (unlike the practitioners of the time) can appreciate the virtues of Turing's approach.

Notes

Thanks to Jack Copeland for his editing of this chapter.

1. See Carpenter, B. E. and Doran, R. W. (eds) (1986) *A. M. Turing's ACE Report of 1946 and Other Papers*. Cambridge, MA: MIT Press. Turing's ACE Report is discussed in detail in Carpenter, B. E. and Doran, R. W. (1987) 'The other Turing machine', *Computer Journal*, 20, 269–79.

2. Amdahl, G. M., Blaauw, G. A., and Brooks, F. P. Jr. (April 1964) 'Architecture of System/360', *IBM Journal*, 8, 87 ('the conceptual structure and functional behavior, as distinct from the organisation of the data flow and controls, the logical design, and the physical implementation').

3. The Annual International Symposia on Computer Architecture, published as issues of the ACM *Computer Architecture News*.

4. There are other criteria in practice—will the architecture sell, be received by customers as sensible or exciting, will the computer engineers be inspired to actually design the computer and the sales folk to sell it? Although such considerations are important we will not discuss them further.

5. We are taking the stance here that the 'level' of a computer architecture is a design attitude rather than any particular architectural features. What is regarded in one era as aiding programming may well be regarded as a hindrance later.

6. Described on p. 78 of Case, R. P. and Padegs, A. (1978) 'Architecture of the IBM System/370', *Communications of the Association for Computing Machinery*, 21, 73–95.

7. At the time of writing (2002).

8. Patterson, D. A. and Ditzel, D. R. (1980) 'The case for the Reduced Instruction Set Computer', *Computer Architecture News*, 8, 25–32. Interestingly, although the goal of a single chip was often justified in terms of performance—shorter paths and cycle time—it was not until 1999 that the fastest commercial computer was a single rather than multiple chip CPU. A single chip CPU is very important for cost reasons, although, again, often the CPU needed support chips anyway.

9. Wallich, P. (August 1985) 'Toward simpler, faster computers', *IEEE Spectrum*, 22, 38. Designers of high-speed computers would have known that the most direct determiner of performance is cycle time (1/clock-rate) and that this is determined by the time taken to perform a useful unit of work, such as 64-bit binary addition. Reducing instruction count or complexity has little effect on this. Making architectures smaller does affect cycle time, but only in the second order.

10. The PDP11 was itself an academic design from Carnegie-Mellon (Bell, G., Cady, R., McFarland, H., Delagi, B., O'Laughlin, J., Noonan, R. and Wulf, W. (1970) 'A new architecture for mini-computers—The DEC PDP-11', *AFIPS Conference Proceedings*, 36, 657–74.

11. Counter to two well-known architectural rules of thumb for fast computers: 'When an instruction is fetched its course of execution should be clear' and 'Avoid instructions that have different courses of execution based on their operands'. The first PDP11s were intended to be low-cost, not-very-parallel machines and the architecture reflects aspects of this intention.

12. Colwell, R. P., Hitchcock, C. Y. III, Jensen, E. D., Sprunt, H., Brinkly, M. and Kollar, C. P. (September 1985) 'Computers, complexity and controversy', *IEEE Computer*, 8–19.

13. Ibid. Of course, microcode was not reasonable for a single chip design because of the space cost of the memory. Apart from the Cray machines, the fastest general purpose computers at the time used microcode, as do single-chip designs today.

14. Patterson, D. A. (1985) 'Reduced Instruction Set Computers', *Communications of the Association for Computing Machinery*, 28, 8–21. Designers would use 'clock skewing' so that the mean stage-length is the determiner.

15. Ibid. Detecting interlocks certainly can occupy a clock cycle but it is in parallel with other work rather than an add-on.

16. Patterson, D. A. and Sequin, C. H. (September 1982) 'A VLSI RISC', *IEEE Computer*, 8–29.

17. Instead of a cache, for design reasons: 'An effective data cache would require a much larger area than our register file'; Patterson and Sequin, 'A VLSI RISC'.

18. Hennessy, J. L., Jouppi, N. P., Baskett, F., Gross, T. R., and Gill, J. (1982) 'Hardware/software tradeoffs for increased performance', *Proceedings of a Symposium on Architectural Support for Programming Languages and Operating Systems (ASPLOS-I)*. Palo Alto: ACM Press, 2–11.

19. Hester, P. D. (1990) 'RISC System/6000 Hardware Background and Philosophies', IBM RISC System/6000 Technology, SA23-2619, IBM Corporation, 2–15.

20. As did the mainframe manufacturers. The mainframes and Intel architectures, although complex, did not have the VAX's intrinsic defects.

21. The adoption of 'MIPS' as a company name avoided having to change its meaning to 'Microprocessor *without* Independent Pipeline Stages'.

22. The same applies to the Pilot ACE, although the architecture was cleaner and had some features that made it even more suited to fast operation with the serial technology—two-operand addresses and a branch address with each instruction.

23. However, Turing certainly pointed out that an architecture different from his own would require 'a very much more complex control circuit'; see his memo to Womersley in Chapter 3.

24. Manchester University Computer Inaugural Conference, July 1951 (proceedings published with the cooperation of Ferranti Ltd.), p. 26.

25. Burks, A. W., Goldstine, H. H., and von Neumann, J. (1946) 'Preliminary Discussion of the Logical Design of an Electronic Computing Instrument', reprinted in C. G. Bell and A. Newell (eds) (1971) *Computer Structures: Readings and Examples*. New York: McGraw-Hill.

26. Wilkes, M. V. (1951) 'The best way to design an automatic calculating machine', Manchester University Computer Inaugural Conference.

27. A good indication that a computer is not intended to be programmed with its machine language is when the individual instructions do not perform useful work by themselves. This applies to low-level machines like ACE, but also, paradoxically, to very high-level computers such as the Burroughs B6700 (at the extreme, a machine language becomes a specification of programmer intentions rather than a sequence of instructions).

28. For example, addition of two floating-point numbers to produce a third (all in standard locations) would be programmed as B, BURY, B, ADD.

29. See p. 277 of Carpenter and Doran, 'The other Turing machine'.

30. An earlier continuing line of architecture, Unisys A-series (Burroughs B6700), also maintained strict control over access to the machine language. All programs had to be written in a higher-level language and translated by a vendor-supplied compiler. In this case the restriction is a key part of the security and integrity of the entire system.

Part III
The ACE Computers

9 The Pilot ACE instruction format

Henry John Norton

A boy in a man's world

I left school at eighteen intending to do my National Service in the RAF before going up to Cambridge to take a Mathematics degree. When I was declared medically unfit for armed combat I had to find employment for a year. A group had been set up at the National Physical Laboratory to make some sort of calculating machine and they were looking for someone to do something called 'programming'. My first day at the NPL was October 1947. Mike Woodger showed me round. In one room there was a solitary man, Cyril Cain, sitting on a lab stool with several vertical copper rods in front of him. He was soldering bits of wire between them according to a diagram on a scruffy bit of paper. 'He's building the ACE', Mike said. In those days Maths Division had a growing library of programs, but no computer.

The head of the group I had joined I knew always as Dr Turing (I never called him Alan to his face). It was only later that I learnt of his fame. He and I used to compete in middle distance races—he always beat me. I have been told that he could often be seen striding the streets of Teddington in bare feet—so as to harden the skin.[1] My main memory of him is his irascibility. On one occasion we were visited by Andrew Booth of Birkbeck College, who later developed rotating drum storage. Booth said something that annoyed Turing, who shouted at him as if he were a naughty little child. This was in front of the whole team. I was amazed that a grown man could behave in such a way. Another thing I remember is that Turing had a faulty chain on his ancient bicycle. Rather than repair it, he had found that by counting the number of rotations of the pedals he could get off and walk a few paces and then remount and ride without the chain jumping off the sprocket.[2] I translated several articles about Turing from Russian journals and sent copies to his mother. In return she wrote at length, telling me that Alan would never have committed suicide.

Turing's second in command was Jim Wilkinson, who later earned fame for error analysis in matrix computation. I always called him Wilkie. Wilkie claimed to get all his inspiration riding his bicycle. He had a fund of stories. Snow lay six inches deep over the whole of Kent when he went for a medical at Fort Halstead at the beginning of the war. During the examination the doctor made him remove his glasses and, pointing to a window that Wilkie could hardly see, asked 'What are those men doing?'. Wilkie made a calculated guess and said 'Shovelling snow'. He passed the medical. He also told me that before his honeymoon he wanted to be as fit as possible and this included a check-up at the dentist. He felt he was justified because the dentist found a tooth that needed a filling. Unfortunately something went wrong with the filling and he had to take aspirin throughout his honeymoon.

Wilkinson told the following story about Turing (the words are Wilkie's own, from a letter).

> During the early stages of the development of the ACE some of the practical work was done at Dollis Hill Research Station some fourteen miles from NPL. The journey from Teddington to Dollis Hill is rather tiresome and one day Turing announced that he intended to run there next time. This was regarded as a joke until the next occasion for the journey came and Turing proceeded to put on his running shorts and set off at a steady pace. I travelled in a more orthodox manner, by train and two buses, and arrived to find that Turing had been there for some time and appeared to regard his behaviour as in no way remarkable![3]

The Pilot ACE Instruction Set Architecture

The Instruction Set Architecture (ISA) is an important abstraction for the interface between the hardware and the low level software. It standardizes instructions, machine language bit patterns, etc. It has the advantage that we may use different implementations of the same architecture on different machines. (One disadvantage is that it acts as an obstacle to innovation.) The modern ISAs include: 80×86/Pentium/K6, Power PC, DEC Alpha, MIPS, SPARC, and HP.

One way of classifying instruction sets is according to the number of operands, as follows:

One-address machine code—as its name suggests, this has a single address defining an operand. If a second operand is involved, it is contained in a special word store called the 'accumulator', with various associated logic and arithmetic input gates. Another part of the instruction, the 'opcode', defines the operation to be performed on one or both operands.

Two-address machine code—this has two addresses for the operands. One of these addresses also indicates where the result is to be stored. The opcode defines the operation to be performed on the operands.

Three-address machine code—as for two-address, except that a separate address is given for the storage of the result.

Four-address machine code—as for three-address, with the addition of a part of the instruction that defines the address of the next instruction.

Some architectures use all these types of instruction format, while others use a restricted set. The Pilot ACE instruction set was different from all of the above. It had three addresses—operand, result, and next instruction—and *no opcode*.

Today the dominant architecture is 80 × 86, which evolved from the 16 bit architecture of the 8086 in 1978 to the Pentium Pro of 1995 with MMX (multimedia extensions) added in 1997. New features were added wily-nily. This architecture has:

- instructions from 1 to 17 bytes long
- one operand acting as both source and destination
- one operand coming from memory
- complex addressing modes (e.g. 'base or scaled index with 8 or 32 bit displacement').

Despite the architecture's complexity, the most frequently used instructions are not too difficult to build and compilers can avoid the portions of the architecture that are slow.

In contrast the MIPS (Multiprocessor with Independent Pipeline Stages) architecture is based on a few instruction types and is known as a 'Reduced Instruction Set Computer' (RISC). Like the Pilot ACE, all instructions are 32 bits long. It is very structured with no unnecessary complications. There are only three instruction formats:

R-type

opcode	rs	rt	rd	shamt	function

The opcode defines the type of operation, such as an arithmetic or logic or shift operation. rs and rt are registers containing the operands and rd is the register where the result is to be stored. 'shamt' is the amount of shift in that type of operation. 'function' defines the type of arithmetic or logic operation, e.g. floating-point division.

I-type

opcode	rs	rt	16 bit address

Since all arithmetic, etc, operations can be performed only on words in registers, there has to be a way of moving words between memory and registers. This is achieved by the I-type instruction. The word in the memory position defined by rt and the 16 bit address is put into, or taken out of, register rs depending on the opcode. This format is also used for conditional jumps.

J-type

opcode	26 bit address

This is used for unconditional jumps.

Of course, the full MIPS instruction set has more facilities than these, but this description gives the flavour of a modern ISA.

The Pilot ACE

The Pilot ACE instruction format is:

NIS	Source	Destination	Characteristic	Wait Number	Timing Number	Go Digit

Each instruction contains the address of the *next* instruction. This address is given in two parts. NIS or 'Next Instruction Source' is the address of the memory unit—a mercury delay line—containing the next instruction (e.g. DL 1). The Timing Number gives the position in that memory line at which the instruction is to be found.

Every instruction is of the same form: Transfer data from Source to Destination. The Source and Destination are always mercury delay lines (except in some special cases such as input and output, multiplication, division, and the alarm). The Characteristic specifies the number of words that are to be transferred to the Destination. Each word consists of 32 bits.

The Go Digit may be used to halt the execution of an instruction. If the Go Digit is 1 the instruction is executed straight away. If the Go Digit is 0 the machine does not execute the instruction until it receives a signal either from the operator at the console, or from the card reader indicating that the card is in position to be read, or the card punch indicating that a row of a card is ready to be punched.

The Wait Number is used to delay execution of the instruction. When the instruction arrives at TS COUNT—which would nowadays be called the *instruction register*—the data that is to be transferred may not yet be available at the Source. By setting the Wait Number to 1, say, the programmer causes the instruction to wait one minor cycle—32 'ticks'—before executing. Programming the Pilot ACE required a detailed knowledge of where each word of data is at each point in time. The programmer's aim is always to position data in the Sources in such a way that the Wait Number is as small as possible—preferably zero.

There is *no* opcode in the instruction. Which operation the instruction will perform is implicit in the Source and Destination. For example, if the Source is S 21 then the data that the instruction transfers is the logical sum of the contents of the two temporary stores TS 26 and TS 27 (also mercury delay lines):

$$S\ 21 = TS\ 26\ \&\ TS\ 27.$$

If the Destination is D 17 then the instruction transfers to that destination the result of adding the contents of temporary store TS 16 to the Source:

$$D\ 17 = S + TS\ 16.$$

Thus the instruction

DL 7	S 21	D 17	1	0	2	1

adds the contents of TS 16 to the logical sum of TS 26 and TS 27 and transfers the result to D 17. Only one word is transferred. Because the Wait Number is 0 and the Go Digit is 1, the instruction is carried out without a pause. The next instruction is the second word in DL 7.

Notes

I am greateful to Jack Copeland for assembling this chapter from various Sources.

1. Thanks to Geoff Hayes for this information.
2. Editor's note. It seems that Turing had the same bicycle at Bletchley Park during the war. He was still riding it after he left the NPL for Manchester University (Tommy Thomas in interview with Copeland (March 2003)).
3. Letter from Wilkinson to Newman, 10 June 1955 (among the Turing Papers in the Modern Archive Centre, King's College, Cambridge (catalogue reference A.7)).

10 *Programming the Pilot ACE*

J. G. Hayes

Introduction

I wrote programs for the Pilot ACE from about the time that it was moved into Mathematics Division in February 1952 until early 1954. (I have never written a computer program since, so I am probably the only one who did a fair amount of programming using solely the machine's first implemented coding system.) I shall give an overview of the programming process against the background of the only one of my programs that has stuck in my mind. This program was for the back-substitution phase of solving a set of linear algebraic equations, with multiple right-hand sides, up to order 32. (The program has stuck in my mind because of the considerable effort that was needed to optimize it.)

Storage

The main features of the Pilot ACE from the point of view of the programmer were the storage, the adder, and the multiplier. There was initially no divider so division had to be achieved by programming. The numbers, or 'words', in the store each consisted of 32 binary digits—4 bytes in modern terminology. (A binary digit is either a one or a zero, these being represented electronically by a pulse or no pulse, respectively.) The main storage consisted of 10 delay lines (DLs for short), tubes filled with mercury down which pulses of sound passed. At one end of the DL the electronic pulses representing the digits in a word were converted into sonic pulses, and these were converted back into electronic pulses when they reached the other end of the DL. The electronic pulses were then fed back into the DL, and the whole process was repeated again and again. The delay due to the low speed of sound (relative to electronic) transmission provided the storage capability of this form of memory.

The DLs each held 32 words. The time of transmission along a DL for a single word was called a *minor cycle* (mc) and the time of transmission for all 32 words in a DL was called a *major cycle* (Mc). One major cycle occupied approximately 1 millisecond.

In addition to the DLs there were a handful of shorter delay lines called Temporary Stores (TSs) and Double Stores (DSs), able to store one word or two words each. The DLs provided the storage for all the program instructions as well as for numerical data to be used in the calculation. The TSs and DSs were there to speed up the computation by saving time that would otherwise be lost waiting for a particular word to come out of a DL.

The following figures give some idea of how all this compares to today's machines.

Pilot Model	TSs and DSs: 40 bytes	DLs: 1280 bytes
PC (2000)	Internal cache: 8160,000 bytes	RAM: 384,000,000 bytes

The two stages of programming

A program was constructed from very basic instructions. Each instruction transferred a number from one place, called the Source, to another, called the Destination. The transfer could be, say, from the input to a storage location, or from a storage location to the adder or multiplier.

The first of the two stages of writing a program was the construction of the 'flow diagram'. In the flow diagram, transfers were written in the form '4–11', meaning 'copy the word from DL 4 into TS 11'. Which word in DL 4 would be copied depended on the timing—on which word popped out of the Delay Line when the transfer took place. So the instruction had in addition to control the timing of the transfer. The instruction also had to say where the next instruction was to be found.

Some of the Destinations were not storage locations, but initiated particular operations. For example, sending a number to Destination 17 would add the number to the one stored in TS 16. Sending a number to Destination 18 would subtract it from TS 16. Sending any number to Destination 19 would multiply the numbers in TS 20 and one half of DS 14 together and give their double-length product in DS 14. The full list of Sources and Destinations is given in Table 1 of Chapter 11.

At the second stage of programming, known as 'coding', the programmer produced the instructions that were to be entered into the computer. At this stage an instruction would have its full form, explained in the preceding chapter:

$$N \quad S \quad D \quad Ch \quad W \quad T \quad G$$

W is the Wait Number and T is the Timing Number. If $W = T = 0$ then, once the instruction was read by the machine, the content of Source S would be sent to Destination D two minor cycles later. During that same minor cycle the next instruction would be read from Delay Line N. In the general case, W and T contributed in the following way: if the instruction was read in minor cycle k, transfer would start in minor cycle $k + W + 2$, and would continue into minor cycle $k + T + 2$, at which point the next instruction would be read from Delay Line N. Thus the transfer would take place for $(T - W + 1)$ mc if $T > W$, but for $(T - W + 33)$ mc if $T < W$. This was so if the Characteristic (or 'Serial Digit') Ch was zero, but when this was unity, the transfer went on for only one minor cycle. The next instruction, though, was still read at $k + T + 2$, allowing it to be located in any minor cycle. Finally, the Go Digit G had no effect when unity, but, when zero, held up the transfer until a signal was received (e.g. from the card reader, indicating that a row on the card had reached the reading head).

In the flow diagram, the instructions would be written in the order in which they were to be carried out, and so were easy to understand. The coding form, though, was a list of instructions in the order in which they were stored in each DL. The allocation of instructions into particular positions in the store—a process known as 'optimum coding'—aimed generally to minimize the waiting time and so reduce the computation time.

When an instruction had been allocated to its particular minor cycle in a DL, the DL's number, with the number of the minor cycle as a suffix, was added in front of that instruction as it appeared in the flow diagram, so as to complete that line of the diagram. The next chapter gives a number of examples.

So now on to a mathematical computation to be programmed.

Programming back substitution

One standard way of solving a set of linear algebraic equations starts by subtracting an appropriate multiple of one equation, called the pivotal equation, from each of the other equations in turn, so as to eliminate from all of them the first variable, x_1, say; then selecting one of these modified equations to be the next pivotal equation so as similarly to eliminate x_2 from the rest of them, and so on. The original equations can then be replaced by the set of pivotal equations, which can be written

$$a_{11}x_1 + a_{12}x_2 + a_{13}x_3 + a_{14}x_4 + \cdots\cdots\cdots\cdots + a_{1n}x_n = b_1$$

$$a_{22}x_2 + a_{23}x_3 + a_{24}x_4 + \cdots\cdots\cdots\cdots + a_{2n}x_n = b_2$$

$$a_{33}x_3 + a_{34}x_4 + \cdots\cdots\cdots\cdots + a_{3n}x_n = b_3$$

$$\cdots\cdots\cdots$$

$$a_{n-1,n-1}x_{n-1} + a_{n-1,n}x_n = b_{n-1}$$

$$a_{nn}x_n = b_n$$

The last equation contains only one variable and so we can immediately provide that variable's value. Substituting that value into the next equation above makes x_{n-1} the only unknown variable in that equation, so readily providing *its* value, and so on up the chain. This is called back-substitution, which was the object of my program.

These computed x-values are stored in order in one of the DLs, put there one at a time as they are computed. The program was aimed at solving a set of equations containing up to 32 unknowns; thus a whole DL was allocated to store the solution. Earlier, on desk calculators, then the main means of computing, it was rare to try to solve more than six equations, though rather more would be solved using the mechanical Hollerith machines. With the advent of the Pilot Model, the aim was naturally much higher. However, the machine's storage was not sufficient to store all the matrix coefficients, the a's in the above equations. The solution process therefore involved reading in these coefficients from Hollerith cards, with one coefficient on each line of a card, in binary form. The aim was to do the required calculation with that coefficient before the next line of the card arrived to be read.

The steps of the program

So let us take an example from the above equations, omitting the row suffix for simplicity:

$$a_3x_3 + a_4x_4 + a_5x_5 + \cdots\cdots\cdots\cdots + a_nx_n = b.$$

At the stage where this equation was being dealt with, all the x-values except x_3 would already have been calculated and put in the chosen DL. The main part of the process would then be to run the Hollerith cards through the reader (with the coefficients in reverse order) and when a coefficient, say a_5, was read in, to do the multiplication $a_5 x_5$, and the associated organization, before the coefficient from the next row of the card appeared on the scene. Before the multiplication was initiated, the value of a_5 had to be sent to the multiplier from the reader and the value of x_5 sent there from its DL. The program had to be able to fetch down any of the x's, from 1 to 32, when required, so a whole major cycle had to be allowed for that before starting the multiplication. The resulting product then had to be added to the sum of the previous products. So we have:

LOOP 1

1. Take the current coefficient (e.g. a_5) from the reader to the multiplier.
2. Take the corresponding x value (here x_5) from its place in its DL to the multiplier.
3. Multiply.
4. Add the result to the partial sum of products already computed.
5. Ask: 'Is that the last product for this equation?'—which here would be $a_4 x_4$.

Then, if the answer to that question is 'no', go back to the beginning of the loop and deal with the next coefficient (a_4); if the answer is 'yes', go on and calculate the unknown for this particular equation, using the next coefficient from the reader, and store it in its proper place in its DL; test if that is the last x to be calculated, that is, x_1; if it is, end the computation; if it is not, go back to start on the next equation.

So now we have:

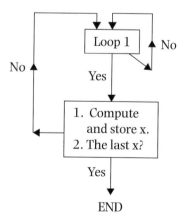

Now let us look at some of the details.

Details

Discrimination

Destination 24 discriminates between positive and negative numbers sent to it, whereas Destination 25 discriminates between zero and non-zero. If you send a number to either Destination, the program will move on to one or other of two instructions stored in adjacent positions in the same DL, depending on the sign of the number in the case of D 24, for example. So the first instruction in a loop and the first instruction after the end of the loop have to be placed in adjacent positions in the same DL.

Counts

To ensure we go round Loop 1 the correct number of times—$(n - 3)$ times to deal with the example equation above—we put that number in a store, say TS 15, before going into the loop. Then, during the loop, we subtract unity from it, and at the end of the loop send the result to D 25, in order to decide whether we go round the loop again or not. If we store $(n - 3)$ in the lowest digits of TS 15, we can use Source 25, which provides P1, a number with one in only the first digit, to help the count. Thus, during the loop, we would have

TS 15–TS 16	Copy the number in TS 15 into 16.
S 25–D 18	Subtract unity from that number.
TS 16–TS 15	Store the count ready for the next time round the loop.
TS 15–D 25	Test if the count is zero.

Multiplication and addition

To compute the product, a_5x_5 say, we need one of the two numbers to be in TS 20 and the other in the upper (most significant) half of DS 14. After multiplying them, their product appears, double length, in DS 14 and then has to be added to the partial sum already computed (preferably stored in DS 12). The instructions are:

S 0–DS 14	Take a_5, now at the reader, to the upper half of DS 14.
S 28–DS 14	Make the other half of DS 14 zero.
S 7–TS 20	Take x_5, stored in minor cycle 5 of DL 7, say, to TS 20.
S 0–D 19	Do the multiplication.
DS 12–DS 13	Add the old partial sum to the new product.
DS 14–DS 12	Store the new partial sum for the next round.

On to multiple right-hand sides

Loop 1, plus any preparations needed before entering it, has to be fitted into the 15 Mc between card rows. As the minimum time to carry out an instruction is 2 mc, the first of the above two blocks of instructions would take a minimum of 8 mc, or 9 mc if the count is non-zero. Those in the second block would take at least 9 mc + the (2 Mc + 1 mc) of the multiplication + 1 Mc to allow the fetching of the x-value (since that may be from any minor cycle in its DL). This totals 3 Mc + 10 mc if carried out as shown and with no time gaps between the instructions.

Therefore there was no difficulty in fitting Loop 1—for a single right-hand side—into the time between card-rows (15 Mc). However, my program was for sets of equations with *multiple* right-hand sides (that is to say, various sets of b-values in the set of pivotal equations displayed earlier). Indeed, the program had to deal with as many of these sets as I could fit into a single passage of the cards. (So saving time: if, for example, two right-hand sides were dealt with in a single pass of the cards, the computer time would be halved.) So steps 2, 3, and 4 of Loop 1 had to be carried out several times, with the same value of a_5 in the example, but several different values of x_5 and different partial sums, before exiting. These steps therefore became part of an inner loop. It was not just a matter, though, of how many times these steps could be done in the 15 Mc, since treating several right-hand sides complicated the program significantly.

Treating four right-hand sides, for example, meant that four solutions for each set of a's would be obtained. Thus four DLs had to be allocated to store the solutions, and, on each of the four passes through the inner loop, an x had to be extracted from a different DL. The instruction to do the extracting therefore had to be altered on each pass: its source number had to have unity subtracted from it. For that one needed P5, a number with unity in just the fifth digit, which was the position of the bottom digit of the Source number in an instruction. Unlike P1, used earlier, P5 was not provided as standard, and so had to be created and stored in a DL. That count number could also be used in the extraction of the different b-values, which also had to be stored in different DLs in the same minor cycle as each other.

Another complication was that DS 12 could not be used to store the accumulating sum of products (as described above), since there were now four of these products. Thus four double-length numbers had to be stored in four different DLs, but in the same positions, say minor cycles 2 and 3. Now, to build up the summation there had to be an instruction to bring down the previous summation to DS 13; and another one to put it back a major cycle later—after the addition of the new product—into the minor cycle it came from. Both these instructions, like the one above for bringing down the *x*-values, had to be modified as the program ran in order to deal with the four different DL storages. One modification needed P5, as before, the other P10, the latter requiring an additional count number of its own.

The three central steps of Loop 1 had therefore become part of an inner loop, with much more complication. With the time needed for fetching the *a*-value, and a few other necessary preparations before starting the inner loop, there was less than 15 Mc available for carrying out this loop the chosen number of times. Since a multiplication takes just over 2 Mc, the minimum time taken by a loop that contained just one multiplication and nothing else would be 3 Mc. The maximum number of right-hand sides one could hope to deal with on one pass was therefore 4.

An additional complication was the fact that the six available DLs all became packed with instructions, making it difficult to place the instructions for optimum timing. It was a great help, though, that other instructions could be carried out while a multiplication was taking place. Even so, the effort it must have taken to succeed in getting the inner loop down to the minimum 3 Mc, and so deal with 4 right-hand sides in one pass, is no doubt the reason that this is the one program that has stuck in my mind these many years.

End of program

I stopped programming in 1954 in order to concentrate on work in statistics. So it is that I claim to be the World's First Ex-Programmer (though I still have not made it into the Guinness Book of Records).

Note

I wish to thank my friends and colleagues of yesteryear, Charles Clenshaw, Frank Olver, and Mike Woodger, for their comments and corrections. My thanks also to Mike for helping me to sort out some of the fine detail. The coding of my back substitution program still exists in complete form; an extract from it, together with a reconstructed flow diagram, are available at
<www.AlanTuring.net/hayes>.

11 *The Pilot ACE: from concept to reality*

Robin A. Vowels

Development of the Pilot ACE from Turing's design

The original conception

Turing's 1945 report 'Proposed Electronic Calculator' set out a detailed design, including logic diagrams, for a stored-program general-purpose computer of the binary serial type. Turing planned for a high clock rate of 1 MegaHertz (i.e. a pulse repetition time of 1 microsecond)—about five times faster than any other computer project at that time.[1] His design specified 200 delay lines of 1024 bits—an extraordinarily large number. A water–alcohol or mercury delay line was proposed, with a preference for mercury. Turing included a detailed mathematical treatise on the practicability of the delay line as a storage medium.

The control unit had two registers; one for holding the current instruction (now called the *instruction register*) and the other for holding the position of the currently executing instruction (now termed the *instruction address register*). The machine would have subprograms—'instruction tables'—to permit the calling of a subroutine and for returning. The return address was stored in a push-down pop-up stack embodied in a delay line; a Temporary Store or TS (i.e. a register) functioned as stack pointer. A word size of 32 bits was proposed, with two instructions per word, a bit for distinguishing instructions from data, and another for the sign. There would be floating-point hardware for dual-word values, one word holding the mantissa and the other holding the exponent. Input and output were to be by means of Hollerith 80-column punch-card equipment, rather than the slow paper tape of other projects. The *pièce de résistance* of the design was the 'discrimination'—now *conditional branch*—which (implemented as software) would permit decisions

223

to be made based on computed values, thus eliminating the need for a human operator to be involved at any of the intermediate steps of the calculation.

It is clear, however, and should have been clear at the time, that the plan for this huge machine—requiring between 6000 and 10,000 radio valves (vacuum tubes)—was impracticable, on account of the considerable complexity of the control unit.[2] The ACE was to have 'an output equal to the total output of all the large machines so far constructed in the U.S.A.'.[3]

Version V

There are no extant records of Turing's intermediate versions prior to Version V (the version current when Wilkinson joined in May 1946; see Chapter 4, 'The Pilot ACE at the National Physical Laboratory').[4] However, considerable refinement of the design must have taken place to arrive at Version V. Instead of the 14 TS (Temporary Stores) principally for internal use proposed in Turing's 1945 report, Version V had only two internal TS, namely INST (the instruction register) and a TS for counting, and moreover involved an entirely different proposal for the handling of conditional branch instructions.[5] It is obvious that Version V was far more economical of hardware.

The lectures delivered by Turing and Wilkinson from 12 December 1946 to 13 February 1947 reveal the progression of development of the ACE design (see Chapter 22, 'The Turing–Wilkinson Lecture Series'). These lectures cover aspects of the design of Versions V, VI, and VII of the ACE. The lectures show that there had been significant progress in the design since Turing's 1945 proposal. Comparing Turing's 1945 report and Version V (as set out in Chapter 22), several striking features are immediately apparent:

1. The mechanism for conditional branching is completely different. In Turing's 1945 design, the conditional branch is implemented in software. In Version V, the mechanism is implemented entirely as part of the control logic, that is, it was to be realized wholly in hardware (see lecture 13). The conditional branch mechanism of Version V is similar to those of the ACE Test Assembly, the Pilot ACE, and DEUCE.
2. The control mechanism has been simplified (see lecture 13). This simplification made possible the new hardware mechanism for conditional branches. Some of the new terms (TIMCI and TRANSTIM[6]) are those used with the Pilot ACE.
3. The 1945 report includes a brief note that a zero word would serve as a useful instruction (permitting an instruction to pass from the card input

device to Delay Line o). Lecture 26 expands this thought, allowing an instruction to pass from the card input device direct to the control unit, and covering the filling of a delay line in an empty machine.

The elusive delay line

The linchpin of Turing's design, namely a *reliable* delay line, was not available in 1946 or 1947. At a meeting held on 21 January 1946, Hiscocks (Secretary to the NPL) reported that the Post Office had successfully kept a number circulating in a delay line for half an hour.[7] A subsequent report in March said that the Post Office was experiencing problems with 'digit pickup' (the appearance of extraneous digits in the delay line).[8] In April 1947 Thomas stated that '... Dollis Hill have after eighteen months' work produced a tolerably satisfactory mercury line'.[9]

Earlier in 1947, frustrated by lack of progress in the construction of hardware, Turing had made a 'breadboard' amplifier circuit and begun to conduct experiments using a drain pipe as a delay line.[10] He placed a loudspeaker at one end of the pipe to generate acoustic pulses and situated a microphone at the other end.[11] According to one witness, 'we were able to catch glimpses of Turing, on the ground floor [of Teddington Hall] and in the basement, sitting with legs off ground, gingerly probing a mass of wires, and fiddling about with drain pipes'.[12]

In Chapter 4, Wilkinson recalled that even as late as 1950, the 'probability of remembering a pulse pattern for as long as a minute was not very high'. In the end, an adequately reliable delay line was not produced until 1951.

Thomas's minimal ACE—a digression

It was decided early in 1947 that the ACE project would be transferred to the Electronics Section of Radio Division under Thomas (see Chapter 3, 'The Origins and Development of the ACE Project'). Smith-Rose, Superintendent of Radio Division, asked Thomas to prepare a plan for the production of the ACE, which he did in a report dated 12 April 1947.[13] Womersley was presented with a *fait accompli* when he received Thomas' report together with a memo from Smith-Rose inviting him to a meeting to discuss staffing arrangements for the transfer.[14]

Thomas' name was not on the list of attendees at the Turing–Wilkinson lectures, held a few months earlier, and it is strange that Thomas should have been given a copy of 'Proposed Electronic Calculator' from which to produce his plan, rather than one of the then current (and simpler) Versions V,

VI, or VII. Was Thomas' plan prepared without the knowledge of the ACE group in the Mathematics Division?[15] The ACE group would not have been impressed by this outsider's gaffes—which included using the wrong name for the ACE: 'Analytical Computing Engine'. To his credit, however, Thomas proposed a prototype with only three delay lines, two temporary stores, and one adder.

Thomas was sceptical about the high speed proposed by Turing (1 MHz), quoting from the EDVAC Report to support the view that pulse rates above 100,000 Hz were not achievable: 'There is a practical lower limit to the pulse time which may be used and this appears to be about 10 microseconds.'[16] Thomas cited problems about wire lengths, when he should have been aware that a simple cathode follower circuit would provide the low impedance needed for long inter-chassis connections.[17] In estimating equipment needs he relied on a figure of 8 valves for each delay line, whereas the Post Office was using 16.[18]

The Test Assembly

Meanwhile, in the Mathematics Division, events were moving somewhat in parallel. In the spring of 1947 Huskey convinced the Maths Division ACE group to build a small machine based on Version V of the ACE.[19] This was called the *ACE Test Assembly* (see Huskey's chapter 'The ACE Test Assembly, the Pilot ACE, the Big ACE, and the Bendix G15').[20] The embryonic computer began to take form from about May 1947.[21] Later that year a report entitled 'ACE Test Assembly' set out specifications that bore a striking resemblance to the Pilot ACE as it was finally built.[22]

The ACE Test Assembly was to have eight 1024 microsecond (μs) delay lines (DL) of 1024 bits, and six 32 μs Temporary Stores of 32 bits. Two of the TS were to be connected together to give 64 bits. Designed as a transfer machine, it had 32 Sources and 32 Destinations. Several Sources provided the convenient constants of 1, 2^{16}, 2^{31}, 0, and ones. Each instruction consisted of five fields: Source, Destination, number of words to be transferred, the Delay Line number, and relative word number of the next instruction. Discrimination instructions provided the means for decisions based on computed values. Input and output devices were included in the design, and incorporated Hollerith punch-card equipment.

A complete and detailed logic design of the control unit was provided, with explanations for the diagrams. The logic diagrams used terms that were retained for the Pilot ACE and DEUCE. Programming details were also provided, including a description of the manner in which instruction tables were to be input, and the way in which the initial orders and the instructions were to be punched on the cards.[23] (The means of entering a new program, while feasible, was not practical or convenient.)

The ACE Test Assembly was to include a multiplier unit. It seems strange that the project, struggling to achieve the bare basics, should be contemplating a multiplier, the logic and construction of which would prove to be as complex as the control itself—in the end, the multiplication unit was not built until 1951. The Test Assembly was a substantial machine, virtually a complete Pilot ACE or DEUCE, when all that was needed was a delay line that worked, a control unit with TS, and an arithmetic unit with TS. And the control was more complex than was needed—a control unit taking instructions sequentially from storage would have been sufficient to demonstrate that the machine was capable of running at the speeds planned by Turing.[24]

Effort was not devoted single-mindedly to the ACE Test Assembly. Just how much the project was thrown off-course is evident in the curious roster dated 21 November 1947.[25] Entitled 'A.C.E. Problems', it assigned responsibilities to Wilkinson and Davies for drawings for a '40–41 digit machine' (Version V and the ACE Test Assembly were 32-digit machines), to Alway for programming, to Wilkinson for a multiplier unit, and to Gill for a divider circuit. Huskey was responsible for setting up the instruction digits from INST (the instruction register) and for the 'short tanks for instructions'. This effort had little to do with the Test Assembly. Apart from Huskey's work, which would be relevant to any of the versions, the rest of the effort was directed toward Version VI or VII of the ACE, and certainly not to a pilot model.

In the end the Mathematics Division Test Assembly was stopped in its tracks by NPL bureacracy (see Chapter 3, 'The Origins and Development of the ACE Project'). The ACE group was left to write a lengthy report[26] (because, Wilkinson said, Womersley 'decided we really must have something to show for our work'[27]).

The Pilot ACE in earnest

The ACE Section devoted 1948 to the development of numerical subroutines but 1949 saw a fresh start on hardware. A request for new funding was made on 16 February 1949[28] and subsequently approved by DSIR. By August 1950

Bullard, the new Director of the NPL, was able to report to the DSIR that the Pilot ACE was complete and could do simple calculations.[29] (Work on the full-scale engine was expected to start in September of that year.) However, the Pilot ACE was still not yet at the stage where it could be used by the NPL, for it had only one DL and one TS with an adder, and lacked input–output equipment. There was no multiplier unit. Moreover, work was needed to make the delay line storage reliable. It was not until 1951 that these enhancements had been effected and the machine was working again.

In 1954 an experimental magnetic drum was installed. Although small in capacity, the drum offered 32 words on each of 32 tracks. 1024 words of storage was more than double the size of the high speed store and there was the promise of more to come—a drum with 128 tracks. Other enhancements included improved facilities for addition, subtraction, and the transfer of double words (whether of 64-bit integers, floating-point values, complex values, or other data 'aggregates').[30]

How much of the Pilot ACE design was Turing's? It has generally been accepted that the Pilot ACE (and therefore the DEUCE) were 'based' on a design by Turing. However, the dependence is much greater than has previously been acknowledged. The Turing–Wilkinson lectures provide definitive evidence that almost all the design was Turing's, with assistance from Wilkinson (although the design of the electronics was largely done by Wilkinson and his team). Version V—which Turing had nearly completed when Wilkinson started at NPL on 1 May 1946—was stripped down for the ACE Test Assembly and this design was subsequently taken forward to the Pilot ACE. The logic of the multiplier was already part of the ACE design (Chapter 22, lecture 16). The only things that were added or changed after Turing left the project were the combination of INST and the count register into a single register called COUNT,[31] and the redesign of Initial Input,[32] including the introduction of the 'Stop' instruction that gave the machine a halt state as well as a run state. (The modifications considerably simplified the input of programs and the interface with the input–output equipment.)

General description of the Pilot ACE

The Pilot ACE was a general purpose computer of the serial type, with a clock frequency of 1 MHz. The word size was 32 bits, and dual arithmetic units were provided. One arithmetic unit supported 32-bit numbers (about 9 decimal digits), while the other was for 64-bit numbers (about 19 decimal digits).

The 64-bit accumulator was also capable of functioning as two independent 32-bit accumulators. A multiplication unit was provided, but no divider. The Pilot ACE had Hollerith 80-column punch-card input and output equipment, the input device running at 200 cards per minute and the output device at 100 cards per minute. Only 32 columns of the 80 available were used for data.

The main store (the delay-line memory) consisted of eleven 32-word DLs, supported by five 32-bit TSs[33] and two 64-bit double-word stores (DS). One DS incorporated an accumulator that operated in conjunction with the multiplier.

Words, minor cycles, and major cycles

Timing considerations necessitated the introduction of terms to represent events. Each mercury delay line of the main store held 1024 binary digits, organized as 32 words of 32 bits each (see fig. 1). Data was held acoustically. Electronic pulses representing the digits were converted to mechanical vibrations or pulses by a crystal placed at one end of a column of mercury. The acoustic pulses were received at the opposite end of the column 1024 microseconds later, by another crystal that converted the pulses back to electrical form. The electrical pulses were amplified and fed back to the opposite end of the mercury column. The acoustic pulses were thus circulated indefinitely until some or all were replaced. The time taken to circulate the 1024 bits was termed a *major cycle*. A group of 32 consecutive bits, representing a useful number, was termed a *minor cycle*. There were thus 32 consecutive minor cycles (or words) stored in a delay line. Any one of them could be accessed by waiting for it to emerge from the end of the delay line.

In contrast, the cycle time of the 32-bit stores was 1 minor cycle (or 32 microseconds); thus there was no waiting time for data from these stores, and their content could be transferred to or from any word of the main stores. The double-word stores were of 64 bits, with a cycle time of 2 minor

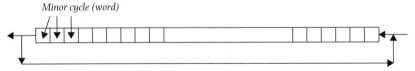

Minor cycle (word)

Fig. 1 A mercury delay line of 1024 bits (microseconds), composed of 32 minor cycles or words. The lower line represents the electronic circulation path outside the delay line.

cycles (64 microseconds). The maximum waiting time to access an individual word of these stores was 1 minor cycle, while the minimum time was 0 minor cycles. The lower 32 bits of these stores could be transferred to an even-numbered word of store, and the upper 32 bits could be transferred to an odd-numbered word of store.

The provision of the double-word stores bridged the gap between the relatively slow access time of the main store and the fast access time of the registers. They were also intended to facilitate double-precision arithmetic.

The architecture of the Pilot ACE

Outline

The Pilot ACE was organized as a series of stores (delay lines) attached to a *Highway* (data bus). Each store had a separate data path from its output terminal to the Highway, and to its input terminal from the Highway. The Temporary Stores were also connected to the Highway. Additional connections were made to the Highway by means of which arithmetic, logical, and other miscellaneous operations were performed. The card reader and punch were attached to the Highway. A separate highway called the *Instruction Highway* connected eight of the delay lines to the control unit and made it possible for instructions to be routed to the control unit while an instruction was being executed. (These highways and connections are set out in fig. 2.) Turing's 'Highway' can be seen in the Unibus of Digital Equipment Corporation's computers[34] and in microprocessor systems in which the input–output equipment and every register and memory are connected to a single bus.

The Pilot ACE differed considerably from other computers. Present-day machines typically have, say, an *add* instruction that consists of an *operation code* that signifies *add*, and which specifies any two registers that contain the values to be summed. As discussed in previous chapters, an addition instruction in the Pilot ACE specified only the addresses of the stores involved. Data was transferred (copied) from a particular store to one specific Destination address; the operation of addition was implied in the Destination address. Thus, although Temporary Store 16 (TS 16) was the store associated with the accumulator, data sent to Destination 17 (D 17) was added to it. Data sent to Destination 18 was subtracted from it. To store data in TS 16, data was sent to Destination 16 (see fig. 3).

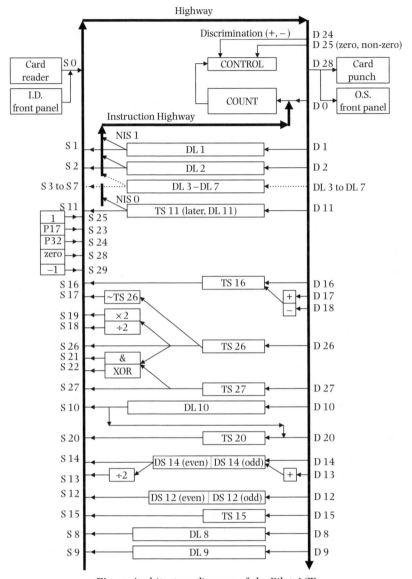

Fig. 2 Architecture diagram of the Pilot ACE.

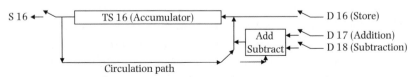

Fig. 3 Organization of the accumulator and TS 16, showing its three Destinations: 16, 17, and 18.

Every instruction specified a Source and a Destination, and the particular operation required was implicit in either the Source address or the Destination address (or both). The Pilot ACE may thus be considered to be a *transfer machine*. An instruction could specify the transfer of from 1 to 32 words from the Source to the Destination. In general, data could be transferred from any Source to any Destination, whether it was a Temporary Store, one of the Delay Lines comprising the main store, or to or from the input–output equipment. Transfers between two delay lines of the main store were restricted to transfers between word(s) in corresponding position(s), because a word could be transferred only when it was emerging from a delay line; the word had to be received by the Destination delay line at that same moment. Some Temporary Stores had more than one input and/or more than one output (we have already seen that the accumulator TS 16 had three inputs: one when it was desired just to transfer (copy) a word into it, one for addition, and a third for subtraction).

For the convenience of the user, words in a Delay Line were distinguished from each other by a system of numbering the minor cycles. Thus the first word was known as minor cycle 0, the second as minor cycle 1, and so on, up to minor cycle 31. Minor cycle 0 of each Delay Line emerged simultaneously, as did minor cycle 1, and so on. To facilitate the writing of programs, minor cycles were identified by Delay Line and position. For example, the first minor cycle of DL 4 was written 4_0, the second as 4_1, the third as 4_2, and so on up to 4_{31}. Minor cycle 0 can be thought of as being at the 'top' of a delay line and minor cycle 31 at the 'bottom'. (The identification of one particular minor cycle as minor cycle 0 was entirely arbitrary.)

Another fundamental difference between the Pilot ACE and other machines of the period lay in the *duration* of the transfer. A transfer of more than 1 minor cycle—a *multiple-word* transfer—enabled consecutive words of a delay line to be copied to another, or enabled the content of a delay line to be cleared (initialized to zero). A transfer of more than 1 minor cycle to the adder terminal of the accumulator enabled multiple copies of a register to be added to the accumulator. In another Temporary Store, TS 26, a transfer of n minor cycles enabled a shift of n places. In a transfer a block of words could be tested to determine whether any was negative or whether all were zero. In the case of the card reader, a transfer of more than 1 minor cycle propagated a copy of the word at the reading station into consecutive words of a delay line. (A transfer of more than 1 minor cycle to the card punch had the effect (curiously) of performing a *logical or* on the words transferred.) Turing's multiple transfer and vector arithmetic may be seen in vector processors from the 1970s.

Optimum coding

This topic has been discussed in previous chapters. Because the storage was serial in nature, instructions in a program were not placed in consecutive words of store. If they were, the next instruction would have passed the output terminal of the delay line before the current instruction could be executed; consequently a complete cycle of 32 minor cycles would be wasted before the next instruction would be ready to emerge again from the delay line. Ideally, instructions were placed so that while the last word of data was being transferred by the current instruction, the next instruction was emerging from a delay line and was being transferred into control.

Because instructions to be executed in sequence could be placed in arbitrary locations in store, each instruction specified the *position* of the next. Positions were relative. Thus each instruction specified the delay line where the next instruction was to be found, as well as the relative position of that instruction in that delay line. The relative position was called the *Timing Number*.

Transfer instructions had to wait for the particular word that was to be transferred to emerge from the delay line. Consequently, each instruction specified a *Wait Number*. Since programmers placed instructions in main memory so as to minimize waiting, the Wait Number was zero or a small number for many instructions. The Timing Number similarly was kept small, or at least greater than or equal to the Wait Number. A Timing Number less than the Wait Number caused a loss of one major cycle, and was avoided except for certain operations (and in cases where the extra time taken was unimportant).

Because each instruction specified the relative location of the next instruction, the system of addressing was called *two-plus-one*: instructions contained a Source address, a Destination address, and the address of the next instruction.

Entering a new program

To resolve the difficulty of entering a new program into an empty machine, Turing made use of the fact that in an empty machine all the stores— including the store in the control unit called INST, which held the instruction currently being executed—contained zero words. (INST was renamed COUNT in the Pilot ACE). A zero word appeared as an instruction having Source 0 and Destination 0. By treating the card reader and the operator's console as Source 0, and INST (or COUNT) as Destination 0, instructions could be passed directly from the card reader to INST (COUNT) when the store was initially empty. Certain of these instructions, called *filler instructions* (now *bootstrap instructions*), enabled the store to be filled with instructions read from punch cards (see below). These were fed into the card reader by the operator.

Conditional branching

A means of testing for zero and non-zero was provided. If the test yielded zero, the next instruction was taken in the usual way. If the test yielded non-zero, however, the instruction in the word *following* the normal next instruction was taken. A similar facility was provided to test for positive and negative. Loop control was handled using one of these tests, or in a comparable manner involving modification of the current instruction.

The starting of input and output equipment and the multiplier was achieved by specifying a particular Destination address. In this special case, no transfer took place (and the Source address was ignored).

Hardware constants

Certain of the Source addresses supplied constants that were needed on a frequent basis, for example, zero, unity, 2^{16} (P17), 2^{31} (P32), and minus 1 (32 ones). (See fig. 4.) Other small constants could be generated by a multiple transfer to one of the accumulators.

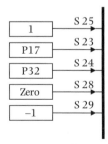

Fig. 4 Sources providing constants.

Control

A glimpse of the way in which the control operates the machine is achieved by considering COUNT. A single Temporary Store (register) like any other in the machine, COUNT would now be termed the instruction register. Before an instruction could be executed, it had to be brought into COUNT (which was Destination 0). While an instruction was stored in this register its Wait Number and Timing Number were each decremented by unity as each and every minor cycle passed. When the Wait Number counted down to zero, the instruction could be executed. When the Timing Number counted down to zero, the next instruction could be brought into COUNT (provided that execution of the instruction was complete or due to complete by the end of the current minor cycle).

In the event that an instruction could not be executed for some reason (e.g. because its Go Digit was 0), the Wait Number and Timing Number would continue to count down (modulo 32) with each passing minor cycle. Once every major cycle (every 32 minor cycles or every 1024 microseconds), the Wait Number counted down to zero, but execution could not take place. Likewise the Timing Number counted down to zero once every major cycle. Counting continued until the inhibition was removed. In the case of a Stop instruction (i.e. an instruction with Go Digit 0), a *Single-Shot* from either the operator or the card equipment released the instruction for execution. The operator gave a Single-Shot by pressing a key on the console of the machine.

There were two paths to COUNT. One was from certain of the Delay Lines via the Instruction Highway and the other was to Destination 0 via the Highway. The path via the Instruction Highway was the normal way in which instructions reached COUNT. The path from the Highway enabled instructions to be sent directly from any store. When an instruction had been modified (e.g. for indexing, arithmetic shifting), and when returning from a subroutine, the instruction would be sent from a register to COUNT.

Figure 5 shows the Instruction Highway linking DL 1 to COUNT. (Delay Lines 2 to 7 and 11 were similar.) The gates (switches) are shown selecting Next Instruction Source 1 (NIS 1), ready to send the next instruction to COUNT. The connection to NIS 1 was at the input side to DL 1, after the switch (*), and not at the exit end. This was to enable a transfer to be made to a DL and to allow the next instruction to be taken from the new content of that same DL. In all other cases, it was convenient to conceptualize the

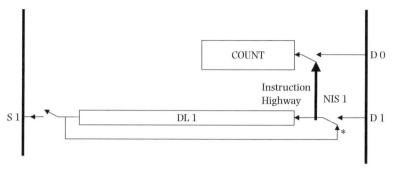

Fig. 5 Instruction Highway (only DL 1 shown; DLs 2–7 and 11 are similar).

NIS as effectively taking the output from the DL in question, because when the circulation path was complete (as shown in fig. 5), the input to (and output from) the DL were identical.

Programming the Pilot ACE

> This process of constructing instruction tables should be very fascinating. (A. M. Turing, 'Proposed Electronic Calculator')

Considering some aspects of programming the Pilot ACE will provide insight into the operations of the machine. The operations may be classified as simple transfers, arithmetic operations, logical operations, multiple word operations, and discriminations. This section gives examples of instruction sequences which perform various typical transfers and arithmetic operations. The next section gives some examples of complete programs. (It will be beneficial to refer to the Pilot ACE architecture diagram in fig. 2 and to the table of Sources and Destinations in Table 1.)

Transfers

One of the simplest operations was the transfer of data from one Temporary Store (TS) to another. To transfer (copy) the content of TS 15 to TS 16, the instruction was written:

15–16.

In transferring a word to either of the two double-word stores (DS) of 64 bits, it was necessary to specify which word of the store was to receive it. The words in a double-word store were referred to as *even* and *odd,* corresponding to an even-numbered minor cycle of a DL and an odd-numbered minor cycle, respectively.

To transfer the content of TS 15 to the even minor cycle of DS 12, one wrote

15–12 (e)

and to transfer the content of TS 15 to the odd minor cycle of DS 12, one wrote

15–12 (o),

where (e) and (o) signified even and odd, respectively. To transfer the content of minor cycle 27 of DL 4 to TS 15, one wrote

4₂₇–15.

Several Sources provided useful constants. S 28 supplied zero, and S 25 provided unity, for example. To clear TS 15 (that is, to initialize it to zero), one wrote

28–15.

To initialize TS 15 to unity, one wrote

25–15.

A special case involved the transfer of an instruction direct to control. The instruction

16–0

transferred the content of TS 16—assumed to be an instruction—to COUNT (control), where it was executed.[35]

Addition and subtraction

To add the content of TS 15 to the accumulator TS 16, one wrote

15–17,

which transferred the content of TS 15 to Destination 17 (which was the adder input to TS 16—see fig. 3). Similarly, to subtract the content of TS 15 from TS 16 one wrote

15–18,

Destination 18 being the input associated with the subtraction facility of TS 16.

A commonly required operation was to double the number in the accumulator TS 16. To do that, one wrote

16–17,

which added a copy of TS 16 to itself, via the adder input, Destination 17.

Logical operations

The logical operations unit was associated with the two stores TS 26 and TS 27. Sources 26 and 27 transferred the content of TS 26 and TS 27, respectively, to any desired location. Source 21, however, supplied the *logical and* of TS 26 and TS 27, while Source 22 supplied the *exclusive or* (XOR) of TS 26 and TS 27 (see fig. 6). To form the *logical and* of the contents of TS 26 and TS 27, placing the result in TS 15, one wrote

21–15.

To form the *exclusive or* of the contents of TS 26 and TS 27, placing the result in TS 15, one wrote

22–15.

Two other addresses were associated with TS 26, Source 18, and Source 19. Source 19 provided double the number that was in TS 26; Source 18 provided the contents of TS 26 shifted down by one place, that is, half[36] (see fig. 7). To shift the number in TS 26 one place up,[37] that is, to double it, one wrote

19–26.

To place double the content of TS 26 in TS 15 one wrote

19–15.

This instruction did not change the content of TS 26. Similarly, to halve the number in TS 26, one wrote:

18–26.

(Note that the shift down, halving, was an arithmetic shift, that is, a division by a power of 2.)

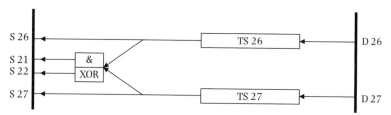

Fig. 6 TS 26, TS 27, and logical operations.

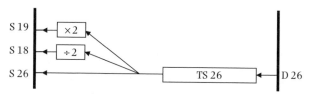

Fig. 7 Shifting using TS 26.

The multiplier

The unsigned integer multiplier was associated with DS 14 and TS 20. To multiply two 32-bit positive integers, the multiplicand (x) was placed in TS 20, the multiplier (y) was placed in DS 14 (odd), and DS 14 (even) was set to zero. The instruction 0–19 started multiplication. The 64-bit product was gradually formed over the next 2 milliseconds in DS 14, and at the end of that time could be used. The low-order 32 bits were held in DS 14 (even) and the high-order 32 bits were in DS 14 (odd).

When signed numbers were to be multiplied, a software correction was required (this is covered in Example 6). The correction was prepared during the multiplication and was added when the multiplication was complete. Note that the multiplier was *asynchronous*, that is, once multiplication had commenced with the 0–19 instruction, any other instructions could be executed provided that those instructions did not modify the content of TS 20 and DS 14.

The following sequence of instructions initiated a multiplication of unsigned numbers. The multiplicand is assumed to be in DL I_0 and the multiplier in DL I_3.

I_0–20	Fetch the multiplicand to TS 20.
I_3–14 (o)	Fetch the multiplier to the upper word of DS 14.
28–14 (e)	Clear the low-order word of DS 14.
0–19	Start multiplication.

The product was available in DS 14 after 2 major cycles (2 milliseconds).

The Double Store DS 14

DS 14 could be used as one 64-bit accumulator or as two independent 32-bit accumulators. When the instruction

0–23 (1)

was executed, DS 14 behaved thereafter as two independent 32-bit accumulators ('(1)' indicating a Characteristic of unity). When the instruction

0–23

was executed, DS 14 behaved thereafter as a single 64-bit accumulator.

Multiple-word transfers

Multiple-word transfers considerably enhanced the utility and speed of the machine. They permitted a wide variety of operations. The following examples use Sources 25 and 28, which supplied the frequently used constants unity and zero respectively. Consider first the instruction

25–17.

This increments the content of TS 16 by unity. Source 25 supplies unity, which when transferred to Destination 17 is added to TS 16. The effect of a multiple-word transfer, using this same instruction, is to add the constant 18 to TS 16. In

25–17 (18 mcs)

the notation '(18 mcs)' indicates that the transfer is to take place for 18 minor cycles. The instruction added unity to TS 16 eighteen times.

The instruction

28–14 (2 mcs)

placed zero in both words of DS 14 (that is, *cleared* DS 14). The instruction

28–1 (32 mcs)

cleared all 32 words of DL 1.

In the subsection 'Addition and subtraction' it was mentioned that the instruction 16–17 doubled the content of the accumulator TS 16. A multiple transfer had the effect of an arithmetic shift (multiplication by a power of 2). For example, to multiply the content of TS 16 by 16 (i.e. a shift up by 4 places) one wrote

16–17 (4 mcs).

A similar operation could be performed in the arithmetic shift register TS 26 as follows:

19–26 (4 mcs),

which shifted TS 26 up by 4 places, while the instruction

18–26 (4 mcs)

shifted down by 4 places (that is, divided TS 26 by 16).

Even DS 14 behaved as an arithmetic shift register. For example, the instruction

14–13 (e, o)

added DS 14 to itself, thus doubling its contents, while the instruction

14–13 (6 mcs)

added DS 14 to itself 3 times, effectively multiplying the double-word contents by 8. By the same token, the instruction

13–14 (e, o)

halved the content of DS 14, and

13–14 (6 mcs)

performed an arithmetic shift down by 3 places, that is, divided DS 14 by 8. (Note that since DS 14 was a double store, the transfer had to be for 2 mcs to double or halve the entire content, and 6 mcs to shift by three places.)

Discrimination (conditional branching)

Following a test for zero, the machine executed the next instruction as usual when the value tested was zero, and executed the instruction in the following minor cycle when the value tested was non-zero (see above).[38] Thus, if the instruction

15–25

normally took the instruction in minor cycle 20 of DL 4 as the next instruction, and the content of TS 15 were zero, then the instruction in 4_{20} was executed next. If the content of TS 15 were non-zero, then the instruction in 4_{21} was executed next. The test was written showing the two paths, one leading to 4_{20} and the other to 4_{21}:

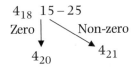

(Here 4_{18} is the location of the instruction 15–25.) Destination 25 was said to *discriminate on zero*.

Destination 24 was said to *discriminate on sign*. The following instruction sequence tests whether the accumulator TS 16 is positive or negative.

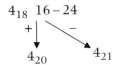

The instruction in 4_{20} is executed next if TS 16 contains a zero or positive value, and the instruction in 4_{21} is executed if TS 16 contains a negative value.

To test whether any word of a group was non-zero, a single instruction was required. The following sequence checks whether both words of DS 14 are zero (e.g. following a double-word arithmetic operation).

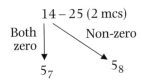

The sequence

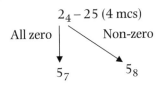

tests four consecutive words of DL 2, namely 2_{4-7}.

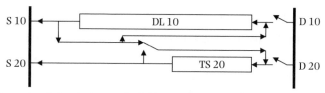

Fig. 8 The normal circulation path of TS 20 is interrupted while connected to DL 10.

Special effects

DL 10 had a special connection to TS 20. Normally this connection was disabled. Executing the instruction

0–21 (1)

enabled the connection. This connected DL 10 to the input terminal of TS 20, disabling the normal circulation path of TS 20 (see fig. 8). A word emerging from DL 10 was thus fed into TS 20, and emerged from TS 20 one minor cycle later. The normal operation of DL 10 itself was unaffected.

This special connection enabled the content of DL 10 to be moved into DL 10 but one minor cycle later. The instruction

20–10 (32 mcs)

caused DL 10_{31} to be moved to DL 10_0, DL 10_{30} to be moved to DL 10_{31}, DL 10_{29} to be moved to DL 10_{30}, and so on. The facility enabled the systematic processing of all 32 words of a delay line, rotating them by one word after each value had been processed.

An alternative use of the facility provided a push-down stack. For example, the following instructions caused the content of DL 10 to move down by one word, leaving room at 10_0 for a new item to be inserted:

0–21 (1)	Enable the connection between DL 10 and TS 20.
20–10_1 (31 mcs)	Move 10_0 through 10_{30} into 10_1 through 10_{31}, respectively.

To cancel the connection, the instruction

0–21

was executed.

Instruction format

As explained in an earlier chapter, the Pilot ACE instruction word consisted of seven parts: the Source, the Destination, the Characteristic (which specified whether one or more words should be transferred), the address of the next instruction (as two components: a Next Instruction Source and a Timing Number), a Wait Number, and a Go bit. They were set out as shown in fig. 9 (the least significant bit is on the left, and the bit positions are given underneath).

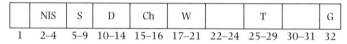

	NIS	S	D	Ch	W		T		G
1	2–4	5–9	10–14	15–16	17–21	22–24	25–29	30–31	32

Fig. 9 The Pilot ACE instruction word.

Notes: NIS = Next Instruction Source (a Delay Line number in the range 0–7; DL 11 is NIS 0); S = Source address; D = Destination address; Ch = Characteristic; W = Wait Number; T = Timing Number; G = Go bit.

If the Go bit was 1 the instruction was executed as soon as possible. If the Go bit was zero, the machine halted (in this case the instruction was said to be a 'Stop instruction'). The machine resumed only upon the receipt of a Single-Shot, which could be given by the computer operator at the console of the machine (also by the card reader, when the next row of a card was ready to be read, or by the card punch, when the next row of a card had reached a position in the mechanism where it could be punched).

The Characteristic specified whether the transfer was to be for one word or for multiple words. If the Characteristic was 1, the transfer was for one word (the normal case). If the Characteristic was zero, the transfer started from the beginning of execution of the instruction and continued until the next instruction was fetched into control; that is, the transfer moved $T-W+1$ words (modulo 32).

If the Characteristic was three, the transfer was for two minor cycles, and was intended to simplify operations on 64-bit values. To transfer the 64-bit value in DS 14 to DS 12 one wrote

    14–12 (2 mcs).

To transfer two words from DL $56,7$ to DS 14 one wrote

    56–14 (2 mcs).

Table 1 lists each Source and Destination address and the store or function associated with that address.

Table 1   Source and Destination addresses

Source address	Function	Destination address	Function
0	Input from the card reader or from the keys at the front panel	0	COUNT (Instruction register)
1	DL 1	1	DL 1
2	DL 2	2	DL 2
3	DL 3	3	DL 3
4	DL 4	4	DL 4
5	DL 5	5	DL 5
6	DL 6	6	DL 6
7	DL 7	7	DL 7
8	DL 8	8	DL 8
9	DL 9	9	DL 9
10	DL 10	10	DL 10
11	DL 11[39]	11	DL 11
12	DS 12	12	DS 12
13	DS 14÷2	13	Add to DS 14
14	DS 14	14	DS 14
15	TS 15	15	TS 15
16	TS 16	16	TS 16
17	~TS 26 (ones complement)	17	Add to TS 16
18	TS 26÷2	18	Subtract from TS 16
19	TS 26 × 2	19	Stimulate multiply (Source is irrelevant)
20	TS 20	20	TS 20
21	TS 26 & TS 27	21	Modify Source 20 via a connection to DL 10
22	TS 26 ≢ TS 27	22	Unassigned
23	P17 (2^{16})	23	0–23 makes DS 14 behave as a 64-bit accumulator; 0–23 (1) makes DS 14 behave as two 32-bit accumulators; (modifies Source 13 and Destination 13)
24	P32 (2^{31})	24	Discriminate on sign
25	P1 (unity)	25	Discriminate on zero
26	TS 26	26	TS 26
27	TS 27	27	TS 27
28	Zeros	28	Output to the card punch or to the front panel lights
29	Ones (−1, or P1 to P32)	29	Buzzer
30	Non-zero on last row of card	30	Stimulate the card punch to punch one card (the Source is irrelevant)
31	Unassigned	31	Stimulate the card reader to read one card (the Source is irrelevant)

The Pilot ACE instruction card

Figure 10 shows an ACE Pilot Model Instruction Card. It is a standard 80-column punch card. The card reader read only the first 32 columns. The twelve rows of the card are identified by the lettering Y, X, 0, 1, 2, 3, 4, 5, 6, 7, 8, 9 from top to bottom in columns 1, 17, 33, 54, 66, and 74. The alphabetic codes are shown in columns 51 to 53. In the first 32 columns, the headings 'N.I.S., SOURCE, DEST., CH, WAIT, TIMING, GO' along the top of the card, together with the vertical lines, helped the programmer to read binary instructions punched on an instruction card. Column 35 has a box in row 9 (called the *nines row*) which marks the place for a hole that would cause the following card to be fed.

Table 2 Next Instruction Sources and their corresponding Delay Line numbers

DL 1	NIS 1
DL 2	NIS 2
DL 3	NIS 3
DL 4	NIS 4
DL 5	NIS 5
DL 6	NIS 6
DL 7	NIS 7
DL 11	NIS 0

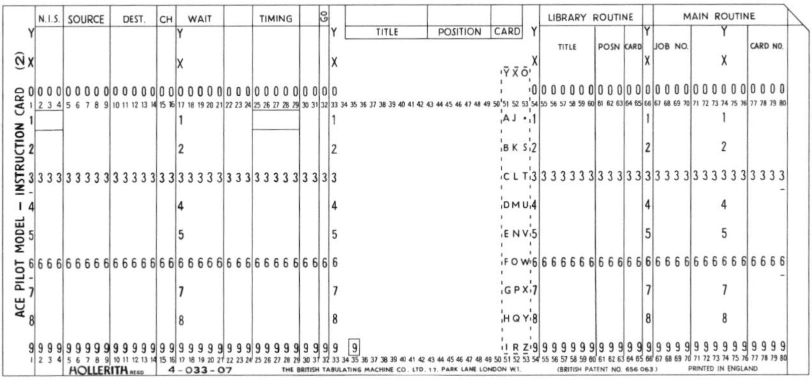

Fig. 10 An instruction card for the Pilot Model ACE (actual size $7\frac{3}{8}''$ by $3\frac{1}{4}''$).

Some programs

Example 1

This example concerns the reading of binary data from the input device. The program forms the sum of n binary integers ($n < 12$). The integer n and the values to be summed are punched on one card.

The card reader is started and one binary integer n is read from a punch card. The instruction 0–16 to read this number n is a Stop instruction; the machine waits for a signal from the card-reader, indicating that the first row of the card is in a position where it may be copied to TS 16, before executing the instruction. The instruction is written 0–16X, the 'X' signifying that the machine must wait (in the corresponding machine instruction, the Go bit is 0). The program then goes on to read in n further binary integers. As each integer is read in, the integer is added to the partial sum being formed in DS 14 (e). (14 (e) refers to the even-numbered minor cycle of DS 14, 14 (o) to the odd-numbered minor cycle, and 14 (e, o) to both minor cycles.) The instructions are as follows.

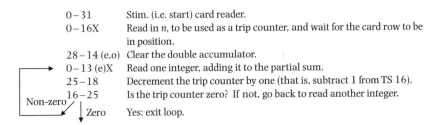

0–31	Stim. (i.e. start) card reader.
0–16X	Read in n, to be used as a trip counter, and wait for the card row to be in position.
28–14 (e,o)	Clear the double accumulator.
0–13 (e)X	Read one integer, adding it to the partial sum.
25–18	Decrement the trip counter by one (that is, subtract 1 from TS 16).
16–25	Is the trip counter zero? If not, go back to read another integer.
Non-zero	
Zero	Yes: exit loop.

The sum is left in DS 14 (e), and the card reader stops automatically after reading the card.

Example 2

This program forms the sum of all 32 words of a DL (DL 5 is chosen). It would be used, for example, in forming a sum check of an array, or to sum the elements of an array or the rows of a matrix.

28–16	Clear the accumulator TS 16.
5–17 (32 mcs)	Add all 32 words of DL 5 to TS 16.

Since all 32 words of a DL would usually be cleared prior to storing values in it, these same instructions would also serve to sum the first n words of a DL.

Example 3

The program in this example multiplies two positive integers (or two signed integers when the product is smaller than 2,147,483,648). The code assumes that the multiplier is in TS 26, and the multiplicand is in TS 20. The multiplier unit requires that the multiplicand be in TS 26, and that the multiplier be in DS 14 (odd). DS 14 (even) must be set to zero prior to starting the multiplier. It is usually convenient to place the multiplier first in a TS before copying it to DS 14 (odd) (it can be copied from *any* minor cycle in any DL).

Location	S	D	
I22	28 – 14 (e)		Clear the lower word of DS 14 (i.e. the word in the even minor cycle).
I25	26 – 14 (o)		Place the multiplier in the upper word of DS 14 (i.e. in the odd minor cycle).
I27	0 – 19 (mc 29)		Stim. (i.e. start) multiplication in an odd minor cycle.
I28	1 – 1 (mc 31)		Waste time.
I30			The result is now available in DS 14.

If the product is smaller than 2,147,483,648 it may be taken from DS 14 (even), because the result does not then exceed the capacity of a word.

It is necessary to include an instruction to waste time, since multiplication takes 65 mcs (or just over 2 milliseconds). The instruction 1–1 transfers 1 minor cycle of DL 1 to itself and thus changes nothing. More than two major cycles elapse in order to allow the result to be formed in DS 14.

Example 4

This example illustrates the multiplication of two signed integers or signed fractions. As the hardware multiplier treats the operands as unsigned, sign correction must be performed. In this program, advantage is taken of the fact that the multiplier is asynchronous. Thus preparation of the correction may be performed while multiplication is in progress. The correction is added after the conclusion of multiplication. The correction is formed in TS 16. Before executing the code, the multiplier is placed in TS 26, and the multiplicand is in TS 20. The product is formed in DS 14.

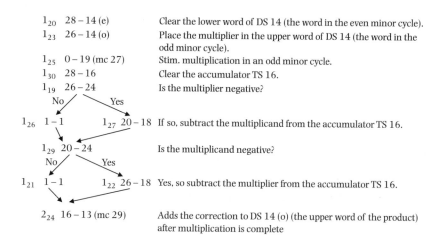

1_{20} 28 – 14 (e)   Clear the lower word of DS 14 (the word in the even minor cycle).

1_{23} 26 – 14 (o)   Place the multiplier in the upper word of DS 14 (the word in the odd minor cycle).

1_{25} 0 – 19 (mc 27) Stim. multiplication in an odd minor cycle.

1_{30} 28 – 16   Clear the accumulator TS 16.

1_{19} 26 – 24   Is the multiplier negative?

1_{26} 1 – 1 1_{27} 20 – 18   If so, subtract the multiplicand from the accumulator TS 16.

1_{29} 20 – 24   Is the multiplicand negative?

1_{21} 1 – 1 1_{22} 26 – 18   Yes, so subtract the multiplier from the accumulator TS 16.

2_{24} 16 – 13 (mc 29)   Adds the correction to DS 14 (o) (the upper word of the product) after multiplication is complete

Example 5

This example concerns the punching of data at the output device. The program punches out on cards the contents of a DL (say DL 4, which contains 32 words). This is useful if the DL contains some results or instructions that the operator wishes to inspect.

A loop is used. An instruction to fetch a word from the Delay Line needs to be modified each time around the loop, in order to fetch successive words of DL 4 (starting with the word in minor cycle 0). This instruction is termed a *pro forma* instruction (also known as a 'quasi instruction'): its Wait Number is incremented by 1 each time around the loop. The loop terminates after 32 iterations because on the thirty-second execution of the pro forma instruction, a carry is propagated from the Wait Number into the Timing Number, increasing its value by 1. For the first 31 executions of the loop, the instruction 4–28 leads to the instruction at 3_4. On the thirty-second execution, the pro forma instruction leads to 3_5 (which is executed next), when the instruction is said to 'spill out' of the loop.

The notation '$Q_{29}(3_3)$' in the instruction '$Q_{29}(3_3)$ 4_{31}–28X (7)' means that the instruction at location 3_3 is executed as if it were stored in minor cycle 29. The instruction is termed 'Quasi 29', abbreviated Q29. The notation '(7)' refers to bits in an unused part of the instruction P22 – 24, where the value 7 is held. The bits of a word are numbered P1, P2, P3, . . ., P32.

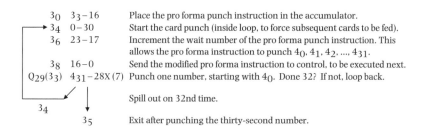

3_0	$33-16$	Place the pro forma punch instruction in the accumulator.
3_4	$0-30$	Start the card punch (inside loop, to force subsequent cards to be fed).
3_6	$23-17$	Increment the wait number of the pro forma punch instruction. This allows the pro forma instruction to punch $4_0, 4_1, 4_2, ..., 4_{31}$.
3_8	$16-0$	Send the modified pro forma instruction to control, to be executed next.
$Q29(3_3)$	$4_{31}-28X(7)$	Punch one number, starting with 4_0. Done 32? If not, loop back.
		Spill out on 32nd time.
3_4		
	3_5	Exit after punching the thirty-second number.

Example 6

The program in this example forms the sum of the last n words of a DL (DL 5 is chosen). The program illustrates the array capability of the Pilot ACE. A single instruction is required to sum the numbers. To prepare for the summation, n needs to be subtracted from the Wait Number of an add instruction, in order that the transfer commences with the appropriate word.

It is assumed that n is in TS 20, that $1 \leq n \leq 32$, and that n is given times P17, that is, as $n \times 2^{16}$—so that n is in a position ready to be subtracted from the Wait Number of the instruction in TS 20.

The line '$Q_{30}(4_{21})$ 5_0-17 (n mcs) (1)' signifies that the instruction held in TS 20 is executed Quasi 30 (i.e. as if it were stored in minor cycle 30). The notation '(4_{21})' means that the instruction was obtained from 4_{21} (see the first line of the program). The notation '(1)' signifies that the instruction at 4_{21} has a digit at P22. (This digit is needed to prevent the subtraction at 4_{22} from altering the Timing Number of the instruction held in TS 20.)

4_{19}	$4_{21}-16$	Place the pro forma add instruction in the accumulator, TS 16, which is 5_0-17 with P22 set (see text).
4_{22}	$20-18$	Subtract n from the Wait Number of the instruction in TS 16.
4_{24}	$16-20$	Copy the instruction to TS 20.
4_{26}	$28-16$	Clear the accumulator to receive the sum.
4_{28}	$20-0$	Send the modified pro forma add instruction to the control unit.
$Q30(4_{21})$	5_0-17 (n mcs) (1)	Add DL 5_{32-n} through 5_{31} to the accumulator TS 16.
4_{31}		The sum is left in the accumulator TS 16.

Example 7

In this example a binary value is converted to decimal, for punching on a card. An image of the decimal part of the card is built up in a DL (DL 7 is chosen) and is then punched.

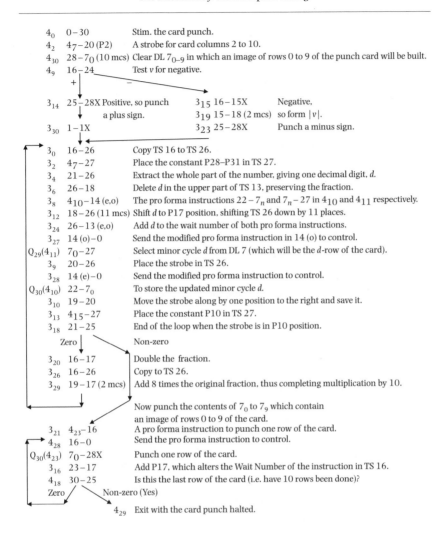

4_0	$0-30$	Stim. the card punch.
4_2	4_7-20 (P2)	A strobe for card columns 2 to 10.
4_{30}	$28-7_0$ (10 mcs)	Clear DL 7_{0-9} in which an image of rows 0 to 9 of the punch card will be built.
4_9	$16-24$	Test v for negative.

$+ \quad \big| \qquad -$

3_{14}	$25-28X$	Positive, so punch a plus sign.	3_{15}	$16-15X$	Negative,
			3_{19}	$15-18$ (2 mcs)	so form $\lvert v \rvert$.
3_{30}	$1-1X$		3_{23}	$25-28X$	Punch a minus sign.

3_0	$16-26$	Copy TS 16 to TS 26.
3_2	4_7-27	Place the constant P28–P31 in TS 27.
3_4	$21-26$	Extract the whole part of the number, giving one decimal digit, d.
3_6	$26-18$	Delete d in the upper part of TS 13, preserving the fraction.
3_8	$4_{10}-14$ (e,o)	The pro forma instructions $22-7_n$ and 7_n-27 in 4_{10} and 4_{11} respectively.
3_{12}	$18-26$ (11 mcs)	Shift d to P17 position, shifting TS 26 down by 11 places.
3_{24}	$26-13$ (e,o)	Add d to the wait number of both pro forma instructions.
3_{27}	14 (o)-0	Send the modified pro forma instruction in 14 (o) to control.
$Q_{29}(4_{11})$	7_0-27	Select minor cycle d from DL 7 (which will be the d-row of the card).
3_9	$20-26$	Place the strobe in TS 26.
3_{28}	14 (e)-0	Send the modified pro forma instruction to control.
$Q_{30}(4_{10})$	$22-7_0$	To store the updated minor cycle d.
3_{10}	$19-20$	Move the strobe along by one position to the right and save it.
3_{13}	$4_{15}-27$	Place the constant P10 in TS 27.
3_{18}	$21-25$	End of the loop when the strobe is in P10 position.

Zero $\big|$ Non-zero

3_{20}	$16-17$	Double the fraction.
3_{26}	$16-26$	Copy to TS 26.
3_{29}	$19-17$ (2 mcs)	Add 8 times the original fraction, thus completing multiplication by 10.

Now punch the contents of 7_0 to 7_9 which contain an image of rows 0 to 9 of the card.

3_{21}	$4_{23}-16$	A pro forma instruction to punch one row of the card.
4_{28}	$16-0$	Send the pro forma instruction to control.
$Q_{30}(4_{23})$	7_0-28X	Punch one row of the card.
3_{16}	$23-17$	Add P17, which alters the Wait Number of the instruction in TS 16.
4_{18}	$30-25$	Is this the last row of the card (i.e. have 10 rows been done)?

Zero $\diagup$ Non-zero (Yes)

4_{29}		Exit with the card punch halted.

The binary value is less than 10 in magnitude and is given to 27 binary places. In order to convert the value to decimal, the individual decimal digits need to be obtained. These are obtained one at a time by multiplications by 10. However, each decimal value thus obtained is in binary; hence the process is termed conversion to *binary coded decimal* (each decimal digit is separately represented by a binary integer). The process begins by extracting the whole number part, then removing the whole number part from the original number, and then multiplying the fraction that remains by 10. The new whole number part is extracted and removed from the fraction, as before. The successive whole number parts are binary coded decimal digits. These values are used to select one minor cycle of DL 7 (decimal 0 to 9 select minor cycles 0 to 9, respectively). Minor cycles 0 to 9 of DL 7 correspond to rows 0 to 9 of the card to be punched. For each decimal digit, a single bit (representing a hole to be punched) is stored in one of these corresponding minor cycles of DL 7.

The program punches a decimal number on a card. The binary value v is assumed to be in TS 16, and $0 \leq |v| < 10$, with 27 binary places. First the sign is punched in the Y-row or X-row of the card, and the absolute value is formed. Then an image of the card rows 0 to 9 is prepared in DL 7. Finally, that image is punched.

The program uses the concept of a 'strobe'. This is a bit used as a marker that will be shifted right (i.e. shifted up) in a register as the conversion progresses. The bit will determine the position of a decimal digit on a card, and will also be used to determine when the last decimal digit has been processed.

Example 8

This more elaborate example illustrates the input of a decimal integer from a punch card and its conversion to binary.[40] A decimal value is punched with the sign in card column 1 (see fig. 10). A single punching in the Y-row indicates a positive integer; a single punching in the X-row indicates a negative integer. Decimal digits are punched in rows 0 to 9 to represent the digits 0 to 9. It is assumed that the integer to be input is punched in card columns 1 to 10, with the sign in column 1, and the digits in card columns 2 to 10. The first row to be read is the Y-row; that is, the sign is read first.

The first instruction in the subroutine copies the instruction that is in TS 16 into I_{30}. This action is termed 'planting a link'. To plant a link is to place an instruction in a specified location in a DL, so that the subroutine can return to the program that called it. The last instruction in the subroutine leads to this link, and so back to the next instruction in the calling program.

A decimal punching check is carried out by the subroutine—that is, a check that there is only one hole in each of the nine columns where decimal digits are expected. For each of the rows 0 to 9, a 'strobe' is used to ascertain whether a column has been punched. The strobe is used to scan a TS that holds a copy of the row. The strobe is compared with each of the bits in the TS in turn to ascertain whether a hole has been punched in a particular column of that row. If the column has been punched, a power of 10 is added to the partially converted binary integer. The power of 10 is initially zero for the ones row, and increases as each digit in the row is processed.

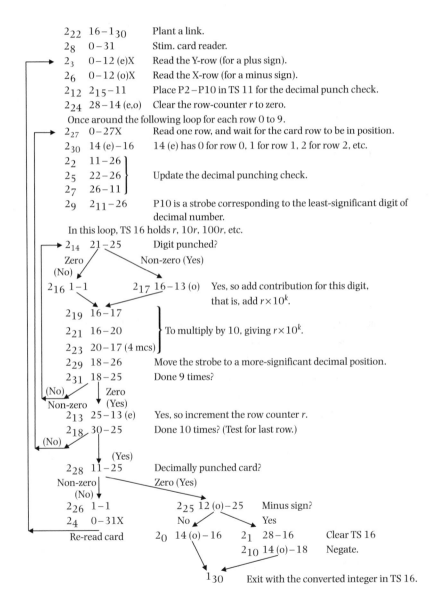

2_{22} 16–1_{30} Plant a link.
2_8 0–31   Stim. card reader.
2_3 0–12 (e)X   Read the Y-row (for a plus sign).
2_6 0–12 (o)X   Read the X-row (for a minus sign).
2_{12} 2_{15}–11   Place P2–P10 in TS 11 for the decimal punch check.
2_{24} 28–14 (e,o)   Clear the row-counter r to zero.
Once around the following loop for each row 0 to 9.
2_{27} 0–27X   Read one row, and wait for the card row to be in position.
2_{30} 14 (e)–16   14 (e) has 0 for row 0, 1 for row 1, 2 for row 2, etc.
2_2 11–26 ⎫
2_5 22–26 ⎬   Update the decimal punching check.
2_7 26–11 ⎭
2_9 2_{11}–26   P10 is a strobe corresponding to the least-significant digit of decimal number.
In this loop, TS 16 holds r, $10r$, $100r$, etc.
2_{14} 21–25   Digit punched?
Zero / (No)   Non-zero (Yes)
2_{16} 1–1 2_{17} 16–13 (o)   Yes, so add contribution for this digit, that is, add $r \times 10^k$.
2_{19} 16–17
2_{21} 16–20 ⎫
2_{23} 20–17 (4 mcs) ⎬   To multiply by 10, giving $r \times 10^k$.
2_{29} 18–26   Move the strobe to a more-significant decimal position.
2_{31} 18–25   Done 9 times?
(No) / Non-zero   Zero (Yes)
2_{13} 25–13 (e)   Yes, so increment the row counter r.
2_{18} 30–25   Done 10 times? (Test for last row.)
(No)   (Yes)
2_{28} 11–25   Decimally punched card?
Non-zero (No)   Zero (Yes)
2_{26} 1–1 2_{25} 12 (o)–25   Minus sign?
2_4 0–31X   No   Yes
Re-read card 2_0 14 (o)–16 2_1 28–16   Clear TS 16
    2_{10} 14 (o)–18   Negate.
1_{30} Exit with the converted integer in TS 16.

Entering a program

The significance of Source 0 *and Destination* 0

It was mentioned earlier that Source 0 and Destination 0 were specially chosen. The significance of this will now be explained.

One of the problems facing the designer of any computer is that of introducing a program into an empty machine. Turing solved it rather simply, by making use of the zero words that would be in every store whenever the stores were cleared (i.e. set to zero) ready for a new program.

He arranged for Source 0 to be the card reader, and for Destination 0 to be INST (as mentioned previously, the instruction register INST was renamed COUNT in the Pilot ACE). A zero word also had to be a useful instruction, namely 0–0 (this instruction says 'copy Source 0 to Destination 0'). Furthermore, a zero word had to be a Stop instruction, otherwise the machine would not wait for the card reader to be ready with a row of a card.[41] (Recall that a Stop instruction had the Go bit set to 0, while an instruction that could be executed immediately had the bit set to 1.) The use of a zero word as a meaningful instruction allowed a new program to be entered into the machine. The next subsection describes how, in an empty machine, instructions were taken from cards in the card reader and executed.

Obeying instructions from the card reader

A completely empty machine contained zero words in every store (except for COUNT, which had zero in all fields except for the Wait and Timing Number fields, which were continually counting down). Considered as an instruction, each zero word specified Source 0 and Destination 0. Since the Go bit was zero, the instruction could not execute immediately it was sent to control, but waited until a row of a card was in position to be read. When the row was in position, the card reader gave a Single-Shot which allowed the machine to execute the 0–0X instruction. A word on a row of the card was then transferred into COUNT, where it was executed. In this way, an instruction could be executed directly from the card reader.

When a new program was to be entered, control and the high speed store were cleared, and a deck of cards that had been placed in the card reader began to pass the reading station. The instruction in control after the store was cleared was 0–0X, with NIS 0, and with Wait and Timing Numbers that were usually unequal. As soon as the first row of the card was ready to be read, the reader gave a Single-Shot and one word of DL 11 was sent to control (the NIS of zero causing DL 11 to be selected).[42]

Because DL 11 was also cleared when all the stores were cleared, it contained thirty-two copies of a zero word—that is, the instruction 0–0X with NIS 0 and with equal Wait and Timing Numbers. One of those words was sent to control and was executed when the Single-Shot for the second row of the card was given.

The instruction in the second row of the card was then sent to control. After the instruction was executed, the NIS specified which of the eight DLs provided the next instruction. Typically this was another zero word, which was executed in the same way. Then another instruction was taken from the third row of the card in the reader, and so on.

Reading in the initial instructions (bootstrap)

This subsection describes how a new program was read in, starting with the initial orders. (It will be helpful to refer to the Pilot Model Instruction Card in fig. 10.)

In order to read in a program on cards and to store the instructions in a DL, a special set of instructions called *filler instructions* was required. A typical set of filler instructions, on the first four rows of the card, were:

Y-row: blank
X-row: 1, 0–1   26 25X   (1 = NIS; 0–1 are S and D;
                                 26 = Wait Number;
                                 25 = Timing Number)
0-row: 1, 0–1 (1) 30 31X
1-row: 1, 0–1 (1) 30 31X

These instructions function as follows. First, a copy of the instruction in the 0-row is placed in all words of DL 1:

1, 0–1 (1) 30 31X.

Then minor cycle 31 of that delay line is replaced with the instruction

1, 0–1 (1) 30 31X

from the 1-row. The instructions in DL 1 are then executed, starting at the first one in minor cycle 0. Each instruction reads in one word from a row of a card. (The remainder of this card is read and two more cards.) The first instruction—the one in 1_0—reads in the word at row 2 of the same card, and stores that word in I_0. That is to say, the instruction that was just executed is itself replaced by the content of row 2 of the card. Then the instruction in minor cycle 1 of DL 1 is sent to control. This causes row 3 of the card to be read and stored in I_1. The instruction in I_2 reads in row 4 of the card and stores it in I_2. And so on. Each instruction in DL 1 reads in a word from a row of a card, and overwrites itself with a copy of that word. The last instruction in minor cycle 31 likewise reads in a word, but is special in that it also breaks the sequence, either to enter the program just read in (as is the case here), or to allow another delay line to be similarly filled.

In detail the process is this. When the first row (all blank) is ready to be read in, a zero word from DL 11 is taken into control.[43] The zero word that is now in control will not be executed until the second row of the card arrives at the reading station (since the zero word is a Stop instruction). When the row arrives, the instruction 0–0X is executed, causing the instruction on the X-row to enter control. When the third row arrives at the reading station, the instruction 1, 0–1 26 25X is executed. It copies the third row into all 32 words of DL 1, because it is a long transfer, that is, of 32 mcs. DL 1 now contains the instruction 1, 0–1 (1) 30 31X in every word. The NIS of the instruction just executed is 1, so the instruction in DL I_{31} is brought into control and is executed when the fourth row of the card is ready to be read. This instruction is 1, 0–31 (1) 30 31X. It causes the fourth row of the card (the 1-row) to be transferred to DL I_{31}, overwriting the instruction that was recently placed there. Since the Timing Number is 31, the instruction in DL I_0 is fetched to control and is executed, bringing the fifth row of the card to DL I_0 (overwriting the instruction previously placed there), and so on for the next 31 rows. Now 32 consecutive rows of three cards are stored in DL I_0 to DL I_{31} respectively, beginning with the 2-row of the first card. Once the last of the 32 rows has been read, the next instruction is taken from I_0, so initiating the execution of the program that has just been read in. (If the first instruction in the program is not in the first minor cycle of DL 1 it is selected by changing the Timing Number of the filler instruction in the 1-row.)

Performance

Some program times for the Pilot ACE (before the magnetic drum was installed) are given in Table 3.[44]

Table 3 Pilot ACE program timings

Program	Order of task	Time taken
Matrix multiplication	14×14 by 14×14	$1\frac{1}{4}$ sec
	47×47 by 47×47	15.36 min using card equipment for interim storage
Solving simultaneous equations	13 equations, 2 right-hand sides, or 10 equations with 10 right-hand sides	A 'few seconds' (less than the time to read the decimal data)
	50 equations	Less than 10 min[45]
Solving equations (Gauss-Siedel iteration)	120 equations with 2000 non-zero coefficients	0.835 min per iteration (the time taken to read the cards; computation was carried out between rows of each card)
Latent roots of matrix	Symmetric matrix of order less than 19	$2\frac{1}{2}$ sec
	Unsymmetric matrix of order 15	2 sec
	Symmetric matrix of order 60	$1\frac{1}{2}$ min per iteration, using cards as intermediate storage
Ordinary Differential Equations[46]	Runge-Kutta (10 simultaneous equations)	$70 + 4T$ ms per step (T = time to evaluate the function $F(x, y)$)
	Lagrangian Formulae, version 1	15 ms per step plus time to evaluate $F(x)$
	ditto, version 2	10 ms per step plus time to evaluate $F(x)$

A comparison with the Manchester University Mark I is given in Table 4.[47] In his *A History of Manchester Computers*, Simon Lavington claims that a 'contemporary benchmarking exercise rated the Mark I at about the same raw power as the NPL ACE'.[48] This claim does not withstand scrutiny. The timings in Table 4 show that the instructions of the Pilot ACE were an

Table 4 Pilot ACE and Manchester Mark I timings

	Pilot ACE	Manchester Mark I	Units
Clock rate	1000	117	KHz
Integer add/subtract	64	1800	microseconds
Integer multiply	2.06	2.16	ms (Ferranti Mark I of 1951)
Integer multiply with fetch	2.12	3.36	ms (Ferranti Mark I of 1951)
Fixed-point divide (software)	32^{49}	1500	milliseconds (ms)
Square root (fixed point)	32		ms
Floating-point add/subtract (software)	8	60	ms^{50}
Floating-point multiply (software)	6	80	ms
Floating-point divide (software)	34	150	ms
Speed (maximum)	15,625	600	Instructions per second
Speed (average)	7800	600	Instructions per second
Input speed	106	150	Decimal digits per second
	1280	750	Bits per second
Output speed	53	15	Decimal digits per second
	640	75	Bits per second

order of magnitude faster than the Mark I machine—the only exception being the multiply instruction, which was about the same speed on both machines. The times reflect the higher clock rate of the Pilot ACE. Even allowing for the serial nature of storage, the Pilot ACE executed instructions at a rate at least three times as fast as the Manchester Mark I. But instruction time was not the only measure of relative performance. In the Pilot ACE, instructions did more work than in other machines. For example, to sum 32 words on the Pilot ACE required only one instruction, which performed 32 additions in one major cycle (1 millisecond). To do the same task on the Manchester Mark I required the execution of 96 instructions. On the Mark I, where the basic instruction took more than a millisecond, the separate additions took 96 × 1.8 ms = 172 ms. The Pilot ACE was 172 times faster at this task.

Notes

The author is grateful to Jack Copeland for many valuable comments and suggestions on this chapter and for much editorial work, and also for providing access to facsimiles of extant correspondence and reports of the NPL through The Turing Archive for the History of Computing <www.AlanTuring.net>. The author also extends thanks to John Webster for his comments and suggestions, and to Michael Woodger for the ACE Pilot Model Instruction Card in fig. 10.

1. The addition rate was 15,625 additions per second (30,300 additions per second for vector addition). By comparison ENIAC had a pulse rate of 100,000 Hz and an addition rate of 5000 per second (Goldstine, H. H. and Goldstine, A. (1973) 'The Electronic Numerical Integrator and Computer (ENIAC)', in B. Randall (ed.), *The Origins of Digital Computers*, 2nd edn, Berlin: Springer-Verlag, p. 335; the addition rate of the Cambridge EDSAC was 700 per second (Lavington, S. (2000) *The Pegasus Story*, London: Science Museum, p. 50; see also the comparison table in Lavington, S. (1980) *Early British Computers*, Manchester University, p. 118).

2. Turing was expecting his machine to solve up to 50 linear simultaneous equations, requiring an estimated 6400 words of storage. No fast storage medium having greater capacity than delay lines was available at the time. The DEUCE circuitry associated with a delay line required 27 valves: 23 for the circulation unit, 3 for the receiver, and 1 for the transmitter, and the ACE circuits would have been similar. Turing's design also needed 32 temporary stores. For 200 delay lines and 32 temporary stores, 6264 valves would have been required. ('Circulation Unit. Unit S. Mark II. D.E.U.C.E.', English Electric Company Ltd. (Nelson Research Laboratories, Stafford), 30 August 1955, 'Receiver Units Type I & II. Mark II. D.E.U.C.E.', English Electric Company Ltd. (Nelson Research Laboratories), 28 May 1956; 'Transmitter Unit Type I & II. Mark II. D.E.U.C.E.', English Electric Company Ltd. (Nelson Research Laboratories), 28 May 1956.) Turing estimated that the control unit would require 1000 valves, thus bringing the total to at least 7264 valves. Wilkinson was later able to solve 192 equations with a machine one-tenth the size of Turing's. (Wilkinson, J. H. (1954) 'Linear algebra on the Pilot ACE', *Automatic Digital Computation: Proceedings of a Symposium Held at the National Physical Laboratory, 1953.* London: HMSO, p. 131.)

3. Womersley, J. R., memo to the Executive Committee of the NPL, E.881, 13 February 1946. The memo accompanied Turing's report E.882. (Digital facsimiles of all NPL and DSIR documents referred to in this chapter are available in The Turing Archive for the History of Computing <www.AlanTuring.net/aceindex>.)

4. Version III is mentioned in Goldstine, H. H. (1972) *The Computer from Pascal to von Neumann.* Princeton, N.J.: Princeton University Press, p. 218. Professor D. R. Hartree gave ' "Circuits for the ACE," third version' to Goldstine between c.4 July and 20 July 1946 (ibid., pp. 218, 219, 233). (The term 'circuits' did not

then refer to 'electronic circuits' but to logic diagrams, as in 'control circuit'— used throughout Chapter 22, 'The Turing–Wilkinson Lecture Series (1946–7)'.) I believe that 'Proposed Electronic Calculator' was Version I (although Woodger says not—see his letter to Copeland (25 February 2003), quoted in Chapter 3, 'The Origins and Development of the ACE Project'). Supporting my view is the time scale: Versions V, VI, and VII were produced from May to November 1946, taking about two months per version. 'Proposed Electronic Calculator' was completed late in 1945. It is reasonable to think that three more versions could have been produced in the four months to the end of April 1946, but four versions seems less likely. Further support is provided by Goldstine (ibid., p. 218) who refers to 'Proposed Electronic Calculator' as 'first version'.

5. The question facing Turing was: how to design a branch instruction that needed a branch address, when the instruction already contained the address of the normal next instruction? The answer was delightfully simple. When a branch was required, add unity to the Timing Number of an instruction when it was in the control unit, thereby allowing control to select an adjacent instruction instead of the normal instruction.

6. TRANSTIM is an acronym for 'Transfer Time', while TIMCI is 'Time to Call the Next Instruction'.

7. Unconfirmed Minutes, Executive Committee Meeting, 21 January 1946.

8. Memorandum, 'Status of the Delay Line Computing Machine at the P. O. Research Station', 7 March 1946.

9. Thomas, H. A., 'A Plan for the Design, Development and Production of the "ACE"', 12 April 1947.

10. Letter from Woodger to Copeland, 21 May 2003.

11. Ibid.

12. *NPL News*, August 1955, p. 3 (quoted in Yates, D. M. (1997) *Turing's Legacy: A History of Computing at the National Physical Laboratory 1945–1995*, London: Science Museum, p. 24).

13. Thomas, 'A Plan for the Design, Development and Production of the "ACE"'.

14. Womersley's memorandum records that discussions took place between the Director, Hartree, and Smith-Rose (Womersley, 'A.C.E. Project', undated, but written after 18 August 1947).

15. Huskey testifies: 'we had no knowledge of any technical activity in the Radio Division' (in a letter to Copeland, 16 May 2003).

16. Thomas, 'A Plan for the Design, Development and Production of the "ACE"', p. 2. In point of fact the ACE group was considering pulses as short as 0.3 microseconds, which would be required for transmission around the machine and for clocking purposes (Memorandum to Director, 'A.C.E. Pilot Test Assembly and Later Development', from R. L. Smith-Rose and J. Womersley, 30 April 1947).

17. Indeed, did Thomas overlook the detailed description of the cathode follower circuit in Turing's report (under the heading 'Use of cathode followers')? Not only does Turing give a circuit; he unequivocally states that it has 'a very large input impedance and a very low output impedance', and points out that the output can have many connections made to it. Turing evidently had considerable knowledge of electronics.

18. 'Status of the Delay Line Computing Machine at the P. O. Research Station'.

19. Version V was by far the simplest of the three Versions V, VI, and VII presented and discussed in the lecture series at the beginning of 1947 and it was natural for Huskey to adopt it as a starting point. Versions VI and VII used vector to vector arithmetic, while the earlier Version V allowed limited vector operations (see Chapter 4). But even Version V needed considerable pruning—for example, it had 1024 Destinations and 1024 Sources. Strictly speaking, it should not have been necessary for Huskey to convince anyone of the need to build a small trial machine, because DSIR had given approval to proceed with a first pilot machine in July 1946 (Memo on behalf of the Lord President of DSIR to Sir Edward Appleton, 2 July 1946).

20. Woodger, M., 'ACE Test Assembly', September–October, 1947.

21. Womersley, 'A.C.E. Project'. (The content of this memorandum cannot be entirely relied upon.)

22. Woodger, 'ACE Test Assembly'.

23. This method was taken directly from Turing's lecture 26 (see Chapter 22). To input a new program from cards the machine had to be switched to 'Stop'. Six punch cards were required to fill a delay line with a program. The instructions and constants (words) for the delay line would not be arranged consecutively on the cards, but were to be punched on alternate rows of the cards, each word preceded by an instruction that would read in the following word to the delay line. After the cards had been read in, the machine would be switched to 'Run' for the program to be executed. There was no provision to stop the processor other than to operate the Stop/Run switch.

24. The successful machines at Manchester and Cambridge consisted of a few stores and an arithmetic unit.

25. Anon., 'A.C.E. Problems', NPL.

26. Wilkinson, J. H., 'Progress Report on the Automatic Computing Engine', DSIR, April 1948.

27. Wilkinson in interview with Christopher Evans in 1976 (*The Pioneers of Computing: An Oral History of Computing*. London: Science Museum).

28. DSIR Minutes, 16 February 1949.

29. Letter from Bullard to DSIR, 4 August 1950.

30. Mathematics Division, NPL, Appendix to 'Programming and Coding for the Pilot Model ACE (1951)', 1 March 1954.

31. Wilkinson, J. H., 'Report on the Pilot Model of the Automatic Computing Engine. Part II: The Logical Design of the Pilot Model', Mathematics Division, NPL, September 1951, Figure 13.

32. Campbell-Kelly credits S. Gill with assisting in the logical design of input–output (Campbell-Kelly, M. (1981) 'Programming the Pilot ACE: early programming activity at the National Physical Laboratory', *Annals of the History of Computing*, 3, 133–68, p. 135). However, Gill's details of the electromechanical circuits and the programming of the input–output equipment follow Turing's design (S. Gill, 'Description of Hollerith Input Output for the Pilot Model', Mathematics Division, NPL, June 1949; see part two.)

33. Initially 10 and 6, respectively. When the magnetic drum was added in 1954, one of the TS was replaced by a Delay Line (see note 39).

34. Levy, J. V. (1978) 'Buses, the skeleton of computer structures', in C. G. Bell, J. C. Mudge, and J. E. McNamara (eds), *Computer Engineering—A DEC View of Hardware Systems Design*. Maynard, MA: Digital Press, p. 277.

35. The Next Instruction Source (NIS) of the instruction 16–0 is ignored.

36. For positive values only.

37. The least-significant bit of a word emerges from a store first, and is displayed on the console and the monitor (VDU) with the least-significant bit on the left. A binary word on a punch card has the least-significant digit on the left. The terms 'up' and 'down' were used to denote shifts that would now be called 'left' and 'right'. However, on ACE, a doubling was actually a shift to the right, and a halving was a shift to the left. Even the components of an instruction were written with the least-significant components on the left, and binary constants were written and punched with the least-significant bit on the left (hence the term 'Chinese binary').

38. Normally an electronic signal, TIMCI (Time to Call the Next Instruction), allowed one instruction into COUNT. However, when the word being tested was non-zero, the signal TIMCI was extended by 1 minor cycle, thus allowing two instructions to enter COUNT. The first was discarded, because the circulation path in COUNT was broken while the transfer took place. This allowed the second instruction to enter COUNT and to be executed next.

39. Initially this store was TS 11 (which also was NIS 0). When the magnetic drum was added in March 1954, TS 11 was replaced by DL 11.

40. In this program, Source and Destination 11 is TS 11, not DL 11 as it became in 1954.

41. The Go/Stop bit was conceived after Turing had left the NPL, quite possibly by S. Gill.

42. In the case where the Wait and Timing Numbers are equal, the next instruction is taken from Source 0, which is the next row to be read by the card reader. This first

row is usually blank, which is the instruction 0–0X with NIS 0, and with equal Wait and Timing Numbers.

43. Should the Wait and Timing Numbers of the instruction in control happen to be equal, then the instruction in control, 0–0X, will cause the Y-row of the card to be read. In any event, a zero word, coming either from DL 11 or from the card reader, is transferred to control when the first row of the card is ready to be read in.

44. Wilkinson, 'Linear Algebra on the Pilot ACE', pp. 129–36.

45. *The Engineer*, Vol. CXC, 8 December 1950, London, p. 560.

46. Fox, L. and Robertson, H. H. (1954) 'The numerical solution of ordinary differential equations', in *Automatic Digital Computation: Proceedings of a Symposium Held at the National Physical Laboratory, 1953*. London: HMSO.

47. Lavington, S. (1998) *A History of Manchester Computers*. Swindon: British Computer Society, pp. 17, 25.

48. Lavington, *A History of Manchester Computers*, p. 26.

49. Could be programmed as 16 ms, if required.

50. The floating-point times are from Campbell-Kelly, 'Programming the Pilot ACE', p. 159, as are the input and output times for the Mark I (ibid., p. 137). In preparing Table 4, the author has written DEUCE subroutines to compute floating-point addition, subtraction, and multiplication. They take 5 ms, 5 ms, and 4 ms, respectively. The times—which are maximum times and include times for post normalization and checks for overflow and underflow—would have been exactly the same for the Pilot ACE. The times compare with those published for DEUCE of 6 ms, 6 ms, and $5\frac{1}{2}$ ms average, respectively.

12 Applications of the Pilot ACE and the DEUCE

Tom Vickers

Practical mathematics

The arrival of Pilot ACE in 1950 had a dramatic effect on practical mathematics.[1] The new wind—it was a hurricane—felled some very fine trees but at the same time scattered far and wide seeds that have now grown into substantial forests.

Up until the Second World War practical mathematics was in a primitive state.[2] My own experience provided ample evidence of this. After graduating in 1940, I was directed to join the External Ballistics Department of the Ministry of Supply, which had taken over the new Mathematics Laboratory in Cambridge. I was a very junior member of a team of about twenty mathematicians. E. T. Goodwin and J. H. Wilkinson were also members of the team. Most of our time was spent calculating the trajectories of various types of shell. The simple differential equation involved was solved by an ancient predictor–corrector method due to Adams–Bashforth. We accepted the method without question. Our lecturers had taught us how to prove the existence of a solution to a differential equation, but finding it was irrelevant unless it could be written down directly in terms of known functions. Indeed, *numbers* were considered to be infra dig, although we got very familiar with the Greek alphabet. Only five years later, Goodwin and L. Fox publicized at least seven methods for the numerical solution of differential equations. These would have allowed the wartime trajectories to be computed by schoolleavers rather than graduates. When Goodwin, Wilkinson, and I moved to the National Physical Laboratory, we were pleasantly surprised to discover the fascination of numbers and errors (although other members of the Cambridge group were glad to escape from the Brunsviga).[3]

A unit such as NPL's Mathematics Division was sorely needed to promote knowledge of numerical methods; 'To develop and apply numerical methods and mathematical statistics' was included in the terms of reference of the new Division. From 1945 to 1950, Goodwin, Fox, Olver, and others flushed out various aids to practical mathematics from obscure publications. Their work helped to define a new discipline, Numerical Methods, and lead eventually to the publication in 1957 of the popular book *Modern Computing Methods*, written by various members of Maths Division.[4] (A second edition in 1961 became a classic, remaining a standard reference for over twenty years.) Later on, Wilkinson's investigation of inherent errors led to the new subject of Numerical Analysis. Maths Division's extensive knowledge of good practical methods was a major factor in the successful exploitation of the Pilot Model ACE.

In 1943, most of my group was relocated to Fort Halstead to join the newly created Armament Research Department. My work was mainly in ballistics; the tasks of my many new colleagues and friends were unknown to me. On occasion I was unaware of the purpose of my own calculations. One bank holiday, Heather Wilkinson and I were required to carry out some very urgent sums. When the V2 rockets were launched shortly afterwards it was obvious that we had been calculating trial trajectories for them. We were never told officially and even now I may be breaching security in revealing it!

When I joined NPL in 1946 I found an entirely different ethos. Womersley had collected a remarkable number of escapees from secret wartime work. We were expected to share experiences and to interact closely with the rest of NPL. My main responsibility was to organize problems submitted to the Laboratory in such a way that they could be solved by our team of school-leavers. (I shared this task with Joan Staton.) Other responsibilities included investigating new computing aids, in particular desk calculators and accounting machines, providing an advisory service on the acquisition and use of calculating machines, and organizing the use of the National Accounting machine for table-making.

Goodwin told the school-leavers: 'At school, if you made an error you were punished by loss of marks. Here, errors will be made all the time. They must not leave the building.' We all became obsessed with checking and I believe that our reputation for accuracy was widely recognized. It is not surprising that the first Quality Manager in NPL was previously the Operations Manager on DEUCE.

My first internal investigation was into a minor difference in the results obtained by new members of staff. 'Did you use sufficient figures in π?', I enquired. Oh yes, 3.14285714—the handy approximation 22/7 taught in schools. It should have been 3.14159265!

Careful scrutiny of submitted problems often identified poor formulation, where the problem could have been solved more easily at an earlier stage. Another common failing was to seek the evaluation of a complicated formula which proved to be the analytical solution of an integral or differential equation, again reflecting the neglect of numerical methods in mathematics degrees. For example, the solution of the simple differential equation $dy/dx = (y - x)/(y + x)$ given in standard texts requires logarithms and inverse tangents, yet it is trivial by a numerical method. In our work it was necessary to strike a balance between the analytical approach and a numerical one, a matter completely neglected in formal teaching.

Our ability to assess new machines was soon noted by both the O and M (Organization and Methods) Unit of H. M. Treasury, responsible for authorizing office equipment in Civil Departments, and also by His Majesty's Stationery Office, who supplied the equipment. Calculators were not in common use then and these departments were relieved to pass the job of technical assessment to me. NPL gained by acquiring new machines. This one-man advisory service was to develop into a major department.

Pilot ACE goes to work

Following the successful demonstration of Pilot ACE in 1950, it was proposed from above that the machine be abandoned in order to give us freedom to concentrate on designing the full-scale ACE. Wilkinson disagreed strongly with this. Following intervention by Goodwin, who was aware of several potential customers, the proposal was rapidly overturned. Not long afterwards Goodwin was appointed Superintendent of Maths Division. We began using Pilot ACE.

My duties changed dramatically. I was told 'Here is a computer. We hope that it will work for a reasonable number of hours. Do what you can with it.' This was no hardship. I had been looking for desk calculators that might reduce a multiplication time from 15 seconds to perhaps 12; Pilot ACE took 1/500 of a second. A new accounting machine that I had been hoping to acquire contained 10 registers rather than 6; Pilot ACE had 350. Ted York— who had been squeezing the pips out of punched-card machines—also jumped at the opportunity to use the ACE. He and I had no doubts that this move would be interesting, challenging and exciting. We little realized, though, the extent of the revolution in computing, communications, and information that was about to take place.[5]

I was asked to develop an operating and programming section from the existing junior staff and future recruits. Since all staff in the Division had received some training in handling punched-card machines, operation of the new computer was easy to pick up. Many new machine-room duties were soon identified—the organization of rosters, the printing of results, the development of a subroutine library and appropriate documentation, and so on. An evening shift taken by junior staff was established and normally ran lengthy jobs. Josie Wright (now Snook), who had already organized the work of the National Accounting machine, quickly rose to the new challenge of ACE. I feel ashamed that her annual salary was easily covered by the income from one good day of ACE. Josie supervised the operation of the machine, prepared suitable jobs, and interacted with the clients, contributing substantially to the smooth running of jobs and to the high quality of the output. She needed to be particularly alert when the job was for Sir Edward Bullard, the director of NPL, who was developing his theory of the origins of the earth's magnetic fields. This work involved the solution of twelve second-order differential equations and was the largest and one of the most difficult pieces of computation tackled by Maths Division up to that time. Bullard lived on site and was liable to appear well before official opening time, clamouring for the results from the night's operation.

Early applications

Table 1 lists some of the more significant applications of Pilot ACE. I will discuss a number in more detail.

Linear equations

Initially, the focus of interest was the use of Pilot ACE for problems where a method of solution was known but lengthy in practice. The solution of linear simultaneous equations is a prime example. A brief history of this area illustrates the difference brought about by Pilot ACE.

- Pre-war—Cramer's Rule was taught at school, mainly to demonstrate the use of a determinant. In theory, Cramer's Rule offers an elegant solution, but in practice is useless other than in the case of three equations.

Table I Use of Pilot ACE and DEUCE at NPL from 1950 to 1960

1950 'ACE Test Assembly' demonstrated at the Physical Society Exhibition (April). Pilot ACE carried out instructions in one delay line (May). Press demonstration—factors of numbers, simple differential equation, ray-tracing (November).

1952 Pilot ACE deemed reliable enough to be moved to Maths Division. Involvement of junior staff for full-time production. Visit by Duke of Edinburgh (April 22). Solution of 129 simultaneous equations. Problems concerning defence, photogrammetry (Ordnance Survey), aircraft flutter, traffic simulation. Report by a Department of Scientific and Industrial Research (DSIR) working party into applying computers to data processing (scientific, industrial, administrative): 'It is already clear that in the organisational and administrative fields, there are, subject to the solution of certain technical problems, possibilities of immense economies in money and man-power, and of the release of human effort from machine-like occupations.'

1953 Symposium at NPL entitled 'Automatic Digital Computation' (March). Twelve second-order differential equations solved for Sir E. Bullard, Director of NPL, in connection with his theory that the centre of the earth is a mass of spinning molten material. Representation of mathematical functions by Chebychev series. Problems for CERN (stability of synchroton orbits), Atomic Energy Authority (AEA), and aircraft industry. Substantial use of the matrix library. The 'mechanization of clerical processes' under active consideration.

1954 Increasing pressure for computer time. Addition of magnetic drum. Investigation of Comet disaster (8 million multiplications were a small part of this problem). Budget Day special: we were all set to produce the new PAYE tables, but in the event the Chancellor of the Exchequer made no change to PAYE that year. Crystallography. Ray-tracing. Warning system in mines for the National Coal Board. Computing train time-tables. Wind tunnel design. Stresses in catapult of Ark Royal (involving 20 nonlinear differential equations).

1955 DEUCEs delivered to NPL, Royal Aircraft Establishment, and English Electric. Agreement to share production of software. New tax tables calculated on Budget Day, 6–11 p.m. The Society of British Aircraft Constructors (SBAC) set up panels to agree on a common formulation of problems in aircraft and engine design; this produced a large demand for matrix programs. Birth of GIP and matrix Schemes A and B. (Scheme A was the first matrix scheme developed. Because of the machine's small store,

Table 1 Continued

	Scheme A relied heavily on cards as backing store. Scheme B was a specialized use of GIP for handling matrices held in the newly added drum stores; for very large matrices Scheme A was still available.) Eigenvalues found up to order 60. Major problems in theoretical physics. Table-making. Mixing of salt and fresh water in the Thames. Freezing slabs of fish.
1956	Pilot ACE transferred to the Science Museum (June). Publication of *Wage Accounting by Electronic Computer*, written by a small group in NPL's Autonomics Division.[6] Comprehensive set of 100 matrix programs. Published tables of zeros of Bessel Functions. Problem for Central Electricity Generating Board on the efficient distribution of power and the optimum siting of power stations. Crystallography programs made widely available by J. Rollett. Study of problem of coding the National Insurance Index (which contained information concerning 30 million people).
1957	Many DEUCEs sold (matrix library a key feature for the aircraft industry). Publication by Maths Division of first edition of *Modern Computing Methods*. Bristol Engines produce TIP for DEUCE. Computing training course at NPL for senior staff from major Government Departments. Analysis of family expenditure survey for the Central Statistical Office. NPL's weekly payroll program demonstrated. Mass spectrometry for British Petroleum.
1958	The full-scale ACE is available on a limited basis for subroutine development (September). Symposium at NPL entitled 'Mechanisation of Thought Processes' attracts wide TV and radio coverage. New approach to error analysis—birth of 'Numerical Analysis'. NPL's eigenvalue programs claimed to be the most comprehensive in the world. Representation of the shape of a ship by orthogonal polynomials. Fast programs for infrared spectroscopy. Automatic data-capture on paper tape.
1959	Maths Division involved with definition of ALGOL. Mechanical translation (Russian to English). Optimization. Self-consistent fields. Large partial differential equations. Vibrational spectra of disordered lattices.
1960	Big ACE in full operation. Development of Autocode. Automated data capture by paper tape from wind tunnel tests and from trials in a weathership. The design of the NPL Standard Interface and BS 4421. Work on character recognition. Magnetic coding of cheques. Stresses in cooling towers. The Computer Users Panel and Data Handling Panel manage interaction with the rest of NPL.

Source: NPL Annual Reports.

- During the 1930s the Mallock machine and other devices were built.[7] On a *good* day the Mallock machine gave three-figure accuracy with 8–10 equations.
- The Second World War—Wilkinson struggles to solve twelve equations.[8]
- 1946—Junior staff at NPL routinely solve six equations with no difficulty.
- 1946—Hotelling gives wide bounds for solutions of large sets of equations. Turing and Fox disagree on the implications of this, but finally agree that good solutions are possible.
- 1947—With considerable effort a set of sixteen equations is solved on desk machines at NPL. The computation took about a week (partly because of malfunctioning of new machines).
- 1948—The solving of eight equations is cited as one of the objectives of the ACE Test Assembly.
- 1952—Pilot ACE solves seventeen equations. The data arrived in the morning post; the input was punched and the answers sent out the same day. (At that time, post was delivered to a central Registry at NPL at 7.30 a.m. but only filtered through to recipients at 10.40 a.m. We insisted on a change and receipt by 9.10 a.m. became the norm. As a result, a rapid turn around on standard problems became perfectly feasible.) Later in 1952 a set of 129 equations was submitted to us by the English Electric Company via John Dennison, their resident contact. He and Ted York worked overnight and succeeded in solving the equations. The following morning we found the room piled high with punched cards, and the reader and punch showing signs of wear and tear. As a result of this episode we realized that punched cards provided a convenient—and potentially infinite—backing store. Remember, Pilot ACE only had 350 words of store for data and program.

Photogrammetry

The use of aerial photography for map-making was in its infancy at this time and depended on extensive manual calculations. The analysis of each photograph took almost a day; roughly an hour was required to complete the readings and several hours to do the sums. A good day's flying could keep staff busy for many months. Pilot ACE took about 1 min to do the necessary iterations. This led to a completely new approach. The bottleneck became the data collection instead of the calculations and automatic stereo-comparators were soon developed to cope.

This work, done for the Ordnance Survey Department, was one of the first external jobs undertaken by ACE. I took it over from Wilkinson, who had started it, and it was my first real program other than subroutine development. I passed the project to Roy Whiteley, the first of my new programming team and probably the first non-graduate in the world to write a computer program. This work evolved throughout the life of Pilot ACE and DEUCE, and continued to run until the NPL DEUCE closed down.

Traffic simulation

An early elegant application by Donald Davies was the simulation of flow through traffic lights—initially for fixed-time signals and later for vehicle-actuated signals, then under development. A random binary number was used to represent a stream of traffic, '1' representing a car and '0' a gap. The four streams of traffic could be observed on the console monitor. Simulation was not new, having been exploited in wartime Operational Research, but application had been restricted because of the time needed to process large samples. With a car or a gap appearing every microsecond, ample evidence was acquired in a very short time. Simulation soon became a booming application.

Matrix programs

The handling of matrices had featured strongly in our research in linear algebra and it was also found that many problems in engineering and statistics could be formulated in matrix form. As a result, a library of good programs became available and attracted a lot of external interest. When a matrix was sent in, say with a request for it to be inverted, we found that it paid to be inquisitive. How was the matrix derived? Could we have saved effort by taking on the problem earlier? What happens to the results? Would a triangular decomposition be more useful? If a matrix was otherwise symmetric we became suspicious if two corresponding elements were 67,429 and 64,729.

The arrival of the drum in 1954 had a big impact on matrix work. Two major schemes, for large and for small matrices, were developed by M. Woodger and B. Munday. There was already a scheme in use at Manchester in which the standard operations such as ADD, MULT, and 'DIVIDE' (INVERT) were resident on the drum. A typical instruction might take the form A B C ADD, where A, B, and C were track addresses and result was left in C. We took a self-assembly approach. We had a large number (eventually 100) of routines known as 'bricks', all conforming to a standard format, which fitted together like LEGO. Bricks were read in with the data. Our instruction format was A B C 6, say, where 6 referred to the sixth brick read in. Suppose we wished to take two matrices from tracks A and B, add them, invert the answer and punch out. We would collect the necessary bricks on triads of cards and number them in the order of assembly: (1) Read, (2) Add, (3) Invert, (4) Punch. Our program would then be

1. o o A 1 (read A).
2. o o B 1 (read B).
3. A B C 2 (Add A to B and place in C).
4. C o D 3 (Invert C and place in D).
5. D o o 4 (Punch out D).

For many standard jobs, only one assembly operation would be needed.

The flexibility offered by the 'General Interpretive Program' or GIP (see Chapter 14, 'The DEUCE—a User's View') was very popular with users in the aircraft industry and some DEUCEs were purchased solely because of it. The program could have been used in contexts other than matrices but I am not aware of such use. A simplified system for dealing with columns, the Tabular Interpretive Program (TIP) was produced by Bristol Engines Co. This had the characteristics of the modern 'Spreadsheet'.[9]

Demonstration programs

Pilot ACE attracted very many visitors, particularly during the annual NPL Open Week (when we also entertained many local schools). It was necessary to find suitable programs catering to a very wide range of interests. At the initial press launch in 1950 several programs were demonstrated. One found the smallest factor of a six figure integer. Another calculated the paths of rays of light through a series of lenses. To our disappointment, this program aroused little interest in the British optical industry, which seemed reluctant to discuss its design methods. (It is perhaps not surprising that German optics were superior to ours at this time.) The popular calendar program was written at very short notice for the Duke of Edinburgh's visit early in 1952. It was tested briefly an hour before his arrival. There was some confusion about the date actually entered for his birthday and the day of the week given on the output lights—the press thought the computer had got it wrong, and this gave us far more publicity.[10]

Programs for the calculation of π and e to 100 decimal places remained popular on all NPL computers. For π, the common form of the subtraction of two inverse tangents was used. For e one of the methods calculated the successive digits more quickly than they could be printed by teleprinter.

Crystallography

Lengthy Fourier analyses were already being performed in NPL's punched-card section for the important research of Dorothy Hodgkin into molecular structure. Her research student John Rollett wrote a program enabling the work to be carried out on Pilot ACE. The calculations concerned the structure of vitamin B12 and, later, insulin. Other groups working in the same field quickly took advantage of the new program.

Fallen trees, new forests

Pilot ACE led to a 'numerical revolution' which had a number of casualties. The desk calculator no longer provided the main aid to mass calculation and the mechanical version soon became obsolete when electronic devices appeared. The production of mathematical tables became unnecessary when the functions could be derived automatically by subroutine. This was a disappointment at NPL, as it was producing some very high quality tables which rapidly became redundant. Analogue machines finally bit the dust (as Comrie had predicted[11]). The large Differential Analyser at NPL could be replaced by a Runge-Kutta subroutine on Pilot ACE. (The hybrid Digital Differential Analyser or DDA did have a further limited life.) Thus the terms of reference for Maths Division, which had included the exploitation of commercial calculating machines and the production of mathematical tables, required substantial revision after only eight years.

On the other hand, experience with Pilot ACE triggered a host of new uses (see Table 1). As has been mentioned, simulation studies became feasible. Developments in linear programming by S. Vajda and E. L. M. Beale made the use of Operational Research a very practical tool. As computer awareness developed throughout NPL, advances in instrumentation enabled experiments to be adapted for automatic data capture on paper or magnetic tape, for example in a wind tunnel or on a sea trial.

Suitable publication outlets were non-existent at first. Eventually, mainly through the efforts of J. Lighthill, Goodwin, and J. Crank, the Institute of Mathematics and its Applications (IMA) was started and this brought together the developing interests in Numerical Methods. At the same time the British Computer Society was formed, again with strong support from NPL.

The Numerical Algorithms Group (NAG) built up a large library of mathematical software. NPL was a major contributor. Initially NAG provided mathematical software for communal use within universities. Now based in Oxford in Wilkinson House—named after Jim Wilkinson—NAG Ltd is a thriving international company. A key contribution to the NAG Library was NPL's comprehensive collection of matrix software.

H. M. Treasury, which had sought NPL's technical advice on desk machines, was even keener for help when other computers came on the market, and a request from Shell for NPL to devise acceptance trials for their Ferranti computer was followed by similar requests regarding other new computers. As large-scale automatic data processing became feasible in civil government it was clear that the part-time efforts of one person (me) were inadequate. I suggested the formation of a specialist group and the Technical Support Unit was formed. This later became the Central Computer Agency, now CCTA to include Telecoms.

I will mention briefly some key developments from 1958 onward.[12] The invention at NPL by D. L. A. Barber and D. V. Blake of the Standard Interface of 12 wires, 8 for data and 4 for control (later to become the more robust British Standard BS4421 with 16 wires), had the effect of uncoupling the I/O from the device, using a well-defined link, and was analogous to the electric plug and socket on which one can rely (at least within a country!). It became a requirement within NPL for all new computers, peripherals, and data-producers to have interface sockets. New computing services were developed such as File Store (Honeywell 516), Scrapbook (or 'Mailbox') on the MOD1, word processing on the PDP11, laser-scan display on the PDP11, on-line terminals on the PDP8. Developments in instrumentation led to special-purpose devices with non-standard input and output. The invention of packet-switching by D. W. Davies led to an experimental network using the Honeywell 516. The standard interface was used to link all computers and services through the network to user terminals and, where appropriate, to experiments. When I retired in 1977 there were roughly 30 computers, 30 exotic peripherals, and 100 VDU terminals all able to interact through the NPL network. Of course, dedicated systems existed in banks and airlines, but there was nothing to rival

NPL's variety of options, except possibly within ARPANET. At about that time, Barber organized a very ambitious link from NPL's network to a conference in Canada via a European network to France and a satellite. He demonstrated a number of computing facilities, but the one that attracted most attention was our experimental link to the BBC's Teletext, because it offered up-to-date scores in the cricket Test. By 1987, the number of terminals had increased to 500. Network protocols were developed.

Ingredients for the success of ACE

I have been able to mention only a few of the outcomes of Pilot ACE, but enough to demonstrate that it was a highly successful computer, in fact earning around £100,000 in its four years of life. This was a sizeable sum— my annual salary was around £300 then. The earnings to NPL compared very favourably with the few pence to test a thermometer or a pound to test a taxi meter. I will identify some of the key reasons for this success.

Hardware

The insistence by Wilkinson that the full double-length result of a multiplication should be provided was unique at the time and essential for developments in linear algebra.

The use of punched cards for input and output proved a boon—although we had been rather apologetic at first, since it appeared to be slow and cumbersome. However, the use of one binary instruction or number per card led to compact compression: the 15 millisecs available between reading each row of the card allowed much computation in parallel. Early warnings that the card would not withstand multi-punching proved to be unfounded and cards were more reliable than many of the contemporary rival paper-tape alternatives.

Many judged the order code (instruction code) of Pilot ACE to be primitive. However, after being tidied up on DEUCE, with memory locations in sequence, it was simple and quite elegant. Every aspect of it was exploited.

Some ingenuity in the hardware (800 valves compared with 3000 in EDSAC) undoubtedly helped reliability.

The input and output available at the console in the form of lights and switches, although primitive, proved very effective in many applications, particularly the solution of eigenvalue problems.

Software

The powerful 'all zero' instruction—'Clear the machine and obey the contents of any store'—enabled a fast start.

There were fast floating-point routines and the concept of 'block floating'.

There was a large support library for the important matrix schemes.

Optimum coding—good or bad? It led to very fast routines but we could overdo it! I recall that when a built-in divider was available on DEUCE, I explored the usual iterative method for finding square roots to replace our digit-by-digit method learned at school. I found that for a floating point number x in the range $\frac{1}{2}$–2, a starting guess of $a + x/2$ gave full accuracy through the range with, if I recall correctly, three iterations. The value for a was around 0·48. By chance, I spotted an instruction which when converted to decimal was about this size, so I used it to save a store. Some time later, a puzzled user had considerable difficulty in understanding my inadequately documented routine.

Peopleware

Pilot ACE became the servant of a very experienced research team with a wide background in numerical methods and supported by experienced junior staff. All staff were dedicated to providing correct answers, in accordance with the rigorous teachings of Comrie and Sadler at the Nautical Almanac Office. We were young and energetic: when I arrived at NPL, the oldest member of Maths Division was 39, whereas in some Divisions, this was almost the youngest age. (We also produced new challenges on the sports field, with Turing coming close to being selected for the Olympics.)

Much credit should go to J. R. Womersley for his efforts in setting up Maths Division and encouraging staff to pursue the very general terms of reference of the Division. Womersley expected us to be outward looking and did not interfere too much. Turing and Fox were at times critical because Womersley was not an experienced numerical analyst, but such animals were rare. Wilkinson and I were very happy with the support he gave us when needed.

The design of Pilot ACE stemmed from Alan Turing. However, most credit for the successful outcome should go to Jim Wilkinson. He was the 'Capability Brown' who assisted in the building of Pilot ACE and master-minded its exploitation, using the very fertile soil provided by NPL staff. In particular, he insisted on double-length results of multiplications, fought for the machine to be put into use, inspired all staff by his enthusiasm, and used Pilot ACE effectively in his researches in error analysis and to develop Numerical Analysis as a new subject. However, he found the red tape of the Civil Service irksome at times. To avoid a boring meeting he would disappear into a hideout and get on with some useful work. He complained that once, when a colleague from the United States had handed him an appropriate number of dollars for a job done, it took him longer to get the money lodged with the NPL Accounts Department than it had to do the job on Pilot ACE.

The ACE attracted many guest workers, all of whom contributed. Some who later became key figures in the computing world were E. L. M. Beale, creator of linear programming, J. van Garvick, who arrived each morning with a new idea for a machine-code, P. Samet, E. S. Page, J. Ord-Smith (he arrived as a student in chemistry and quickly switched horses), C. S. Strachey, A. van Wijngaarden, B. Randell, and J. Howlett.

The decision in 1945 to situate the proposed Central Mathematics Station at NPL proved to be a very good one.

Notes

Thanks to Jack Copeland for his work on this chapter.

1. Recollections over fifty years can get blurred and I apologize for any errors in what follows. Too few records remain of the fertile 1950s. Thanks to Mike Woodger for reading the manuscript.

2. Its primitive state is described in Croarken, M. (1990) *Early Scientific Computing in Britain*. Oxford: Clarendon Press.

3. The Brunsviga was a desk calculating machine.

4. NPL, *Modern Computing Methods*. HMSO (1957; revised edition, 1961).

5. My association with Pilot ACE and its successors was a long one, lasting from 1950 until 1977.

6. NPL, (1956) *Wage Accounting by Electronic Computer*. London: HMSO.

7. See Croarken, *Early Scientific Computing in Britain*.

8. As described by Wilkinson in his lecture delivered on receipt of the Association for Computing Machinery Turing Award in 1970—Wilkinson (1971) 'Some comments from a Numerical Analyst', *Journal for the Association for Computing Machinery*, 18, 137–47.

9. There is a good introduction to GIP and TIP in Campbell-Kelly, M. (1981) 'Programming the Pilot ACE', *Annals of the History of Computing*, 3, 133–62.

10. I believe that Wilkinson is in error when he gives 1950 as the date of the calendar program on p. 110 of his (1980) 'Turing's work at the National Physical

Laboratory and the construction of Pilot ACE, DEUCE, and ACE', in N. Metropolis, J. Howlett, and G. C. Rota (eds), *A History of Computing in the Twentieth Century*. New York: Academic Press.

11. See Croarken, *Early Scientific Computing in Britain*.
12. Further details can be found in Yates, D. M. (1997) *Turing's Legacy: A History of Computing at the National Physical Laboratory 1945–1995*. London: Science Museum.

13 *The ACE Test Assembly, the Pilot ACE, the Big ACE, and the Bendix G15*

Harry D. Huskey

The ENIAC

In 1943, after graduating from Ohio State University, I accepted an Instructorship in Mathematics at the University of Pennsylvania in Philadelphia. My salary was not all that large so I applied for part-time work in the Engineering Department. They had various wartime projects. After some time I was 'cleared' and found I was to work on a huge electronic computer called the ENIAC, which was to be used to compute firing tables for the military. I joined the ENIAC project in 1944 and started working on the punched-card reader and punch that served as the input and output of the computer. I soon graduated to doing the operation and maintenance manuals.

The ENIAC consisted of electronic versions of 20 desk calculators—called 'accumulators'—plus control units which would use certain of the accumulators to do multiplication, division, square root, and input–output. The ENIAC could store up to 20 numbers of 10 decimal digits each. Each digit was stored in a ring counter consisting of 10 flip-flops. Digits were transmitted from unit to unit as a string of 0 to 9 nine pulses on 1 of 10 channels. Addition was accomplished by the pulses stepping the rings, with appropriate carry occurring.

An accumulator occupied a full panel and had about 550 vacuum tubes (valves). The whole computer used about 18,000 vacuum tubes (compared to less than 1000 in the Pilot Model ACE).

The EDVAC

Some of the ENIAC's shortcomings were realized before it was completed. Five hundred and fifty vacuum tubes to store one ten-digit number and to do addition and subtraction was too many tubes! Thus members of the project were considering ways of storing information more efficiently.

Mercury delay lines had been used for range measurements in radar. A pulse was 'fired' down the mercury line at the same time that the radar signal was transmitted. The time for the echo to return was compared with the fixed time for the pulse to transit the mercury line. J. Presper Eckert, chief engineer of the ENIAC, suggested using mercury delay lines to store numbers. A mercury line 5 ft long could store a train of 1000 pulses operating at a pulse rate of one megacycle. The output from the line could be amplified, standardized, and re-inserted into the line, storing the equivalent of 32 numbers of ten decimal digits each. The associated amplifying and standardizing circuits required less than ten vacuum tubes. Thus, one mercury line would store 32 times as many numbers as the 550-tube ENIAC accumulator.

Sometime in the spring of 1945 Eckert described this re-circulating memory to me. My first question was: 'How do you control the operation of the computer?' On the ENIAC, the program for solving a problem required a large number of patch panel interconnections. How does one do this for the mercury lines? Mechanical control from punched cards or punched paper tape takes milliseconds, which is much too slow compared to the rate at which data are available from the mercury lines. Magnetic wire or tape were not yet developed at that time, and would in any case offer other problems, such as start/stop.

Eckert said: 'Store the instructions in the delay lines just like numbers'. Of course, the answer is obvious! Why didn't I think of that? It was the only way that instructions could become available at rates comparable to the data rates. That was the stored-program computer! This idea gave a tremendous step in flexibility, making it possible to process programs; this led to assemblers, compilers, and languages like FORTRAN, ALGOL, COBOL, etc. A project was started to design a computer called the EDVAC using this memory.

John von Neumann, as consultant to the military at Aberdeen Proving Ground, visited the EDVAC project and, after numerous discussions with the project personnel, produced the *First Draft of a Report on the EDVAC,* dated 30 June 1945. The report, although not formally published, was widely distributed both in the United States and England. In 1947 the War Department sent an 'official' copy to the Patent Office; included was a statement of von Neumann saying 'Reference to specific classified devices and components was avoided...'. This was so that distribution of the report would not be limited for security reasons.

The *First Draft of a Report on the EDVAC* was of little help to those of us working on the project, because von Neumann approached the topic from the more theoretical point of view. His report was of little value to the computer designer. We looked at hardware and tried to decide what could be constructed in a reasonable time. However, the von Neumann report was of immeasurable value in acquiring government financial support for computer development.

Who was responsible for the concept of the stored-program computer? The von Neumann report gives no credit to other individuals and citations were never completed; as a result von Neumann has generally received credit for the stored-program concept. John Mauchly, who had been interested in numerical applications in meteorology, joined with Eckert, who had the hardware experience, and they developed a proposal for Aberdeen Proving Ground which led to the ENIAC. The limited memory of the ENIAC led Eckert, with his experience of delay lines for radar ranging, to propose delay lines for computer memory. Storing instructions like data in the delay lines was the only way that the instructions would be available at the required speed. I think Eckert deserves credit for the stored-program concept.

In 1946 the ENIAC was completed and visitors were invited to see and—in some cases—to use the machine. One such visitor was Professor Douglas Hartree of Cambridge University. He was interested in solving two-point boundary value problems. Sometime in the spring of 1946 I asked him about opportunities to work on computers in England. He seemed non-committal.

In the spring of 1946 several senior people left the EDVAC project. Eckert and Mauchly, unhappy with the University's approach to the commercial development of their ideas, resigned and started their own company, planning to build mercury delay-line computers (the highly successful UNIVACs). Arthur Burks and Herman Goldstine joined von Neumann's computer project at Princeton. I became the senior person on the EDVAC project and was offered the job of directing the activity. However, I was still an Instructor in the Mathematics Department and the chairman of Mathematics blew his stack. The offer was withdrawn. I was mad and resigned from the University.

We moved back to Ohio and for a brief time I taught at Ohio State University. Then in July I received a cable offering me a one-year visiting appointment at the National Physical Laboratory.

At the National Physical Laboratory

In the aftermath of the war, transatlantic travel was difficult to arrange. Finally, with the help of the British Embassy, we obtained passage on the SS *United States* in December. We sold our car to pay for the tickets. Things were still in short supply in England, so on the advice of the British Embassy we arranged to take with us two cases of corned beef. In preparation for our trip someone had given us a book *How to Like an Englishman*. It said the English were very reserved and that it would be difficult to make friends. At Southampton we transferred to the boat train. In our compartment we were joined by an English couple. He looked at our baggage and said 'From the States, eh?'. He talked all the rest of the trip.

I reported for work at the National Physical Laboratory and found that Mathematics Division was in two buildings outside the main NPL grounds—Teddington Hall and Cromer House. Mr Womersley, the superintendent, introduced me to Jim Wilkinson and Mike Woodger. Turing was at a meeting in the United States.

With the possible exception of Turing, no one in Mathematics Division had any electronic experience. The expectation was that hardware would be made somewhere else. The main theme in the ACE was a design that encouraged the programmer to locate data and instructions so as to execute in minimum time. The team concentrated on logical design and did extensive programming to test ideas.

Mr Womersley took advantage of my presence to learn about the status of other computer activities in the United Kingdom, sending me to visit each project. (My original report on these visits is in Part V.) I found Wilkes at the University of Cambridge circulating data in a mercury line. His system tended to gain pulses, leading to thoughts of shielding or increasing signal levels. Williams at the University of Manchester had stored two lines of data on the face of a cathode-ray tube, using a dot–dash pattern. The change of charge at the gap could be detected before the beam refreshing the pattern reached the gap. This could be done sufficiently early that the beam could be turned off, maintaining the gap.

Coombs of the Post Office Research Branch at Dollis Hill had both long and short mercury delay lines working, but not reliably. At that time I knew nothing of the wartime work on Colossus by the Post Office, Turing, or Newman (initiator of the Manchester computer project). On later trips to Dollis Hill some of us raced Turing—he was an Olympic class long distance runner. Wilkinson rode his bicycle and I took the train. The distance was 16 miles and I had to catch three trains. I carried a satchel with Turing's clothes and arrived at the Post Office gate house a few minutes before him. He changed from his shorts and we went to our meeting.

That spring Turing gave a series of lectures at the Ministry of Supply in London (see Part V). Turing, Leslie Fox (who worked on relaxation methods), Wilkinson and I would take the train to London. On the way we talked of computer design problems. Once we argued about the rounding of floating-point numbers and Turing became so upset that he could hardly give his lecture.

At this time Turing was interested in artificial intelligence. I once asked him what he was working on. He said, 'How would you chastise a computer?' It was much later that I appreciated the significance of that question. We all knew of Turing's 1936 paper 'On Computable Numbers', but the machine discussed there was so elementary compared to what we were working on that no one, I think, thought of it as a precursor.

The ACE Test Assembly

There was no hardware effort at the NPL and plans for acquiring hardware at other places, such as the Post Office, were not moving ahead (see Chapter 3, 'The Origins and Development of the ACE Project'). So our group proposed to Womersley that we build a prototype and we started work on a much simpler ACE that came to be called the ACE Test Assembly. All of us except Turing worked on the Test Assembly. Turing was never in favour of the project; he preferred that we all work on the Big ACE. Turing participated to the extent of answering questions. For example, when we were designing the mercury delay lines the question arose: What about reflections of the sound waves in the mercury lines causing spurious pulses? I had expected to set up a test line to see. But not Turing! He spent more than a day working out mathematically the probability of spurious pulses.

Late in 1947 Sir Charles Darwin decided, perhaps rightly, that mathematicians did not know enough electronics to build a computer. The Radio

Division was to be involved. A whole new group had to learn about computers. The Test Assembly was scuttled. I prepared to return to the United States and Turing went to Cambridge—morale was low! The Mathematics and Radio Divisions went on to design and build the Pilot Model ACE, which was similar to the Test Assembly.

The 12-bit ACE

With the demise of the Test Assembly I had time to work on other things. As an exercise I designed a 12-bit ACE. My goal was the simplest design that would work and illustrate the principles of Turing's ideas. My design used one long delay line. There was only 1 Timing Number in the instruction. Instructions came from the instruction register, a 12-bit delay line, at the same time that transfer occurred. This was very restrictive, making the system unsuitable for general computation, but it was sufficient to illustrate the principles.

As shown in fig. 1, instructions came from the instruction line and were complemented at gate *a*. During TRANSFER gate *b* opened passing a new instruction to the half-adder. At the same time gate *g* was inhibited erasing the old instruction. Pulse P7 was added to the Timing Number of the instruction as it passed into the instruction register and during each word time thereafter. At the end of TRANSFER gate *h* passed a pulse which cleared the

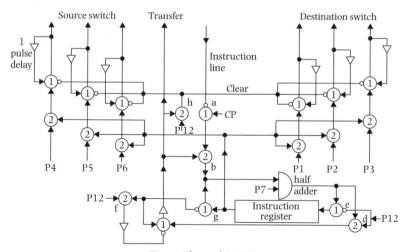

Fig. 1 The 12-bit ACE.

Table 1   Sources and Destinations for the 12-bit ACE

	Source	Destination	Special	Remarks
0	Input	Output		
1	DL	DL		
2	TS & P12	TS		Sign, accumulator
3	TS	+TS	Bell	Accumulator, add
4	P6	Special	Read	
5	P1	—	Punch	'1'
6	—	Instruction		
7	I/O ready	—		

Source and Destination flip-flops. Immediately (at P1 time) the new instruction emerged from the delay line and, via gate g, circulated through the adder. For subsequent word times, the first 6 bits of the instruction were gated by P1 to P6 to set the Source and Destination flip-flops (they were set each word time, which had no effect after the first). When the count overflowed into the P12 position, gate d set the transfer flip-flop and the process continued. Sources and Destinations are shown in Table 1. DL is the single long delay line and TS is the 12-bit register.

The SEAC

The National Bureau of Standards (NBS) established an Institute for Numerical Analysis (INA) on the campus of the University of California at Los Angeles. While in England I had applied for a job at the INA, to start in January 1948.

I was to spend six weeks in Washington to learn how the Bureau operated. It became almost a year. My job was to monitor the US computer projects. These were the Eckert–Mauchly UNIVAC in Philadelphia, the EDVAC at the University of Pennsylvania, the von Neumann project at the Institute for Advanced Study at Princeton, WHIRLWIND at MIT, and a data reduction computer being made by Raytheon for the Navy.

During the year in Washington I initiated a project in the Electronics Division of the Bureau to build a computer. I proposed to use Turing's ideas in order to build a fast delay-line computer. However, the Bureau hired Sam Lubkin from the EDVAC project and it was decided to build a copy of the EDVAC. This became the SEAC (National Bureau of Standards Eastern Automatic Computer).

I proposed to build a parallel computer at the INA using Williams tubes for memory and in October 1948 my proposal was approved. In December I moved to Los Angeles.

The ACE

In the meantime the group in the Mathematics Division at NPL transferred to the Radio Division and worked on a design which became the ACE Pilot Model (see Chapter 3, 'The Origins and Development of the ACE Project'). The success of the Pilot Model and the DEUCE led the team at NPL to continue with the original ACE design. The result was a computer with sixty-four Sources and Destinations and the same type of control system.

The Big ACE had twenty-four long (32-word) delay lines, five 4-word lines, four 2-word lines, and seven 1-word lines. There were four 32K magnetic drums. The primary input–output was punched cards in row binary mode. The timing of events in processing a card was handled by the program. There were two magnetic tape drives with a magnetic core buffer.

As in Turing's design there were two Source switches and a powerful function box. (The Test Assembly and Pilot Model had a single Source address and there was no function box.) The word length was 48 bits and the instruction used 47 bits. Table 2 (from the 1960 'ACE Programming Manual') shows the instruction format. (The wait W and next instruction time T are no longer relative: they refer to the actual time of the event.)

The functions were arranged in eight groups of eight as shown in Table 3. The LOGICAL operations are of the form D = A op B. Shifts are of the form D = A shifted B places (or shifted B + 48 places). SHIFTS may be cyclic, up (more significant), or down. ADDITION/SUBTRACTION works on single-length, double-length, or mixed operands. The CLEAR group works with A = A op B zeroing one or two words of A. If A is a 1- or 2-word register, a single instruction can sum many values of B. In instruction MODIFY an integer from B is added to W. Carry out of W is added to J and to some combination of A, B, or D. With an appropriate initial value in J, branch-out of a loop does not require extra instructions. MULTIPLY and DIVIDE initiate the process and other computation may be done while they are underway. Interlocks delay instructions which would disturb multiply or divide results. EXTERNAL handles input and output and much of the input/output timing is handled

Table 2   Instruction format for the Big ACE

Bits	Group	Effect
1–5	W	Wait—word time in which operation will start
6–11	A	Source for A operand
12–17	B	Source for B operand
18–23	F	Function (8×8 operations)
24–29	D	Destination
30	St	Stop bit (if console key depressed)
31–35	N	Next Instruction Source (1–32)
36–40	J	No effect, counting area
41–45	T	Next instruction time
46–47	Ch	Characteristic bits
48	—	Not used

Table 3   Groups of functions in the Big ACE

Group	Function
0	LOGICAL
1	SHIFT
2	ADD/SUBTRACT
3	CLEAR and ADD/SUBTRACT
4	INSTRUCTION MODIFY
5	EXTERNAL, STANDARDIZE, MULTIPLY, and DIVIDE
6	MULTIPLY by small integers
7	ADD/SUBTRACT and DISCRIMINATE

by the program. Instructions in group 6 do addition or subtraction with 1, 2, 4, or 8 times the operands. DISCRIMINATION or transfer of control involves taking the next instruction 1 word later than was normal in the Test Assembly and Pilot Model. (Destinations 60–63 give discrimination branches if the result of A op B is zero, non-zero, positive or zero, or negative.)

The SWAC

At the INA in 1949 I started out with an empty room. I hired people, bought equipment and supplies, and my Williams tube computer began to take shape.

The computer was dedicated in August 1950—the same year as the SEAC—and was named the SWAC (National Bureau of Standards Western Automatic Computer). Punched card input and output was added to the SWAC, and a magnetic drum auxiliary memory.

In 1953 the Bureau was asked to test a battery additive. It reported that the additive was of no value. A complaint by the manufacturer led to an investigation of the Bureau's activities. The result was a scaling back of the Bureau. Projects done for the Defense Department were moved to military laboratories. The INA became a mathematics project in UCLA's Department of Mathematics and the SWAC was transferred to the College of Engineering.

The Bendix G15

In 1954, I accepted a joint associate professorship in Mathematics and Engineering at the University of California at Berkeley. Here I resurrected my ACE-type computer design, prepared a description and drawings, and showed them to companies who might be interested in building the computer. Figure 2 shows a sketch I prepared, indicating how I expected

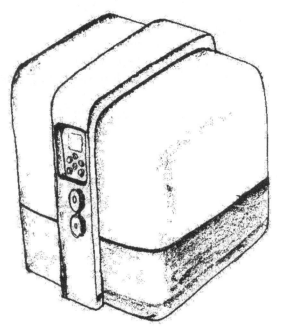

Fig. 2 Preliminary sketch for the G15.

the computer to look. It was to be about 4 ft square and 6 ft high with a separate typewriter serving as input and output. No air conditioning was involved and normal 220 V electric supply was adequate. The Bendix Corporation made the best offer. I sold them the rights and became

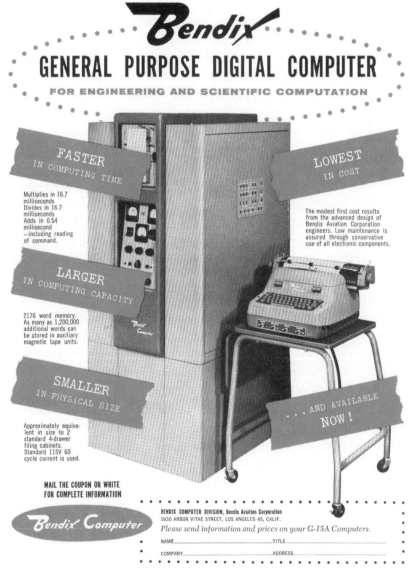

Fig. 3 Advertisement for the Bendix G15.

a consultant to their Computer Division in Los Angeles. (Bendix lent me one of the G15s and it was installed at my home in Berkeley. This machine is now on permanent display in the Smithsonian Museum in Washington DC.)

The problem of locating instructions and data and determining the various Timing Numbers in order to obtain an optimized program was difficult with the ACE-type computers. Bendix produced a program called POGO for the G15 which placed instructions in optimal locations. However, it was never popular with customers.

A G15 customer from the Humble Oil Company produced a floating-point interpretive system which addressed memory using consecutive integers, hiding the memory line and Timing Number structure. Consecutive addresses were spread cyclically over eight lines and the only instruction to the user was that placing instructions in low memory addresses and data in high would produce faster programs. This was called INTERCOM 101. I produced INTERCOM 500 which used the first 100 locations in memory lines for program or data and the positions beyond 100 for index registers, increments, and limits for the index registers. A double precision version was produced called INTERCOM 1000. These two INTERCOMs were more popular with customers than the 101. Each instruction address was converted to binary with each execution. Thus, checking code did not involve conversion from binary.

A digital differential analyser (DDA) was added to the G15. The Northrup company had produced the MADDIDA which imitated the behaviour of the analogue differential analysers. Variables were 'rate' functions which ranged in value from zero (000. . .) to one (111. . .). One-half would be represented by a string of alternate zeros and ones (101010. . .). A problem with the Northrup design was the difficulty of input and output. That part was more complicated than the DDA itself! Bendix bought the rights to the Northrup DDA and were making their own version. It was natural to add the DDA circuitry to the G15 and use the general purpose computer to control it and to handle input and output.

Comparison of the Test Assembly, Pilot Model, and G15

None of these computers had a separate arithmetic unit in the Eckert–von Neumann style. There was a Source switch which connected memory to a 'Highway' (called a 'bus' in the United States) and which connected via a Destination switch back to memory. Arithmetic and control actions were

Table 4 Structure of the ACE Test Assembly (TA), the Pilot Model (PM), and the G15

Sources or Destinations	TA	PM	G15
Delay lines or tracks	L 1–L 7, L 26	L 1–L 7	L 0–L 19
4-word lines			Q 20–Q 23
2-word lines		D 12, D 14, D 26	D 24–D 26
1-word lines	T 8–T 12, T 14	T 8–T 10	T 28 (=AR)
AND	T 9 & T 10	T 9 & T 10	Q 27 (=(Q 20 & Q 21)
			∨ (~Q 20 & AR))
NOT	~T 11	~T 9	
Pulses	P1, P16, P32	P1, P17, P32	
Units for:	1, W, Sign	1, W, Sign	
Precession	L 1 via T 11	L 26 via T 11	Characteristic[a]:
(i.e. delay by 1 mc)			via AR
ADD	T 12	D 14, T 16	T 29, D 30
SUBTRACT		T 16	Characteristic[a]

[a] A 2-bit characteristic determines whether the number(s) from the source switch are negated, changed to absolute value, or delayed by 1 mc via T 28—see fig. 4.

determined by certain positions on the Source and Destination switches. Tables 4 and 5 show that the Source, Destination, and memory line struc-ture, as well as the instruction format, were similar for the three computers. (Although there were some differences in the terminology used and in the parts of the instruction assigned to various functions.) Two components of the Pilot Model (PM) instruction not present in the Test Assembly (TA) were the Characteristic bit and the Go bit described in previous chapters.

The memory of the G15 consisted of circulating loops on a magnetic drum. During each revolution all information was erased and re-written. Short loops for 1-, 2-, and 4-word tracks occupied space between the erase and write heads for the long tracks. Logically they corresponded exactly to the mercury delay lines. The TA and PM each had eight mercury delay lines storing thirty-two 32-bit words (numbers or instructions), one 2-word store (DS), and five 1-word stores (TS). The G15 had twenty 108-word tracks, four 4-word tracks, three 2-word tracks, and one 1-word track.

Many of the Sources and Destinations provided logical or arithmetic values or caused other actions. For example, Source 17 of the TA gave TS 9 & TS 10

Table 5 Instruction format of the ACE Test Assembly (TA), the Pilot Model (PM), and the G15

TA		PM		G15	
Bits	Function	Bits	Function	Bits	Function
1–2	Spare	1	Spare	1	Single/Double
3–7	Source	2–4	Next Instruction	2–6	Destination
6–12	Destination	5–9	Source	7–11	Source
13–15	Next Instruction	10–14	Destination	12–13	Characteristic
16–18	Spare	15	Serial/Deferred	14–20	Next Instruction
19–24	Transfer time	16	Spare	21	Breakpoint
25	Spare	17–21	Wait	22–28	Transfer time
26–31	Wait time	22–24	Spare	29	Block/Deferred
32	Spare	25–29	Timing		
		30–31	Spare		
		32	Go Digit		

('&' is boolean conjunction). This became Source 21 in PM. Source 31 in the G15 gave Q 20 & Q 21. One word addition involved sending a number to Destination 13 (+TS 12) in the TA, to Destination 17 (+TS 16) in the PM, and Destination 29 in the G15. The precession instruction enabled one to cycle through a list (which was less than Turing wanted in his BURY and UNBURY—see his 'Proposed Electronic Calculator' in Part V). Pulse P_n (at the nth position of the word) could be used to 'count' in the SPARE positions of the instructions. Overflow spills into the Timing Number and changes the effect of the instruction.

Figure 4 shows the structure of the G15. Commands with Destination 31 controlled input and output to the typewriter, paper tape, magnetic tape, and punched cards. Punched paper tape was mounted on the front of the computer in a magazine when used.

Although the state diagram of fig. 5 (which is from a Bendix publication) uses the terminology of the G15, all three computers had similar state diagrams. During a specified word time the instruction or command is transferred from memory to a control register (READ COMMAND). If the command is SERIAL in the case of the PM (determined by bit 15) or BLOCK in the case of the G15 (determined by bit 29), it moves immediately from the READ COMMAND state to the TRANSFER state. ('Serial/Deferred' in the PM

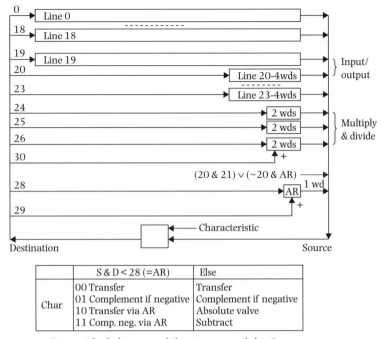

	S & D < 28 (=AR)	Else
Char	00 Transfer 01 Complement if negative 10 Transfer via AR 11 Comp. neg. via AR	Transfer Complement if negative Absolute valve Subtract

Fig. 4 Block diagram of the structure of the G15.

Fig. 5 State diagram for the G15.

corresponds to 'Block/Deferred' in the G15; see Table 5.) The transfer Timing Number determines when the TRANSFER state ends. For DEFERRED instructions the transfer Timing Number determines when the word or double word is transferred.

14 *The DEUCE—a user's view*

Robin A. Vowels

Introduction

This chapter describes the origins and development of the English Electric
DEUCE (Digital Electronic Universal Computing Engine), the production
machine derived from the ACE Pilot Model. The DEUCE was an outstanding success. This is attributable to the DEUCE's high speed, huge program
and subroutine library, fast magnetic drum, enhanced peripheral equipment,
and extraordinary reliability. (The reliability of the DEUCE was the result
not only of the quality of its engineering but also of a rigorous schedule of
preventative maintenance.) Other factors in the commercial success of the
DEUCE included the availability of a backup maintenance team and a user
group sponsored by the manufacturer.

The first DEUCE was installed in early 1955.[1] Most DEUCEs saw a decade
of service. Approximately 20 were still operating in 1965, some continuing to
the end of the decade.

The origins and distribution of the DEUCE

The English Electric Company became interested in computers as early as
1949, when it assisted the National Physical Laboratory in building the Pilot
Model ACE. The Company decided to build the DEUCE in 1951. Sir George
Nelson, chairman of English Electric, offered to build an engineered version
of the Pilot ACE for the token sum of £5000, because he wanted 'to see English
Electric getting into the field'.[2] DEUCE was developed in the Nelson Research
Laboratory (NRL) of the English Electric Company.

Transforming the experimental Pilot ACE into the highly reliable production model required a modified electronic design and an entirely different
functional organization and physical construction.[3] The DEUCE had one

extra delay line of high-speed storage, two additional quadruple stores, an automatic divider, and a larger magnetic drum store. These improvements were the result of experience with the Pilot ACE. The DEUCE was similar in programming and use to the Pilot ACE.

English Electric continued to develop the machine after the first DEUCEs were delivered. Some of the smaller enhancements were offered at no cost. The speed of the card input and output was doubled by means of equipment that read and punched twice as many columns of the card. Some DEUCEs used new card equipment that read all 80 columns of the card, buffered the transfer of data, and performed binary to decimal conversion. Other optional extras included a magnetic tape backing store, paper tape input and output, and more main memory. A kind of parity check was added to the magnetic drum store in about 1961.

Approximately 33 DEUCEs were built. They were used in universities, industries, and government, in work ranging from scientific research and technical applications to commercial data processing. DEUCEs were installed at Queen's University in Belfast, Liverpool University,[4] Glasgow University, and the New South Wales University of Technology (Kensington, Australia). Six machines were used for aircraft design, two at the Bristol Aircraft Company at Filton, two at Bristol Siddeley Engines at Patchway, and two at the Royal Aircraft Establishment at Farnborough. The British Aircraft Company at Warton near Preston had two DEUCE Mark I machines. DEUCEs purchased for government use included three operated by the Ministry of Agriculture, Fisheries, and Food (one a Mark II), one at the NPL at Teddington, one at the Department of Scientific and Industrial Research in Glasgow, one at English Electric's Main Works at Stafford, and one at their Nelson Research Laboratory at Stafford, and there were several more employed on reactor and atomic weapons design, including at least one at the Atomic Weapons Research Establishment at Aldermaston and one at the Atomic Energy Authority at Capenhurst. The English Electric Company's Kidsgrove Data Centre had a Mark I and Mark IIA machine side-by-side, and its London Computing Service had at least one Mark I machine and probably a second. English Electric at Luton had at least one DEUCE, and there were two at the English Electric Whetstone site, servicing the Mechanical Engineering Laboratory and the Atomic Power Division. There was one at the Central Electricity Generating Board (CEGB), one at Short Brothers and Harland in Ireland, and one at British Petroleum at Aldgate in London (used for seismology work). There was at least one DEUCE in Oslo, owned by the Norwegian Government.[5]

Overview: storage, instructions, input–output

The high speed section of the DEUCE consisted of a main storage of twelve mercury delay lines, each containing 32 words. They were written as DL 1 to DL 12. Each delay line was separately addressable, as were the individual words stored in each delay line.

Three levels of registers, also mercury delay lines, were provided:

1. Four 32-bit Temporary Stores, denoted TS 13 to TS 16. TS 13 was an accumulator in which integer addition and subtraction could be performed. Logical operations were associated with TS 14 and TS 15, while TS 16 was associated with the integer multiplier and divider unit. TS 14 could also be used as an arithmetic shift register.

2. Three Double Stores, denoted DS 19 to DS 21, each consisting of two 32-bit words. DS 21 could be used as a 64-bit accumulator (arithmetic register), and the integer operations of addition, subtraction, multiplication, and division were carried out. It could also be used as an arithmetic shift register. Mixed mode arithmetic could be performed using DS 21 (a 32-bit integer could be added to or subtracted from a 64-bit integer in DS 21, and the 32-bit integer would be automatically extended to 64-bits, achieving sign extension). DS 21 could be switched to a second mode, in which it behaved as two separate 32-bit accumulators, and the operations of addition, subtraction, and shifting could be carried out in both. DS 21 was automatically switched to this second mode whenever division commenced.

3. Two Quadruple Stores, denoted QS 17 and QS 18. Both consisted of four 32-bit words. Initially, however, no arithmetic facilities were provided for these registers.

Figure 1 shows the organization of the main memory and registers of the DEUCE.

The hierarchy of storage was intended to provide a more effective interface between the relatively slower main storage (maximum access time 1024 microseconds (μs)) and temporary stores (registers with access time 32 μs). The design struck a compromise between speed, the number of available addresses, and the cost of equipment. Each TS could be accessed in every machine cycle; a word in each DS could be accessed every 64 μs, and a word in each QS could be accessed every 128 μs. Single words could be transferred to and from each of the 18 registers. Multiple words could be transferred

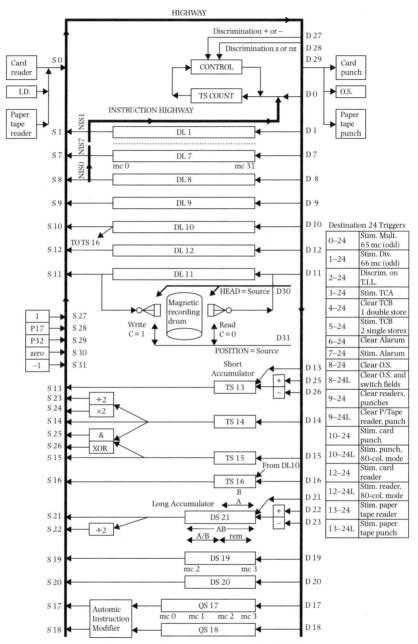

Fig. 1 Architecture diagram of the DEUCE.

to and from the quadruple and double registers by executing a single instruction.

A single DEUCE instruction was capable of transferring from one to 33 words to or from any of the registers and main storage. Specific provision was made to transfer pairs of words (64 bits) to facilitate double-length and quadruple-length arithmetic. Thus all 32 words of one DL could be transferred to another by a single instruction. Likewise all 32 words of a DL could be cleared (set to zero) by a single instruction. Probably the most-frequently used array operation was the summation of all the words in a DL, again with a single instruction.

The DEUCE input–output equipment was Hollerith punch-card,[6] adapted from tabulating and gang punch equipment of the day. One person could barely lift one end of the equipment, which had a heavy cast-iron frame.

Backing storage was realized in a compact magnetic drum having separately movable reading and writing heads. Mounted in a cast-iron frame, it weighed about 25 Kg overall.

The organization of the DEUCE was almost the same as that of the Pilot Model ACE, although different Source and Destination addresses were used for some of the same registers.

Preventative maintenance

The high reliability of the DEUCE was brought about through a régime of regular preventative maintenance, and through the daily proving of the machine under 'marginal conditions'. When a DEUCE was operated under high or low margins, a voltage was simultaneously injected into the circuit of every logic gate.[7] These voltages would take the machine one-third of the way toward the point of failure. The DEUCE had to function correctly under these conditions. The premise was that if the circuits worked correctly under adverse conditions, they would certainly function under normal conditions, with the guarantee of a considerable margin of safety.

A gradually deteriorating circuit would be identified through the maintenance régime of recording and monitoring safety margins of the circuit of each logic gate. This was carried out progressively over a period of three months. Circuits that deteriorated more rapidly would be detected earlier, when the machine failed a marginal test and before the circuit had a chance to cause an intermittent or permanent fault.

The design of the DEUCE enabled a fault to be isolated in one of six sub-circuits in any of the seven bays of electronic chassis. Other fault-finding facilities enabled engineers to explore the range over which an individual logic gate would continue to operate correctly, and to change voltages at sensitive points throughout the machine by means of a single key at the console.

Enhancements of the DEUCE

The first DEUCEs had 80-column punch card input–output equipment (a 200 card-per-minute reader and a 100 card-per-minute punch). However, only 32 columns of the 80 could be used for data. Shortly after the first production machines were introduced in 1955, it was realized that modifications could be made to the input–output equipment to double throughput—not by increasing speed, but by doubling the number of card columns read and punched from 32 to 64.

To make use of this increased throughput, improved instruction modification was needed. Although DEUCE had one single and one double-length accumulator (which could be used as three separate 32-bit accumulators), those arithmetic facilities would be employed in reading or punching the 64 columns, leaving nothing spare for performing conversion between decimal and binary. Improved loop control was needed as well, since performing all incrementing and decrementing through the one register TS 13 would have meant swapping the trip count in and out. (Because the DEUCE had no index register, one of the accumulators—usually TS 13—was occupied for this purpose.) Thus the Automatic Instruction Modifier (AIM) was conceived. With it, the programmer could perform the functions of indexing and trip counting (incrementing or decrementing for loop control), or both combined, all in the one register.

Following a survey of usage of the double and quadruple stores in programs, the English Electric Company chose the quadruple stores QS 17 and QS 18 for the AIM, which luckily provided eight registers in which automatic incrementing and decrementing (now called *auto-increment*, *auto-decrement*) could be done. When used for automatic indexing, the content of the register was incremented/decremented as it was leaving the register en route to the control unit, the updated content being written back into the register. When auto-incrementing/auto-decrementing was required, the transfer to the control unit was suppressed. Automatic increments and decrements could be made to the Source address of an instruction, the Destination address, the Wait Number (effectively the index), and to an unused part of the instruction to assist in loop control.

Another important advantage which came with the AIM was that the control of the most common form of loop required a single instruction, instead of three instructions and typically a constant under the old method.

The AIM became standard on all DEUCEs (and was retrofitted to all pre-existing DEUCEs except the first).

The DEUCE magnetic recording drum had separate reading heads and separate writing heads. The 16 reading heads were all on the one access shaft, and the 16 writing heads on another access shaft. Although the heads could move independently, they could not be moved together on early DEUCEs. The electronics for the drum interlock were completely redesigned to allow the heads to move simultaneously, and to allow other non-conflicting magnetic drum operations to proceed simultaneously, thus speeding up programs that used the drum. This change influenced programming, because it reduced the necessity to anticipate data transfers between the drum and memory.

While DEUCE was initially constructed for scientific use, a second version, *Mark II*, having completely different IBM input–output equipment, was built for the commercial market. The original Hollerith card reader and punch were separate units. They were never intended to be operated simultaneously and ran at different speeds. The new input–output unit was a combined card reader and punch. The reader and punch mechanisms could be locked together to run at 100 cards per minute and to use the same Single-Shot to indicate that a row was ready to be read and punched. Such an arrangement enabled an update of the card just read to be punched on another card simultaneously with reading the next card. In the original DEUCE, the card reader would have had to be stopped after each card, to allow the corresponding card to be punched. Not only would repeated stopping damage the equipment, but the effective speed of each device would have been about 25 cards per minute instead of 100. In the combined unit the reader could still be operated separately at 200 cards per minute, and the punch separately at 100 cards per minute.

With the new IBM input–output equipment came buffered input and output for all 80 columns, with automatic conversion of alphabetic and numeric characters to and from binary, stored as 6-bit codes, thus freeing the processor for computation during the entire card cycle. In 80-column mode, the ten decimal digits, the 26 upper-case alphabetic characters, and punctuation characters could be represented. The IBM equipment also read programs and data in 64-column mode, and was compatible with the earlier Hollerith equipment.

To increase the size of main memory, seven extra delay lines could be installed, giving an additional 224 words of store. Machines having the extra high-speed store had the letter 'A' appended to the model number, as in 'Mark IIA'. The extra delay lines became available in 1959 and could be installed in any DEUCE.

An additional magnetic drum, Decca magnetic tape units, an English Electric paper tape reader, and Creed paper tape punch equipment could also be fitted to the DEUCE. The Decca magnetic tape units read and wrote six-channel half-inch tape at 100 inches per second. Character density was 80 characters per inch, giving a transfer rate of 8000 characters per second (cps).[8] The paper tape reader could read five, seven, and eight-channel paper tape at 850 cps. The paper tape punch could punch five- and seven-channel paper tape at 25 cps.

During the production life of the machine, consistent engineering effort went into improving the basic DEUCE, with circuit changes to enhance its reliability. Evidence of this is to be found in the engineering changes periodically circulated to each site.[9] Furthermore, considerable development of new equipment for the DEUCE and the associated electronic control circuitry undoubtedly continued to make the DEUCE an attractive proposition to users.

The DEUCE program library

There is no doubt that the commercial success of the DEUCE was due in part to the fact that Turing and his team continued to write programs even when construction of the ACE was in the doldrums. The legacy of this work was the hundreds of programs that ran on the Pilot ACE and which could, after minor translation, be run on the DEUCE. Software included subroutines for solving simultaneous equations, differential equations, double- and triple-precision fixed-point arithmetic, and the like.

The DEUCE program library was an extensive collection of at least 738 programs plus over 280 subroutines. Many were contributed by DEUCE users and by the staff of English Electric, and were published by the DEUCE Library Service, a department of English Electric. In addition some specialist programs were distributed by individual sites.

The programs were distributed to each site with operating instructions, technical details of the algorithms used, flow diagrams, and detailed coding sheets. Each site was provided with master and sub-master (backup) copies of the punch cards for each program and subroutine. Perusing the flow diagrams of DEUCE library programs and subroutines often helped the programmer to improve his or her coding of programs generally.

One important publication of the DEUCE Library Service was *DEUCE News*, a regular bulletin of the latest programming tips and techniques, news about sites, hardware upgrades, peripheral timing details, corrections to library programs, programming standards, and the like. Of special importance were the new techniques for programming the AIM and the 64-column read and punch facility. The *News* was essential reading. At least 63 issues were published.

Machine language

The New South Wales University of Technology DEUCE (withdrawn from service at the end of 1966) was opened by the Governor of New South Wales on 11 September 1956 and was fully installed by October 1956 (see fig. 2). My first recollection, in 1961, of UTECOM—as the machine was known—is of being given the *D.E.U.C.E. Programming Manual*[10] and setting out to learn machine language. They said it would take three months to learn thoroughly. And it did.

The *flow diagram*, the first stage of producing a program for the DEUCE, was more than a flowchart—it was the complete machine-language program in two-dimensional form. There was more work to perform even after the flow diagram had been refined and finalized, however. Memory locations had to be allocated to each of the instructions and constants, using a check list to ensure that no location was used twice. Only then could the instructions be written in decimal on a coding sheet. On account of the 'two-plus-one' address system of the DEUCE, the writing of each instruction had to include the address of the next instruction. The Wait and Timing Numbers had to be specified. It will be recalled from previous chapters that the Wait Number specified when the transfer could take place.[11] It was the time when the data was emerging from a storage unit and thus was capable of being transferred to another part of the machine. The Timing Number specified the position in a given delay line of the next instruction, which would also be the moment when the instruction to be executed next was emerging from a DL. (Both the Wait and Timing Numbers were relative to the position of the current instruction.) As with the Pilot Model ACE, instructions had to be placed in memory in such a way as to minimize waiting time both for the

Fig. 2 The UTECOM.

The left-hand side of the operator's console had three main rows of lights. The top-most row of 13 white lights displayed the current instruction (NIS, Source, Destination). Beneath that, 32 white lights displayed Destination 29 (when the card punch was not running). On the next row beneath were 32 white lights showing the state of the 32 digits of Source 0 (when the card reader was not running); the row of 32 dark keyswitches beneath them were used to set individual bits of Source 0. The row of whitish toggle switches toward the bottom of the left side of the console were used during maintenance to set digit patterns for Source 0. Other switches included Single-Shot, Request Stop, and Program Display. The left-hand monitor (VDU) displayed the contents of all the short stores, while the other displayed one DL. A rotary switch beneath the right-hand monitor permitted the selection of any of the twelve DLs and TS COUNT. The top rows of push-buttons and lights switched the machine on and off and showed the status of the power supplies. A telephone dial (obscured by the operator's head) emitted up to ten Single-Shots.

Directly behind the operator's console is part of the mainframe. The large device to the left of the console is the Hollerith card reader. Punch cards were placed in a race at the top and were kept steady by a weighted metal plate. The cards slid sideways one-by-one down a vertical chute, ending up stacked horizontally in the niche half-way down the front of the machine. To the right of the console is the card punch. Blank cards were placed face down in a hopper (hidden from view by the top ledge). Under program control the card in front was fed past a punching station while the next card was fed to the ready position. The two vertical columns of rotary switches on the lower front panel of the card punch are labelled 'CARD NUMBER' (left) and 'JOB NUMBER'. The eight job number switches were set manually, causing a job number to be punched on each card. The four card number dials rotated automatically as each card was punched and the number displayed on the dials was punched on the card.

Source: Photograph by Keith Titmuss (reproduced with his permission).

execution of the current instruction and the fetching into control of the next instruction. (An ideal program was one in which each instruction had zero waiting time—as each instruction completed execution, the next one was already in control ready to be executed in the next machine cycle.) Much use of the DEUCE logic diagram of fig. 1 was made in the early stages of learning DEUCE programming.

The final stage of programming (just as difficult to master) was the hand-translation of the decimal values to 'Chinese' binary (binary with the least-significant digit on the left) and finally to punch each instruction in binary on cards. (The instructions of my first program, incidentally, were correct, but the instruction to punch the binary result unfortunately lacked the bit[12] to synchronize the punch with the control unit (CPU).)

Debugging

The creators of machine-language programs for the DEUCE usually debugged the programs on the machine themselves. Debugging machine-language programs was slow and often difficult.

Probably the most common cause of malfunction of a program was a wrong Timing Number. The Timing Number gave the relative position of the next instruction. In other words, every DEUCE instruction was a branch instruction. If a Timing Number was wrong, the normal sequence of instructions comprising the program was 'broken', and anything could happen, such as:

- branching to a location containing a zero word or a positive constant
- executing an entirely different sequence of instructions
- sticking in a loop.

The first of these was easy to diagnose, because the machine 'dropped out' (i.e. stopped). If this was the first 'rogue' instruction executed, it was usually the easiest to find. A preliminary diagnosis could be made by examining the contents of registers and memory on the visual display. If this did not help, a 'Post Mortem' could be taken. Post Mortem was a program that punched out the contents of all registers and memory (with the exception of one word), and the entire content of the magnetic drum if required. Punching out the contents of the entire machine took 4 minutes. The cards produced by the Post Mortem program could be examined at leisure away from the machine.

Often it was easiest to re-run the program up to a certain point, stopping it at a given machine instruction by using the hardware 'Request Stop' facility. The 'Program Display' hardware facility would then be used. To use Program Display, the user set the machine to 'Augmented Stop',[13] and then pressed the 'Program Display' key on the console of the machine. The machine would run in 'slow motion'. As each instruction was executed, it would be punched out by the card punch. Twelve instructions per card were punched out, and quite soon a sizeable pack of up to several hundred cards could be obtained before the program halted. Typically, however, several dozen cards were sufficient.

If the program became stuck in a loop, the visual display would be used to see whether one of the registers was counting up or down. The machine would be placed on 'stop' and instructions executed one at a time, in order to determine whether the sequence of instructions was one that existed in the program, or if not, whether it contained some instructions that did exist (in which case, the rogue instruction could be identified). Sometimes it then became necessary to use Program Display or Post Mortem. The former could help considerably in identifying long loops.

When an entirely different sequence of instructions was executed, it was usually difficult to find out what went wrong, because with each 'rogue' instruction executed, the contents of more and more registers became corrupted. There was the danger, too, that even the content of main memory could change.

Such errors might be found by running the program up to a certain point—again using the Request Stop facility—and then with the machine on 'stop', executing instructions one at a time using the Single-Shot key on the console. After each instruction was executed, the contents of registers and the high-speed store might be examined on the visual display. If this strategy did not succeed, a Program Display would be taken, followed by a Post Mortem.

Another cause of malfunction of a program was a loop count being 'one off'—that is, being executed one more time than it should or one fewer than it should. These kinds of error are not the kind we know today (where a simple counter is used, decremented at each iteration). In the DEUCE, an instruction to fetch consecutive words from a delay line (a block of 32 words of high-speed store) would do so by having its Wait Number modified (the machine not having an index register). The loop would terminate when the Wait Number exceeded 31, causing a carry bit to be propagated into the Timing Number field of the instruction. The Timing Number would thus increase by 1, and instead of the normal next instruction being taken, the one in the location after it would be taken instead, thus terminating the loop. If the programmer miscalculated when this carry occurred, the loop would terminate prematurely or too late. This type of error was best discovered by using the Program Display facility, but sometimes could be located from an examination of the high-speed store, which might indicate that the last result in a DL had not been computed. A more laborious method would be to run the program, using the Request Stop facility to stop the program at a particular instruction in the loop, and noting the number of times that the loop was executed. Alternatively, an instruction in the loop would be changed to a Stop instruction (by temporarily plugging the corresponding hole in the punch card) and noting the number of times the stoppered instruction was executed (to proceed, the operator would give a Single-Shot from the console).

Other common causes of program malfunction included a wrong Wait Number (resulting in the wrong minor cycle of a register—the wrong memory word—being fetched or stored), violation of the timings of the card reading and punching equipment, and less commonly, accessing a product or quotient before the result was ready (such a fault would show up only when the machine was run at normal speed, but would not show up when the machine was run on 'stop' and the program was given a succession of 'Single-Shots' (i.e. executed one instruction at a time from the console) or run under Program Display. This was because the multiplication or division would have ample time to complete.

One helpful 'tool', used both in the normal running of the machine and during regular maintenance and program testing, was the loudspeaker. This was driven by an amplifier connected to one of the instruction Source digits. In normal operation a program would generate musical notes. The pitch and duration of these depended on the rate at which a particular bit in each instruction changed as instructions were executed. Each segment of program (and each program) thus had its own musical 'signature'—a series of squeaks and musical notes of rapidly changing pitches. During normal operation, the signature would indicate to the experienced operator whether a program was running normally, or whether it had stopped or failed. During preventative maintenance sessions, test programs were run to check the safety margins of circuits. If the signature changed, it indicated that the program had failed. If a program became stuck in a loop during program testing, the speaker would sound a continuous high-pitched note. In each of these circumstances, the loudspeaker provided the programmer with an immediate indication of the state of the machine.

Fig. 3 The DEUCE playing noughts and crosses.

Source: Photograph of the UTECOM by Keith Titmuss (reproduced with his permission).

A number of demonstration programs was available. One played noughts and crosses (see fig. 3). Another played popular tunes, including Christmas carols. Of note was an animated cartoon of the nursery rhyme 'Hickory Dickory Dock'. This displayed on the monitor (VDU) at the operator's console. The monitor showed the mouse running up the clock, the clock striking one, and the mouse running down again. Lacking a bell, the DEUCE sounded its buzzer for the clock striking one o'clock.[14]

Program cards

Each program had to consist of an initial punch card for synchronization of certain parts of the machine, followed by one or more *triads*, each triad consisting of three cards. Each triad typically contained three filler (i.e. bootstrap) instructions and 32 instructions and constants comprising all or part of the program. (A *filler* is a sequence of instructions that will read in instructions and data from cards into main storage. Usually, filler instructions caused 32 or 64 instructions and constants to be read in and stored in one or two delay lines of main memory.) A typical initial card is shown in fig. 4.

The DEUCE card reader read card columns 17 to 80 (in 64-column mode) and card columns 21–52 (in 32-column mode). In 64-column mode, card columns 17 to 48 were called the alpha field, and columns 49 to 80 were

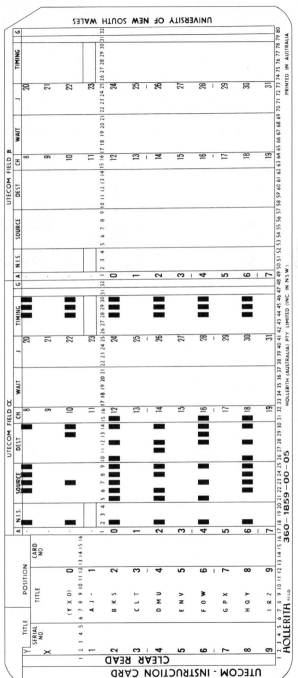

Fig. 4 A DEUCE initial card.

The twelve rows of a punch card were called Y, X, 0, 1, 2, 3, 4, 5, 6, 7, 8, 9. The names of the rows were significant only for decimal data. When decimal values were punched, the 0 row to the 9 row would be used for digits. Thus, the digit 8 was represented by punching a hole in the eighth row. A plus sign was represented by a hole in the Y row, and a minus sign was represented by a hole in the X row.

called the beta field. Card column 1 (column 54 in 32-column mode) was also read by the DEUCE, a hole causing the reader to stop after reading the current row (any remaining rows of the card passing through the equipment unread). In the DEUCE instruction card in fig. 4, the alpha and beta fields are both partitioned by vertical lines to define the fields of the instruction; the names of the fields appear along the top margin. The numbers, letters, and rulings were to help the programmer read the instructions on the card and had no effect on the card reader, which responded only to holes in the card.

The purpose of the initial punch card of the program was to synchronize the double-length accumulator with the odd and even cycles of the machine. The multiplier and divider unit were associated with the double-length accumulator. The low- and high-order words of the double accumulator corresponded to the odd and even addresses of a delay line, respectively. The machine cycles were numbered 0–31, corresponding to the addresses of the words in a bank of 32 words of main memory. Transfers between the double-length accumulator and a delay line required that the low-order word of the accumulator be transferred only to an even-numbered address in a delay line, and the high-order word be transferred to an odd-numbered address. When the DEUCE high-speed store (including the control unit) was cleared ready to accept a new program this relationship could be broken, since any minor cycle of a delay line could then be established as minor cycle zero. The initial card served to ensure that an even-numbered minor cycle of a DL corresponded with the low-order bits of the double store accumulator.

Basic assembler language

Once the basics of machine language were mastered, it then became routine to use a primitive assembler program called ZP43. DEUCE instructions were presented in decimal. As well as converting the supplied components of a given instruction to binary, the details of the next instruction (the Next Instruction Source address and the Timing Number) were calculated by the assembler after processing the next assembler statement. Immediately after all the source statements had been read in, the executable binary program was punched out, ready to run. The assembler, primitive though it was, helped considerably in reducing errors in converting to binary and in punching the seven binary components of an instruction. It also alleviated the great difficulty of altering a program, because alterations to a few instructions (or the allocation of new locations for instructions) could be made in a matter of minutes, and a quick re-run of the assembler produced a new executable. The much longer alternative of modifying the executable was often done with the aid of a *reproducing punch*, capable of omitting rows of a card at the flick of a switch (an instruction occupied a row). Minor alterations were carried out by plugging up holes in a card with 'chips' (of which there was always an abundant supply from punching holes in other cards), and then reproducing the entire card on the reproducing punch. (One user in Tasmania, who had no access to a reproducing punch, re-punched the entire card by hand if an alteration was required!)

Reading integers from cards

Programs (a) and (b) illustrate two ways in which 32 binary integers could be read in from punch cards, storing the integers in DL 10. In example (a), the loop is controlled by a counter in TS 13. In example (b), the loop is controlled using the AIM. (In the read instruction $0-10_0X$, the letter X signifies that the DEUCE must wait until the card row is in position to be read. The instruction is pronounced 'nought to ten nought stop'.)

A 'pro forma' instruction is one that will be modified as the program runs (see the examples in Chapter 11, 'The Pilot ACE: from concept to reality').[15] In example (a) it is the Wait Number of the pro forma instruction that is modified.

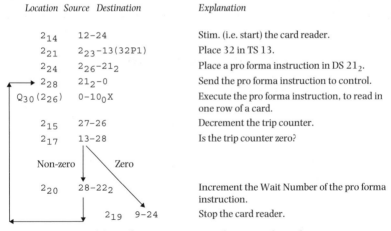

Location	*Source*	*Destination*	*Explanation*
2_{14}	12–24		Stim. (i.e. start) the card reader.
2_{21}	2_{23}–13(32P1)		Place 32 in TS 13.
2_{24}	2_{26}–21_2		Place a pro forma instruction in DS 21_2.
2_{28}	21_2–0		Send the pro forma instruction to control.
Q30(2_{26})	0–10_0X		Execute the pro forma instruction, to read in one row of a card.
2_{15}	27–26		Decrement the trip counter.
2_{17}	13–28		Is the trip counter zero?

Non-zero   Zero

2_{20}	28–22_2	Increment the Wait Number of the pro forma instruction.
2_{19}	9–24	Stop the card reader.

(a) Reading in 32 integers from punch cards

Notes: In the DEUCE, the bits of a word are numbered P1, P2, P3, ..., P32, where P1 represents the least-significant bit of a word and P32 is the most-significant. P1 thus corresponds to unity (or 2^0), P5 corresponds to 2^4, P17 corresponds to 2^{16}, P22 corresponds to 2^{21}, and P32 to 2^{31}. Q30 denotes *Quasi* 30, which means that the instruction is obeyed *as if it were stored in minor cycle* 30.

313

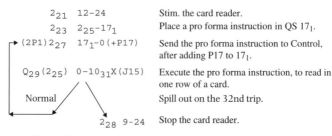

2_{21} 12-24   Stim. the card reader.

2_{23} 2_{25}-17_1 Place a pro forma instruction in QS 17_1.

(2P1)2_{27} 17_1-0 (+P17)   Send the pro forma instruction to Control, after adding P17 to 17_1.

Q_{29}(2_{25}) 0-10_{31}X (J15)   Execute the pro forma instruction, to read in one row of a card.

Normal   Spill out on the 32nd trip.

2_{28} 9-24   Stop the card reader.

(b) Reading in 32 integers from punch cards, using the AIM

Notes: The instruction 17_1-0 adds a P17 to the content of 17_1 and also sends the updated instruction to control, where it is executed in minor cycle 29. As it is the *modified* instruction that is executed, the instruction executed is 0-10_0X. The next time the instruction 17_1-0 is executed, the contents of 17_1 become 0-10_1X, which is then executed, and so on. The thirty-second time it is executed, the instruction is 0-10_{31}X. The Wait Number, previously 31, becomes 0, with a carry of 1 to the P22 position. The P22 position—or 'Joe field'—holds 15, so the carry is propagated to the P26 position. (The Joe field was a spare field of the instruction. Spare fields of an instruction were not sensed by control, and could therefore be used for various purposes by programmers. The Joe field was often used in connection with loop control.) Thus the Timing Number of the instruction is incremented by 1, so that the next instruction executed is at 2_{28} instead of 2_{27}. The term 'normal' in the flow diagram refers to the path taken when the instruction 0-10_{31}X is executed for the first (and subsequent) times. The term 'spill out' refers to the thirty-second execution of that instruction, when the loop is exited.

Optimum coding: examples

Every 32 minor cycles the same word in a DL would reappear at the 'entrance' to the DL. Consider the following four instructions to sum the contents of TS 14, TS 15, and TS 16, using the accumulator associated with TS 13, and leaving the result in TS 16.

14–13   Copy TS 14 to TS 13.
15–25   Add TS 15 to TS 13.
16–25   Add TS 16 to TS 13.
13–16   Store the sum in TS 16.

If these four instructions were placed in the first four minor cycles of DL 1, they would be encoded thus:

	S	D	W	T
1_0	14 –	13	0	31
1_1	15 –	25	0	31
1_2	16 –	25	0	31
1_3	13 –	16	0	31
1_4	. . .			

where S and D are the Source and Destination, W and T the Wait and Timing Numbers.

The first instruction has a Wait Number of 0, meaning that the earliest time that the instruction can be executed is in minor cycle 2 (each instruction takes two minor cycles to enter control and to be set up ready for execution). But by the time the instruction can be executed, the next instruction, which is in minor cycle 1, has already passed the 'entrance' to DL 1, and the machine must wait a complete 'revolution' for it to come around again before it can be copied to the control unit for execution. Similarly for the instruction in I_1. It is executed in minor cycle 3, but as the next instruction is in minor cycle 2, another revolution must elapse before the next instruction can be accessed. To avoid such inordinate delays, these instructions would be placed in alternate minor cycles of DL 1, thus:

	S D	W	T
I_0	14–13	0	0
I_2	15–25	0	0
I_4	16–25	0	0
I_6	13–16	0	0
I_8	...		

At the same time as the instruction 14–13 is being executed in minor cycle 2, the instruction in I_2 is being copied to the control unit, ready to be executed in minor cycle 4, when the instruction in I_4 is available. This arrangement gives the fastest possible execution speed for these instructions.

The aim of optimum coding is to ensure that the Wait Number is as small as possible, and that the Timing Number is equal to the Wait Number or slightly greater than it. A large Wait Number and/or a large Timing Number (or a Timing Number smaller than the Wait Number) indicates that a substantial amount of time is being wasted.

A program for counting the bits in a word will give a better feel for optimum coding.[16] TS 14 contains the word whose bits are to be counted. TS 13 is used for the tally. In this example the instruction in location I_{11} has a Timing Number of 22 so that the next instruction is taken from I_3 (zero path). The instruction in I_6 is coded with Timing Number 1 rather than 0. This is because of the constraint that the instruction in I_7 can lead at the earliest to an instruction in minor cycle 9. In this short loop, each path around the

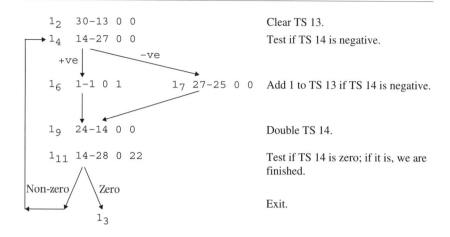

1_2 30–13 0 0		Clear TS 13.
1_4 14–27 0 0		Test if TS 14 is negative.
1_6 1–1 0 1	1_7 27–25 0 0	Add 1 to TS 13 if TS 14 is negative.
1_9 24–14 0 0		Double TS 14.
1_{11} 14–28 0 22		Test if TS 14 is zero; if it is, we are finished.
		Exit.

loop takes 1 major cycle, and so the instructions could have been 'spread out' in the delay line without increasing the time taken for the loop.

Optimum coding, then, involved allocating locations for instructions and data so as to minimize waiting. Instructions were not placed in consecutive locations. For example, if only the single-length registers (32 bits) were used, instructions would be placed in alternate locations in main store. Instructions using the double stores might be placed in even or odd locations in main store, or in every fourth location. Sometimes the programmer coded instructions in a different order so as to reduce waiting time and thus to increase speed. Great attention was given to loops.

Almost inevitably the size of a program exceeded available storage. Consequently, instructions were re-used, and in some cases constants doubled as instructions, in order to cut down on storage requirements.

When a task involved extensive reading of cards or punching of cards, as much computing as possible was done between each card in order to reduce the overall time for the job. If a task required considerable use of the multiplier or divider—in a long loop, for example—then the programmer overlapped other computation with the multiplication or division. Overlapping was possible because multiplication and division were asynchronous operations. Once multiplication or division had been started, all the facilities of the machine were available for use, except registers 16 and 21 (in which multiplication and division were carried out).

Magnetic drum operations also were asynchronous; since these took considerable time (a head shift took 35 ms, reading and writing a track took 15 ms), head shifts were initiated ahead of time (if possible) and reading and writing a track were often overlapped with computation.

A more complex program

The program shown on the next page will read in 16 positive 32-bit integers from punch cards and place the integers in storage, compute and store their squares as 64-bit integers, and finally punch the squares on cards (as 64-bit integers). For clarity, separate loops will be used for each of the three parts. The AIM is used for loop control (so keeping loop control instructions to a minimum) and also for instruction modification. Each row of a punch card contains one integer.

Symbolic assembly language

A considerable improvement over hand-coded machine language arrived with the publication of STAC in 1959. The STAC (<u>ST</u>orage <u>A</u>llocation and <u>C</u>oding Programme) was a symbolic macro assembler—though the extent of symbols and macro facilities was minimal. The symbolic instructions (i.e., instruction mnemonics) extended only to the 'superinstructions' MULT and DIV, transfers between the magnetic drum and main storage, and automatic linking to DEUCE library subroutines.[17]

The principal feature of STAC was the automatic allocation of storage to instructions and data (in a form that was generally as good as could be done by a programmer), along with the computation of Wait and Timing numbers and the next instruction source. It also had facilities for merging binary object code for one or more subroutines, and for reading binary and decimal fixed-point constants. It produced an executable object program. Designated sections of instructions could be given priorities, and STAC would code blocks of high-priority instructions first. This would enable STAC to give the best locations to critical sections of code (loops, especially inner loops), coding each instruction so as to minimize the amount of waiting before it could be executed.

Once the advantages of STAC became appreciated—the elimination of most of the hand allocation, a reduction in the amount of typing (it being no longer necessary to type most of the instruction locations), the ease of changing, inserting, or deleting instructions, the automatic allocation

Read in 16 positive binary integers.

2_0	12–24	Stim. read (i.e. start card reader).
2_2	2_{18}–17_2	Set up loop control.
(5P1) 2_{28}	17_2–0 (1)	+P17 + P22
Q30 (2_{18})	0–10_{31}X	Read in one integer; wait for the row of the card to be ready to read. Loop back to 2_{28}; exit on 16th trip to 2_{29}.
2_{29}	9–24	Clear read (i.e. stop reader).

Form the squares of the integers in storage.

1_{18}	1_{22}–17_2 (2 mcs)	Set up loop control.
(2P1) 1_{20}	17_2–0	+P17
Q30 (1_{22})	10_{31}–16	Fetch one integer (the multiplicand).
1_{19}	16–21_3	Copy the integer to the multiplier register.
1_{21}	30–21_2	Clear 21_2.
1_{25}	0–24 (mc 29)	Start multiplication.
1_{28}	1–1	Waste time, to allow mult. to complete.
(7P1) 1_{27}	17_1–0 (1)	+P18 + P22
Q29 (1_{23})	21_2–11_{30} (2 mcs)	Store one 64-bit square. Loop back to 1_{20}; exit on 16th trip to 1_{21}.

Punch the results.

1_{21}	10–24	Stim. card punch.
2_{19}	1_{26}–17_2	Set up loop control.
(7P1) 2_{21}	17_2–0 (1)	+P18 + P22
Q30 (1_{26})	11_{30}–21_2 (2 mcs) J=15	Fetch a 64-bit square.
2_{22}		Normally executes 2_{23} next; exit on 17th trip to 2_{24}.
2_{23}	21_2–29X	Punch the 32 least-significant bits; wait for the card punch to be ready to punch the row.
2_{25}	8–24 (1)	Clear the output staticizers ready to
2_{27}	21_3–29	punch the 32 most-significant bits. Loop back to 2_{21}.
2_{24}	9–24	Stop the card punch.
2_{30}	12–24	Start the card reader.
1_0		Finish, with the reader called for the next program.

DEUCE program to square 16 positive integers

Notes: P1 corresponds to unity (or 2^0), P5 corresponds to 2^4, P17 corresponds to 2^{16}, P22 corresponds to 2^{21}, and P32 to 2^{31}. The notation Q30(1_{26}) signifies that the instruction originally stored in 1_{26} (and now also in 17_2) is obeyed as if it were stored in minor cycle 30. The instruction 2_{21} 17_2–0 (1) +P18 +P22 means that the content of 17_2 is sent to control to be executed, and that in doing so the content of 17_2 is incremented by P18 and P22. The P18 increments the Wait Number by 2, while the P22 increments the Joe field and serves as a trip counter. On the seventeenth time that the instruction in 17_2 is executed, the P22 that is added causes a carry to the P26 position, which increments the Timing Number, and hence causes the next instruction to be taken from 2_{24} instead of 2_{23}, thus terminating the loop.

of instructions, and the elimination of simple translation and keyboard errors—its use increased.

General Interpretive Programme

The General Interpretive Programme (GIP), with its large suite of numerical algorithms, was one of the most important programs in the DEUCE library.[18] GIP was developed at the NPL in 1955. One of the difficulties of machine language was the long time that it took to write programs. The GIP scheme enabled programs to be prepared in a few minutes to hours (instead of weeks to months). GIP programs ran as fast as any hand-written in machine code.

The GIP scheme consisted of self-contained programs called *bricks*. Each was capable of performing some numerical algorithm, such as matrix inversion or solving simultaneous equations. The bricks, being written in machine code, were optimized for the DEUCE. They were supported by several dozen bricks for reading decimal values and for punching decimal results. In all some 230 bricks were available in the DEUCE library.

To make a GIP program it was first necessary to decide which operations were required (e.g. a program to invert a matrix would require bricks to read in a matrix (in decimal), to invert a matrix, and to punch a matrix in decimal). The *codewords* required to direct the operations would then be prepared.

GIP codewords were of the form:

Source address 1, Source address 2, Destination address, Operation.

An *address* is a track address of the magnetic drum (in the range 0–8191). The codewords for the above operations could be:

$a1$	$a2$	$a3$	op	
0	0	0	1	Read a matrix in decimal to track 0.
0	0	36	2	Invert the matrix at track 0, placing the result at track 36.
36	0	0	3	Punch the matrix at track 36 in decimal.

Next, copies of the punch cards for these three bricks and the GIP control program were made, by means of a Hollerith card reproducer. The user would assemble a deck of cards with the GIP control program followed by a card containing three (the number of bricks), the three bricks (in the order: read decimal matrix, invert matrix, punch decimal matrix), followed by a set of three cards containing the three codewords (in binary). The decimal data would be placed last.

The first card of the data contained the number of rows and columns of the matrix. The bricks took the sizes of the matrices from dimensions stored with each matrix, and thus any size of matrix was automatically handled, the maximum sizes depending on the spaces reserved by the user and by the sizes of the bricks. It was the user's responsibility to allocate areas on the magnetic drum for storing data and for intermediate results. (Some of the drum, from track 8191 downwards, was reserved for GIP and its bricks and could not normally be used for storing data.)

One feature of the GIP scheme was a sum check carried out by each brick on any arrays that it used. Whenever a brick produced an array, a sum check was stored along with the dimensions of the array.[19] GIP was one of the few schemes to incorporate self-checking, despite the fact that the DEUCE hardware did not provide parity checking. Even if parity checking had been provided, the sum check would still have been employed, because the sum check helped to protect against the inadvertent (partial) overwriting of one array with another.

The GIP language included mechanisms for looping, loop control, and conditional statements. It had debugging facilities that were employed from the front panel, including the ability to stop before executing each codeword, to 'request stop' on a given codeword, to replace a codeword with another, to inject a new codeword into the program, to continue from any given codeword, to obey a brick read in by the card reader, and the like. If a brick failed (either because of invalid data, a hardware fault, a surge of the power supply, or a brick under test having a programming error), a 'restore control' card could be executed from the card reader, and any of the above-mentioned debugging facilities could be used to continue. If desired, the failed codeword could be repeated, or partial results punched out. There was even a program to re-synchronize the drum with the high-speed store, should a power supply surge cause the magnetic drum to lose synchronism.

Bricks available for use with GIP included those for solving linear simultaneous equations, for solving differential equations, for matrix multiplication, matrix inverse and transpose, term-by-term matrix operations (addition, subtraction, multiplication, division, sine, cosine, square root, logarithm, reciprocal, power), latent roots, determinants, residuals, linear programming, regression analysis, statistical tabulation, interpolation, decimal and binary input and output, and various utility bricks and programs including extraction of a submatrix, attaching rows and/or columns to a matrix, and eliminating columns from a matrix.

It is interesting to compare GIP with the contemporary Matrix Interpretive Scheme of the Ferranti Pegasus computer. On the Pegasus, each matrix instruction contained the addresses of the matrix operands as well as their dimensions. GIP codewords, on the other hand, specified only the addresses of the matrices. Thus GIP was a form of object-oriented programming. This accounts for its usefulness—once written, a program ran unaltered for differently sized matrices. The Ferranti scheme required the program to be changed whenever the size of matrix changed. Under the Pegasus Matrix Interpretive Scheme it took 42.4 seconds to multiply a 10 × 12 matrix by a 12 × 16 matrix.[20] GIP took 26.6 seconds.[21]

GEORGE

GEORGE was a translation scheme using an addressless language. GEORGE was conceived and developed by Charles Hamblin at the NSW University of Technology. The GEORGE language had all the elements of a modern programming language, in a simple form—including loop control, conditional operations, subroutines, arithmetical and trigonometrical built-in functions, and the ability to deal with scalars, vectors, and matrices.[22] A compiler was available in 1957.

GEORGE relied on a push-down pop-up stack, arithmetic being carried out on the value at the top of the stack, or the top pair of values as appropriate. For example, a read instruction caused one decimal number to be read from the card reader and placed in the accumulator (the top of the stack). An add instruction would remove the top two values on the stack and place the result on the top of the stack. Using the name of a variable, for example k, would bring the contents of that variable to the top of the stack. The reverse operation, a *store*, was indicated by using the variable name in parentheses, for example (k), which would copy the value on the top of the stack and place it in storage. (An exception to this rule about parentheses was the keyword '(punch)'.) The semicolon was used to remove the value on the top of the stack, that is, to cancel the accumulator. Several other operations were available to manipulate the stack, such as **dup** (to take a copy of the top of the stack and push it on the top), and **rev**, which would interchange the top two values on the stack.

The following example of a GEORGE program forms the sum of n values (keywords are shown in bold face, and the right square bracket indicates the end of the loop):

read (n) ;	Read in one number, store it in variable n and then cancel the accumulator.
0,	Initialize the accumulator to receive partial sums.
1, n **rep** (k)	Loop n times with k taking the values 1 to n.
read $+]$	Read one number and then add it to the partial sum; end of loop.
(punch)	Punch the final sum.
;	Cancel the accumulator.
stop	

Expressions were written in reverse Polish notation (suffix notation). For example, to evaluate $y = ax^2 + bx + c$, one writes:

x **dup** $\times a \times bx \times +c + (y)$.

When the elements in this expression are processed from left to right, the expression is evaluated and the result stored in y. The steps of the evaluation are as follows:

Symbol	Meaning
x	Bring x into the accumulator.
dup	Make a copy of x, leaving the copy on the top of the stack.
$\times$	Multiply, giving x^2.
a	Bring a into the accumulator.
$\times$	Multiply by a, giving ax^2.
b	Bring b into the accumulator.
x	Bring x into the accumulator.
$\times$	Multiply, giving bx.
$+$	Add bx to ax^2, giving $ax^2 + bx$.
c	Bring c into the accumulator.
$+$	Add, giving $ax^2 + bx + c$.
(y)	Store the result in variable y.

The following GEORGE program performs the operation of the DEUCE machine-language program given earlier, namely to read in 16 values, to compute their squares, and to print out the results. All arithmetic is performed in single precision floating-point, including loop control.

1, 16 $\mathbf{R}_I\,(a)$	Vector Read into a_I to a_{16} (does not use the accumulator).
1, 16 **rep** (j)	Loop 16 times, with j taking the values 1 to 16.
$j\,\vert\,a$ **dup** $\times j(a);]$	Replace a_j with its respective square.
1, 16 $\mathbf{P}_I\,(a)$	Vector Punch the values of a_I to a_{16} (does not use the accumulator).

stop

Note that a subscripted variable is written with the subscript first, and is separated from the variable name by a vertical bar (|). Thus a_j is written $j\,\vert\,a$. The sequence $j\,\vert\,a$ in the third line of the above example means fetch a_j into the accumulator; **dup** will make a copy of it and push it on the top of the stack; $\times$ forms a_j^2, while $j\,\vert\,(a)$ stores that result in a_j. The semicolon cancels the square a_j^2 from the accumulator ready for the next iteration.

After writing out the program in the above form, the programmer would encode the symbols using a look-up table. The resultant codes, in decimal, would then be punched on a card. Figure 5 shows the codes for the above program; each symbol of the program is represented by the pair of numbers beneath it. In reality, alphanumeric input was not provided for GEORGE programs. Consequently, each operation, name, etc, was encoded in decimal. Later versions of GEORGE still had this form of input, but could print a copy of the program in alphanumeric form, on standard paper tape (or, in the case of the NSW site, via an on-line Siemens M100 teleprinter running at 75 Baud (10 cps), installed in 1963).

An interpreter and a compiler were provided for GEORGE. The interpreter permitted the machine to be halted after executing each code (such as **rep**, **dup**, $\times$, $+$, $-$, and so on). The accumulator and stack could be viewed on

1		1	6	R_1	(a)					
1 | 2	1 | 0	1 | 2	6 | 2	8 | 15	0 | 6					

1		1	6	rep	(j)					
1 | 2	1 | 0	1 | 2	6 | 2	8 | 1	9 | 6					

	j	|	a	dup	$\times$	j	|	(a)	;	]
	9 | 4	0 | 1	0 | 4	12 | 0	7 | 0	9 | 4	0 | 1	0 | 6	2 | 0	5 | 1

1		1	6	P_1	(a)					
1 | 2	1 | 0	1 | 2	6 | 2	9 | 15	0 | 6					

stop										
14 | 15										

Fig. 5 Decimal codes for the GEORGE program to square 16 values.

the video display as the codes were executed. The compiler did not permit program testing, as the codes were compiled into machine instructions.

The 'GEORGE pack', which consisted of about 100 cards, was first read in and stored on the magnetic drum (taking half a minute). A special initial card followed by the GEORGE program codes was then read in. As long as the special initial card was used, any number of GEORGE programs could be read in without having to re-read the GEORGE pack. It was thus ideal for running a batch of programs.

GEORGE was probably the first program in which a software push-down stack used a hardware push-down stack. The GEORGE commands **dup** and **rev** used the hardware stack. The two machine instructions for **dup** that used the stack were 3–24 and 16–10$_n$.

Alphacode

Alphacode, developed by English Electric, provided comparable facilities to GEORGE and could be used to solve similar problems. An interpreter was available by 1958 and a compiler by 1959.[23] Instructions were given one per card, with each instruction consisting of an operation and up to three operands. In no way did the source resemble the more modern algebraic forms of GEORGE and other later high-level languages. Alphacode did, however, contain all the characteristics of such languages, including mechanisms for loop control, conditional operations, subroutines, and arithmetical and trigonometrical built-in functions. It could deal with scalars and vectors (but not matrices).

The basic forms of an arithmetic instruction were: $A = B + C$, and $A =$ function B. Other types of instruction (including input/output) took the form $a\ b\ c$ function d.

Alphacode had the advantage that there were two translators. One was an interpreter that enabled a program to run virtually immediately after it had been read in and translated. The other was a compiler, which could be used for a very long job to substantially reduce execution time. Unfortunately, an interpreted Alphacode program needed to run for more than hour before compilation became worthwhile, as compilation took from 30 to 75 minutes. The compiler itself was huge, consisting of 22,000 instructions, and had to be read in for each compilation. Just reading in those cards—a stack about half a meter high—took 10 minutes.

Even so, running an Alphacode program using the interpreter required two stages. In the first stage, the Alphacode translator read in the Alphacode source statements and punched out a translated binary program on cards. In the second stage, the interpreter program cards were read in, which in turn read in the translated binary program, and then interpreted it.[24]

The GEORGE program to form the squares of 16 numbers (see above) is as follows for Alphacode.

r	R	Field 1	Field 2	Field 3	Field 4	
23			0016	DATA	X0001	Read 16 DATA into X1 to X16.
0		N0001	0000	MOVED		N1 = 0. Set the initial value for loop control.
12	01	N0001	N0001	MODIFY	N0001	Modifies the address of X0001 in corresponding fields on the next line.
3		X0001	X0001	MULTIP	X0001	X1 = X1 multiplied by X1, subscript N1 in each case taken from the previous line.
10		N0001	UP TO	0016	R01	Count N1 up to 16 jumping to R01 on completion.
23		0016	RESULT	X0001		Punch 16 results X1 to X16.
18			FINISH			

DEUCE Alphacode program to square 16 numbers

Notes: The 'r' field is a decimal function number, identifying the operation (such as 'MOVED'). The alphanumeric function shown under the heading 'Field 3' was optional, but if supplied had to agree with the 'r' field. The 'R' field is a statement label. '01' is the destination of the branch address R01 given in the 'UP TO' statement. Variables have pre-defined types and the name 'X' is a pre-defined array name.

Speed

The DEUCE was a fast machine for its time. Its clock rate was about three times faster than its contemporaries. The Ferranti Pegasus was almost the same price but ran at about one-third of the speed of the DEUCE. Single-precision addition time on the DEUCE was 64 micoseconds, double-precision addition time was 96 microseconds, array addition 33 microseconds per word for 32 words, integer multiplication and division about 2 milliseconds. As to software floating-point arithmetic, addition took 6 ms, multiplication took 5.5 ms, and division took 4.5 ms.[25] Floating-point square root took 3 ms to 15 ms, using Newton's method.[26] Table 1 gives additional details.

Table 1   Comparison of the DEUCE and the Pegasus

	DEUCE	Pegasus[a]	
Clock rate	1000	333	KHz
Word size	32	39	bits
Addition, single precision	64	315[b]	microseconds
Addition, double precision	96		microseconds
Addition (mixed double and single precision)	64		microseconds
Array addition, per word (32 words)	33		microseconds
Integer multiplication	2080	2000	microseconds
Integer division	2112	5400	microseconds
Floating-point addition	6	18	milliseconds
Floating-point multiplication	5.5		milliseconds
Floating-point division	4.5		milliseconds
Square root	3–15	40	milliseconds
Block floating-point addition	1		milliseconds
Block floating-point multiplication	3		milliseconds
Maximum input rate			
Cards	426		cps (decimal)[c]
	213		cps (64 column)
	266		cps (80 column)
	2560		bits per second
Paper tape	850	200	cps
Maximum output rate	4250–6800	1000	bits per second
Cards	106		cps
	1280		bits per second
Paper tape	25	25	cps
	125–175	125	bits per second

[a] Lavington, *The Pegasus Story*, pp. 31, 33, 36.

[b] Average of two additions.

[c] Vowels, R. A. (1963) *LR23BM General Decimal (Matrix) Read to Drum*. Kidsgrove: English Electric-Leo Computers Ltd. Based on 64 signed integers per card (128 characters including signs and digits). Typical 8 signed integers per card (72 characters including signs) gives 240 characters per second (cps).

Notes

The author wishes to thank Jack Copeland for valuable comments and editing and John Webster for valuable comments.

1. Three DEUCEs were demonstrated at the English Electric works on 17 February 1955 ('"DEUCE" A computer for solving complex problems at high speed', *Engineering*, 11 March 1955, 313).

2. Memorandum from I. G. Evans to C. Jolliffe of 15 December 1951 (a digital facsimile is in The Turing Archive for the History of Computing <www.AlanTuring.net/evans_jolliffe_15dec51>). The amount is exclusive of input–output equipment. A memorandum from E. S. Hiscocks to Jolliffe on 10 December (<www.AlanTuring.net/hiscocks_jolliffe_10dec51>) confirms the offer, but without quoting an amount.

3. *The English Electric Journal*, 14(8), Dec. 1956, 49.

4. Removed to Stafford Technical College in 1964.

5. Thanks to Brian Randell, Garry Tee, Keith Titmuss, Jeremy Walker and John Barrett for information about DEUCE sites.

6. Manufactured by the British Tabulating Machine Company Ltd.

7. Specifically, the control grid circuit of the valve associated with the gate, which was the most sensitive part of the circuit.

8. 'Guide to the Use of Magnetic Tape Equipment on the N.P.L. DEUCE', June 1959 (a digital facsimile is in The Turing Archive for the History of Computing <www.AlanTuring.net/deuce_magnetic_tape_jun59>).

9. Of the many, I recall only three: an improved earthing system to prevent electrical disturbances from altering the contents of the delay lines (termed 'digit pickup'), an improvement to the magnetic drum interlock system to prevent a magnetics instruction from being ignored (termed 'sneak CMI'), and replacement of a certain type of capacitor that was prone to breakdown.

10. '"English Electric" D.E.U.C.E. Programming Manual', Report NS-y-16, May 1956. Kidsgrove: English Electric Co. Ltd.

11. A DEUCE instruction consisted of seven main components: Data was transferred from a Source address to a Destination address. The address of the next instruction was given in two parts: a Next Instruction Source (NIS) being the address of a DL, and the relative position of the next instruction in that DL (the Timing Number). The moment when the data could be transferred was specified by the Wait Number. Another field specified whether two or more words were to be transferred. The remaining field specified whether the instruction was to be executed immediately or was to wait for peripheral equipment or for the computer operator. The term 'two-plus-one address' referred to the need to specify the address of the next instruction, in addition to the Source and Destination addresses. The structure of an instruction is illustrated in the following diagram, the first line of

which applies to all instructions not using the AIM, while the second is for those using AIM. The 'extra delay line' bit is of significance only when the seven extra delay lines are fitted.

extra delay line	NIS	Source	Destination	Characteristic	Wait	Joe	Timing		Go
Director (4 bits)		Source	Destination	Characteristic	Wait	Joe	Timing		Go
1 bit	3 bits	5 bits	5 bits	2 bits	5 bits	4 bits	5 bits	1 bit	1 bit

The least-significant bit of an instruction is on the left. Although the Joe bits and the bit between the Timing and Go fields were not used by the control unit, they could participate in instruction modification.

12. Termed a 'stop bit'; explained in the section titled 'Debugging'.

13. A keyswitch on the console had three settings: NORMAL, STOP, and AUG. STOP (Augmented Stop).

14. An extensive programmer's and operator's manual was issued in 1959: Burnet-Hall, D. G. and Samet, P. A. (April 1959) *A Programming Handbook for the Computer DEUCE*. Farnborough: Royal Aircraft Establishment. A simulator for the DEUCE is available from the author.

15. I use the term 'pro forma' to refer to an instruction that has been placed in a register to be executed 'as is' and in modified form. Burnet-Hall and Samet used the term 'basic instruction', English Electric used the term 'planted instruction'. I cannot recall any particular name being used at UTECOM other than 'quasi instruction'. In fact, none of these terms is altogether appropriate, as the content of the register need not even be an instruction. (Burnet-Hall and Samet, *A Programming Handbook for the Computer DEUCE*, p. 39; ' "English Electric" D.E.U.C.E. Programming Manual' (see note 10), p. 32.)

16. This example is for illustrative purposes only. A more efficient method relied on the multiplier, and required 2.5 ms, as follows: 30–13; 30–21₃; 14–21₂; 0–24 (even mc); 21–26 (32 mcs); 22–26 (32 mcs); 13–25 (16 mcs). The example using the loop required up to 32 ms.

17. Birchmore, A. and Gilmour, A. 'DEUCE STAC Programming Manual', DEUCE News No. 38, Report K/AA y 1, June 1959. Kidsgrove: DEUCE Library Service, Data Processing and Control Systems Division, English Electric Co. Ltd.

18. 'DEUCE General Interpretive Programme', 2nd edn., *DEUCE News* No. 63, Report K/AA y 32, c.1962. Kidsgrove: English Electric Co. Ltd.

19. The first four elements at the specified drum address were, in order, the sum check, number of rows, number of columns, and number of binary places. Then followed the elements of the matrix in row-major order.

20. Lavington, S. (2000) *The Pegasus Story*. London: Science Museum, p. 35.

21. 'DEUCE General Interpretive Programme' (see note 18). Times are quoted for the brick LMo8B. This was the slower version. A faster version, LMo8B/1, was produced.

22. Hamblin, C. L. (1957) 'Computer languages', *Australian Journal of Science,* 20, 135–9; reprinted in *Australian Computer Journal,* 17 (1985), 195–8.

23. Denison, S. J. M., Hawkins, E. N., and Robinson, C. (1958) 'DEUCE Alphacode', *DEUCE News* No. 20, Report NSy 87. Kidsgrove: English Electric Co. Ltd.

24. For information on Alphacode programming see Denison, Hawkins, and Robinson 'DEUCE Alphacode'. For information about the translator see Duncan, F. G. and Hawkins, E. N. (1959) 'Pseudo-code translation on multi-level storage machines', *Proceedings of the International Conference on Information Processing,* UNESCO, Paris, London: Butterworths, p. 144; and Duncan, F. G. and Huxtable, D. H. R. (1960) 'The DEUCE Alphacode translator', *The Computer Journal,* 3(1), 98–107.

25. 'General Description of "DEUCE" Digital Electronic Universal Computing Engine', August 1958, English Electric Co. Ltd.

26. Hamblin, C. L. 'GEORGE Flow Diagrams', unpublished, *c.*1957. (Author's collection.) The square root was computed to 32-bit precision, and truncated to 22 bits.

15   *The ACE Simulator and the Cybernetic Model*

Michael Woodger

The ACE Simulator

During the early days of the ACE work, when digital processing was novel, a demonstration machine was built as an aid to the visualization of binary operations. (Now that computers and binary arithmetic are commonplace, it is perhaps less easy to appreciate the need that there was at the time for such a machine.)

Constructed out of Post Office relays and lamps, the ACE Simulator occupied a 6-ft rack. Five panels made of an early form of white plastic showed constituents of the ACE. The panels could be interconnected by means of wander-plugs. Behind the rack was a hinged frame containing banks of Type 3000 relays, also a box which held a stepping uniselector, and another holding a thyratron valve with potentiometer for speed control. A 'Westat' power supply gave +50V DC stabilized for the relays, and 2V AC for the lamps.

The lamps displayed the flow of binary digits in the ACE. The problem of showing a flow of digits without confusing transient and static states was partially solved by using white lamps for the positions of digits and extra green lamps for digits in transit between one position and the next. The basic pulse time had four parts. In the second and third parts the white lamps would remain lit while the green lamps came on. Then these would be extinguished at the same moment that the white lamps following would be lit. An impression of flow resulted.

Control of a one minor cycle transfer was illustrated by holding the 'one-shot control' key down until a lamp lit. At that moment, the transfer trigger went on and remained on until the lamp lit again. The transfer signal could be plugged to the transfer control socket of a delay line, when one minor cycle of bits from the input of the delay line would displace the existing content of the minor cycle. To fill the delay line from the input dynamicizer switches, one connected the source dynamicizer socket to the input of the transfer gate, and the output from the gate to the input of the delay line. To 'read' the contents of either delay line one connected its output to the output staticizer, where the bits remained until cleared by the key. Use of the adder input of the accumulator showed binary addition in action, including any carries produced during the process. Two manual input sockets could be used for a constant stream of bits, or could be operated 'on the fly' to perform tricks such as accumulating via the 'exploded' adder and the delay line to build up the Fibonacci numbers.

The ACE Simulator was designed by D. W. Davies and myself in the winter of 1949/1950, and I wired it up and demonstrated it on 30 January 1950 as part of the NPL Jubilee demonstrations to the Royal Society at Burlington House. It was also shown at the Physical Society exhibition at City and Guilds College from 29 March to 5 April 1950. At the time of writing it stands in the computer gallery in the Science Museum, next to the Pilot ACE.

The Cybernetic Model

The Cybernetic Model was constructed in May 1949, before even the first chassis of the Pilot Model ACE had been delivered. 'Cybernetics' is the term introduced by the mathematician Norbert Wiener to mean 'control and communication in man and the machine'. The Cybernetic Model was built in order to explore some of Turing's ideas about learning. (It had nothing to do with the development of the ACE.)

The Cybernetic Model contained only a modest amount of equipment. It comprised six free-standing units of aluminium, each with a lamp on top, connected by wander-plugs to the control unit. Each unit represented a binary digit in a 1-bit store; the lamp was lit for 1 and was extinguished for 0.

For each unit one could set any boolean function on the controls. The inputs to the function could be plugged from any of the units. As the machine passed through a four-point timing cycle, representing one pulse time, the six functions would be applied to the current states of the units to compute six new states, which then became the new current states of the units. The machine could be stepped one pulse at a time, or allowed to run indefinitely.

Publicity for the ACE work led to visits by television teams, and I proposed demonstrating the Cybernetic Model during a BBC program called 'How the

Brain Works'. The program, which went to air on 13 November 1950, was the first in a series of Reith Lectures given by the distinguished biologist J. Z. Young. Young showed how an octopus would learn not to attack a crab when a white plate was exposed followed by an electric shock. I used the Cybernetic Model to mimic the octopus's learning. The following extracts are from my draft script for the program.[1]

> My colleague Mr Davies and I made this machine to study some suggestions of Dr Turing about the possibility of imitating learning. It shows the various patterns of behaviour that can arise from the interaction of things such as electric relays or nerve cells which are always in one or other of two conditions—on or off—excited or quiescent.
>
> It consists mainly of six identical units with a switchboard to set up connections between them. The lamps go on when a unit is stimulated and off when the unit is inhibited.

The draft script indicates that I went on to emphasize the essential simplicity of the units of which the machine is composed, and the fact that the complication lies in the wiring. I then moved to the specific example of the octopus.

> Four units [are] arranged to illustrate nerve cells in an octopus given an electric shock to 'teach it' not to devour a marked crab.
>
> ... This light represents the tendency of the octopus to seize the crab. These two lights represent cells in the brain. So long as their state is not changed, the octopus will continue in its tendency to seize the crab. (slow motion) When however their state is disturbed as for example when this fourth light is brought into play—it could be said to represent the stimulus of pain—the state of two cells in the brain is changed. (slow motion) With the result that, after the first painful experience, the octopus no longer seizes the crab, i.e. the light no longer goes on. (slow motion; repeat at speed to show sequence of events)

The set-up I used in this demonstration of 'learning' consisted of two units representing 'memory neurons' (denoted 1 and 2 in the table below), a third unit (3)—a 'motor neuron'—representing attack, and a 'perceptual neuron' (4), the stimulation of which represented the receipt of a shock after seeing a white plate. I supplied the stimulus to unit 4 manually at the right moment, and then discontinued it.

The function settings and behaviour were as follows:

Unit	Function	Time								
		0	I	2	3	4	5	6	7	8
I	2	0	0	0	0	0	I	0	I	0
2	I + 4	0	0	0	0	I	0	I	0	I
3	Not (I + 2)	I	I	I	I	I	0	0	0	0
4	Manual	0	0	0	I	0	0	0	0	0

At time 3, I gave the 'shock'. Up to that time units I and 2 were copying each other and staying in state 0, so unit 3 stayed in state I (the octopus was happily attacking the crab). At time 4, 2 has got the message, but it takes another moment to affect 3. At time 5, 3 has stopped the attack, and I and 2 continue to 'remember' the shock. 3 will not come on again.

At the end of January 1951, Norbert Wiener and Howard Aiken visited NPL to see the Pilot ACE. Aiken was particularly impressed by the Cybernetic Model, saying 'That is the way of the future'.

The Cybernetic Model was demonstrated at the NPL 'open day' on 23 May 1951.[2] The machine was later dismantled because it was considered a distraction from the work on ACE.

Detailed descriptions of the ACE Simulator and the Cybernetic Model are available on the Internet at www.AlanTuring.net/woodger_ace_sim and www.AlanTuring.net/woodger_cyber.

Notes

1. The draft script is in the Woodger Papers, National Museum of Science and Industry, Kensington, London (catalogue reference N 27).
2. As reported in *Nature*, June 23, 1951, 167, p. 1006.

16 *The Pilot Model and the Big ACE on the web*

Benjamin Wells

Preserving the Pilot ACE

A National Physical Laboratory memorandum dated 30 January 1956 broached the topic of the final resting place of the Pilot Model, by then superseded by the NPL's DEUCE.

Disposal of Pilot ACE

Pilot ACE has now reached the stage where the amount of maintenance it requires precludes it from being used economically as a computer, and in the circumstances we have had informal discussions with the Science Museum about the possibility of it being handed over to the Museum for display there.

The Museum have informed us that they will not have room for the complete equipment even if we could let it go but they would be able to spare enough room to make a display which would be intelligible to visitors and for this purpose they would require the large rack of valves and electronic circuits together with the control desk and an example of a mercury delay line.

Authority is sought to transfer to the Science Museum these and any other sections of Pilot ACE which they may be able to accommodate.[1]

A large fragment of the Pilot ACE is on permanent display in the London Science Museum. The Pilot ACE now also enjoys the functional reincarnation that many outdated and outmoded machines experience: it has been emulated.

The USF ACE emulators

An emulator is a computer program that makes the machine it is running on capable of resembling or mimicking the target machine. An emulator should not only simulate the running of programs by the target computer, but also provide a simulation of the interface and controls of the target.[2]

The first known software emulator of the Pilot ACE was built in Visual BASIC by Donald Davies in the early 1990s. In a file accompanying his program, he indicated a wish that it be rewritten in a more modern language, but he died before he could accomplish this task. In spring 2001, two graduate students at the University of San Francisco, Athena Huang Shih-Yun and Nicola Rugai, ported Davies's code to Java as part of their Master's Project, directed by Greg Benson. Huang wrote additional Java code and improved and extended the interface. The improved emulator was written for The Turing Archive for the History of Computing (directed by Jack Copeland and Diane Proudfoot) and was inspired by the desire to offer a Pilot ACE emulation in a platform-independent format. I served as proxy client and mentor. Huang, Rugai, Benson, and I consulted a distinguished panel: Mike Woodger, David Clayden, Harry Huskey, and Roger Scantlebury.[3]

As this book makes clear, the architecture of the Pilot ACE differed profoundly from today's architectures. Moreover, there was no high-level language divorced from the internal workings of the machine. The students, like the original users, had to grasp the use of functional ports and mercury delay line timing before they could write programs for the emulator, much less write the emulator itself. And they had to be mindful that the Pilot ACE's input and output were punched cards—something they had never seen.

The Pilot ACE control panel could display the contents of two 'fast memory' delay lines directly. Davies included a facility in his emulator for viewing any two additional delay lines. The USF emulator goes a step further by providing optional views of all 11 delay lines at once, so showing the next instruction location and allowing easier memory modification. A speed control allows the simulation to proceed at close to the original pace.

Documents cited on the emulator webpages are helpful for understanding how to program the Pilot ACE. Huang wrote several sample programs to flex the emulator and to demonstrate some subtle features of the architecture. These are tracked in a PowerPoint slide presentation on the website.

Huang subsequently adapted the Pilot ACE emulator to the architecture of the Big ACE. The Big ACE emulator is also available on the website.

The Pilot ACE and Big ACE emulators are located at: **<www.AlanTuring.net/ace_emulators>**.

Notes

1. Memorandum from Hiscocks to the DSIR, 30 Jan 1956 (PRO document reference DSIR 10/275; a digital facsimile is in the Turing Archive for the History of Computing <www.AlanTuring.net/hiscocks_dsir_30jan1956>). I am grateful to Jack Copeland for drawing Hiscocks' memorandum to my attention.
2. For more on simulation versus emulation, see A. Mulder, <www.mediamatic.net/cwolk/view/3139>.
3. Scantlebury collaborated with Donald Davies on an early paper about packet-switching networks (Davies coined the term 'packet'). With Davies ill, Scantlebury demonstrated Davies' emulator at the ACE 2000 Conference (organized by Copeland and held in May 2000 at the London Science Museum and the NPL).

Part IV
Electronics

17 *How valves work*

David O. Clayden

What is a valve?

Very little design work using valves has been done in the last thirty years and it is reasonable to expect that many readers of this book will have had no experience with valves. I provide some background for those chapters which describe valve circuits.[1]

A valve consists of an evacuated glass envelope containing a number of electrodes. These are connected to the outside by wires passing though special seals.

The innermost electrode is the *cathode*. This consists of a metal tube coated with a material that emits electrons when it is heated. The heat is provided by a tungsten wire, situated inside the cathode and connected to a 6 or 12 V supply.

In the simplest form of valve, called a *diode*, the cathode is surrounded by a metal cylinder called the *anode*. If the anode is connected to a voltage that is positive relative to the cathode, the anode attracts electrons from the cathode, and a current flows. Current cannot flow in the other direction.

Triodes

In the type of valve called a triode, a *grid* of fine wires is inserted between the cathode and the anode. A voltage on this grid (normally a few volts negative relative to the cathode) can control the flow of electrons to the anode. This enables the valve to be used as an amplifier, the current depending on both the grid voltage and the anode voltage.

Pentodes

The pentode has five electrodes. As well as the anode, cathode, and grid, there are the *screen* and the *suppressor*. The screen consists of fine wires (like the grid). The screen surrounds the grid and is normally connected to a fixed voltage of about +200V, which causes it to attract electrons through the grid. Most of these electrons pass though the screen and aim for the anode. The anode current is much less dependent on the anode voltage than it is in the triode.

The fifth electrode, the suppressor, is another grid of wires, located between the screen and the anode, and normally connected to the cathode. Alternatively the suppressor can be connected to a variable voltage, giving it the ability to control the proportion of electrons reaching the anode. (This method of control is not as effective as using the grid, however, for reasons explained in my chapter 'Circuit Design of the Pilot ACE and the Big ACE' (Chapter 19).)

Valves as amplifiers and switches

Valves were initially developed as linear amplifiers. When a valve is used in this way, the anode current is related to the grid voltage. For small signal amplitudes, the distortion is small. Valves are used either as radio frequency or as audio frequency amplifiers.

When television started in the 1930s, it was an easy step to use valves as on–off or change-over switches. When the valve is used in this way, the current is either *on* (at some controlled level) or *off*. (The current was usually controlled by negative feedback in some form.) This is how valves were used in digital computers.

Power consumption

The ACE Pilot Model used small valves designed to dissipate 2 watts of power. Larger valves (e.g. the KT66) were used for power supply stabilizers and for driving magnetic drums.

Designing for variations

Within their power limitations, valves vary considerably from their specification. Much of this variation is due to mechanical tolerances. Cathode temperature is another variable. Also the coating of the cathode deteriorates over the life of the valve. In time the hardness of the vacuum deteriorates. Because of these sources of variation it is desirable to employ design methods that will work over a range of valve characteristics and prolong the useful life of the valve as much as is possible.

Note

1. Numerous text books describing valve circuits exist. A famous one of the period is Terman, F. E. (1943) *Radio Engineers Handbook.* London: McGraw-Hill.

18 *Recollections of early vacuum tube circuits*

Maurice Wilkes

The birth of electronics

A remark made by Donald Davies in the Foreword to this book brings back to me very vividly the climate of the late 1940s when the first stored-program digital computers were being designed. Davies said that Harry Huskey's circuits for the ACE Test Assembly were anathema to Ted Newman, who preferred to design the real Pilot ACE in his own way.

The early radio engineers were concerned with sine waves of various frequencies—radio, intermediate, audio—and nothing else. By the 1930s cathode ray tubes were coming into use and bringing with them new and strange wave forms, particularly time bases and strobes. Primitive analogue computing devices were also appearing. A new term, 'electronics', was coined for the new technology.

Electronic techniques were much to the fore in ionosphere research and in television. They were vigorously exploited during the war for radar and other applications and, by the end of the war, knowledge of electronics had become widespread.

The designers of the early digital computers felt entirely confident that electronic techniques would meet the challenge. In fact, electronics offered them an embarrassingly wide range of alternative techniques to choose from. The first thing they had to do was to decide on the best way to realize gates and flip-flops and to evolve a consistent set of principles for putting them together to make a computer. There was not time for a careful and exhaustive appraisal, and each designer made his choice largely on the basis of personal preference. Although their experience in other applications of electronics stood them in good stead, computer designers soon found they had to learn a few new tricks, such as how to handle non-repetitive wave forms.

Design choices

There were three main approaches to the design of trees and gating circuits. One was to use a form of threshold logic, consisting of a simple resistor network feeding an amplitude discriminator. This was essentially an analogue approach. The circuits were sometimes referred to as 'Kirchoff circuits', the reference being to Kirchoff's laws which were widely taught at the period in question. Another approach was to make use of pentodes with independent inputs applied to the control grid and to the suppressor grid. Third, use could be made of diodes.

Obviously vacuum tubes would be used for amplifiers and this seemed straightforward enough. However, the output was at a much higher voltage than the input, and the interstage coupling circuits had to allow for this. The designer could either use capacitors or pulse transformers for interstage coupling, with diodes for zero restoration (otherwise called clamping), or he could use a resistor chain, perhaps with capacitors for frequency correction.

Having made his choice, every designer was firmly convinced that his way was the best. This was only natural. I myself was no exception to the rule. I would stand up stoutly for the superior merits, as I saw them, of the EDSAC design philosophy. Likewise, it was inevitable that Ted Newman, an ex-EMI man and a disciple of Blumlein, should have no time at all for Harry Huskey's ENIAC-style circuits.

Yet in spite of all the strong feelings, it was found, when the chips were down, that all the early computers worked with much the same degree of reliability. It was not that the doubts which had been expressed about pattern sensitivity, stability and so on were not well founded. What experience showed was that, if the engineering were carefully and competently done, most schemes could be made to work.

Table 1 illustrates the great diversity that existed in the way selected circuit functions were implemented in the first wave of computers. It was constructed partly from memory and I make no great claim for its accuracy. Not all the functions required in a computer are included in the chart; for example, there is no mention of control logic.

Events moved fast in the first few years. Threshold logic dropped out and pentode gates became unpopular. Germanium diodes, which were not available when the EDSAC design started, soon came along. At first, there were doubts about their reliability and recovery time, but confidence was soon established, and the SEAC made free and elegant use of them. The merits of parallel architectures became recognized, one being that they opened the way to DC interstage connection. Finally, when all seemed set for a great future with vacuum tubes, transistors came along and we were all back at square one.

Table 1 First generation computers

	ENIAC	SSEM	EDSAC	Pilot ACE	SEAC	SWAC	IAS
Interstage coupling							
Threshold				■			
Capacitor and DC restorer			■	■			
Pulse transformer and DC restorer					■		
DC							■
Trees							
Threshold	■			■			
Diode			■		■		
Adder							
Threshold		■					■
Pentode	■	■				■	
Triode				■			
Diode			■		■		
Flip-flops							
Static	■	■	■	■		■	■
Dynamic					■		

The computers were: ENIAC—Electronic Numerical Integrator and Calculator (University of Pennsylvania); SSEM—Small-Scale Experimental Machine (also known as the 'Baby Machine', Manchester University); EDSAC—Electronic Delay Storage Automatic Calculator (Cambridge University); Pilot ACE (National Physical Laboratory); SEAC—Standards Eastern Automatic Computer (US National Bureau of Standards); SWAC—Standards Western Automatic Computer (US National Bureau of Standards); and the IAS computer (named after the Princeton Institute for Advanced Study where it was built).

Blumlein and the long-tailed pair

The great designer Alan Blumlein died early in the war (see the next chapter) and we can only speculate as to what his approach to the design of digital computers would have been. He is famous for his insistence that a circuit should be designed on paper, with the expectation that it would work first time. This used to puzzle me, until I realized that he must have been referring to analogue circuits. How right he was! Anyone who has worked with such circuits will have found that to proceed without working out a properly toleranced design in advance is a good way to hang oneself!

Blumlein would have approved of one feature in the design of the EDSAC, namely the use of cathode-coupled amplifiers. These are essentially long-tailed pairs, a special favourite of Blumlein's (see the next chapter). If the tail is not made too long they have very good clipping properties and they do not invert the pulses. For this latter reason the EDSAC contained no inverters.

Note

An earlier version of this chapter appeared in the bulletin of the Computer Conservation Society, 'Computer Resurrection' (issue 24, Autumn 2000).

19 *Circuit design of*
the Pilot ACE and the Big ACE

David O. Clayden

Introduction

The ACE Pilot Model project started in the NPL's Radio Division with a team of mathematicians and engineers drawn from Mathematics Division, Radio Division, and Electricity Division (see Chapter 3, 'The Origins and Development of the ACE Project'). The engineers of Radio Division had long experience of radio receiver and transmitter design. In 1947, the team was joined by Ted Newman from EMI (Electric and Musical Industries) Research Laboratories. At EMI Newman gained extensive experience of radar and television camera design. I followed Newman from EMI in September 1947. At that time the logic design of the Pilot Model had been settled by the members of the Mathematics Division, including Alan Turing, Jim Wilkinson, and Donald Davies, and a decision had been made that the main store would be mercury delay lines.[1]

The EMI connection

Newman, the main architect of the circuit design of the Pilot Model, brought to Radio Division some of the practices common at EMI and due to the brilliant electronic engineer Alan Blumlein. Before the war Blumlein produced 128 patents, many concerning the EMI television system. Newman worked on radar with Blumlein during the war. The Blumlein style circuits that Newman designed for the ACE Pilot Model were advanced for the time.

Alan Dower Blumlein

Alan Blumlein[2] was born in 1903 in London.[3] His father, a mining engineer, was a naturalized British subject from Germany. Alan was awarded a first class BSc degree in 1923 following education at Highgate School and City and Guilds College. He joined International Western Electric Corp and Standard Telephones and Cables in 1924, and was involved in long distance telephonic communication in Europe, receiving a prize for his contribution at the age of twenty-three.

In 1929 Blumlein started work for the Columbia Graphophone Company. Within a short time he produced an advanced moving coil wax cutting machine for cutting gramophone record masters. He also designed a moving coil microphone. Around 1930 he became interested in stereophonic recording and reproduction. His work resulted in an outstanding patent on the subject and a new method of cutting stereo disks.

About this time Blumlein developed some circuit design principles aimed at making the behaviour of circuits less dependent on valve characteristics. In those days valves came from the manufacturer with parameters varying up to 30 per cent from their nominal values. (The current passed by a valve with a fixed bias varied so much from valve to valve that sometimes on the production line valves were selected for particular characteristics.) Blumlein's philosophy was that if a circuit is properly designed then it will work without any need to adjust component values on test, and fundamental to this philosophy was the notion of circuits in which the current is controlled within close limits with little dependence on valve characteristics. Blumlein developed the principle of defined current circuits in which the valve's current was defined either by a cathode resistance to a considerably lower potential, or by employing negative feedback. These principles were patented in about 1936. Upon these principles were built other circuit ideas including the wide-band DC couplings described below. During the war these principles survived the rigours of quantity production in a factory under military specifications and, after the war, were used in the design of the ACE Pilot Model, the DEUCE, and the Big ACE.

Before the opening of the London television station at Alexandra Palace in 1936, Blumlein's activities were devoted to the engineering of the 405 line television system. In 1931 the Gramophone Company (HMV) merged with the Columbia Graphophone Company to form EMI. The merger initiated major research into the design and manufacture of vacuum cathode-ray tubes. This research lead to the development of the iconoscope tube for cameras and was crowned by the outside televising of the Coronation procession in 1937. The cameras were linked by an eight mile cable to the transmitter. A new type of aerial was required and Blumlein invented the resonant slot aerial.

In 1939 EMI developed a 60 MHz radar patented by Blumlein and E. L. C. White. Blumlein was involved with a visual display for the stereo sound locators. By 1940 this system was delivering airborne interception radar for Beaufighters. In 1941 work began on H2S, a radar system to help bombers find their targets. It was during the testing of this equipment that Blumlein, Browne, and Blythen of EMI Research Laboratories were killed in a disastrous air crash. In spite of this, H2S became operational in 1943. Some units were allocated to the detection of enemy submarines and H2S made a substantial contribution to the Battle of the Atlantic.

Newman joined EMI Research Laboratories at about the beginning of the war, and after Blumlein's death became responsible for the further development of H2S and other military projects. He was later involved in the development of television cameras. I joined the Laboratory in 1941 and like Newman absorbed Blumlein's design principles.

Other techniques invented by Blumlein included current steering logic, using common cathode circuits (equivalent to emitter coupled logic), and negative feedback. The method used to define the valve current was to use 'long tails', in which each valve's cathode current is determined by a resistance, typically 10k ohms, to a negative voltage, typically −100V. The cathode settles at a voltage which produces the right bias for the grid at this current (10 milliamps). If the bias varies by a few volts from the nominal value, this makes little difference to the current. At that time resistances with a tolerance within 5 per cent were obtainable, and these determined the current.

Another problem with valve circuits at that time was parasitic oscillation at very high frequencies. Various ad hoc methods of dealing with this were used, but EMI's practice was to give every grid a grid-stopper of about 47 ohms so that the oscillation never occurred. This practice was copied in the ACE Pilot Model and we had no trouble from parasitic oscillation.

The long tailed pair

Conventional valve amplifiers traditionally had a bias resistor connected between a valve's cathode and earth (oV), the grid of the valve being biassed to earth. The value of this resistor, a few hundred ohms, was chosen so that the cathode worked at a few volts above earth, thus providing an appropriate grid bias voltage corresponding to the valve's advertised characteristics at the desired working current. (Calculating the value of the resistor involved Ohm's Law.) An undesirable feature of this arrangement is that the valve current can vary from one valve to another by a considerable margin, depending on the characteristics of the particular valve. This feature can be avoided by increasing the value of the resistor to a few thousand ohms and connecting it, not to earth, but to a negative voltage of, say, −100V. The current through the resistor and the valve is then determined by this voltage and the value of the resistor, the valves operating clear of grid current. If the resistor is 10,000 ohms then the current is close to 10 milliamps, the characteristics of the valve having little influence. This high-value resistor is known as a *long tail*.

This principle can be extended by having two valves (triodes or pentodes) working side by side, with their cathodes connected together and sharing the current from one long-tail resistor. Their anodes are connected to some suitable positive voltage, say +200V. The way in which the two valves share the current is then controlled by the relative voltage of the two grids. The input signal is connected to one of the grids, while the other one is typically connected to a fixed potential. If one grid is 10V above the other, then all the current goes to its anode. Thus it is possible to switch all the current from one valve to the other (or, at smaller signal amplitudes, to create a balanced push–pull amplifier). This arrangement is the long tailed pair. One double triode valve may be used in place of two triodes (see fig. 1).

Blumlein patented the long tailed pair in 1936.[4] There are two distinct functions of the long tailed pair. It was first developed as a small signal amplifier producing a well balanced push–pull pair of outputs from the two anodes. The other function is as a switch where all of the cathode current is switched to one or other anode (the anode current of one valve being reduced to zero).[5] As an amplifier it was used at both audio frequency and radio frequency. As a switch it was used in television and radar circuits.

These functions were used extensively in the ACE Pilot Model and its decendants.

Switching advantages of the long tailed pair

An important advantage of the long tailed pair in digital circuits is its switching speed. In general the switching speed of a circuit is a function of the valve's maximum power rating, the anode load resistance and the stray capacitance. For high switching speed the anode load resistance needs to be as low as possible subject to producing an adequate signal size for operating the next valve. The valve current needs to be as high as possible subject to not exceeding the valve's maximum power rating. The 'miniature' double triodes available in the 1950s were ideal for such applications, running at about 10 milliamps with 20V signals and producing switching times of about a tenth of a microsecond. The ACE Pilot Model ran at a speed of a million bits per second, and its successor, the 'Big ACE', which also used this technology, ran at a speed of 1·5 million bits per second.

To achieve an AND gate it was possible to feed the anode current of a valve to the cathode of another, the lower valve having its cathode at about −200V and being fed current from −300V. This provided two input grids supplying signals which were used for various logical operations.

A contending circuit technology used in some computers at that time was the pentode with both the grid and the suppressor being used as signal inputs. Although this circuit provides an AND gate in one valve there are two important disadvantages of the circuit. First, the action of the suppressor is to divert the cathode current from the anode to the screen. Unfortunately in general the maximum screen dissipation is much lower than that of the anode, in the range 20–30 per cent of the anode rating. As a consequence the cathode current has to be limited to about a quarter of the valve's normal current, and the anode load resistance has to be increased by a factor of 4 to achieve the same output signal size. This increases the switching time in proportion. The second disadvantage is that the suppressor provides a much lower gain than the first grid so that its signal amplitude needs to be several times larger. The extent of this depends on the valve type, but the machines of the 1950s which used this technology ran at a rate of about a tenth of a million bits per second.

Figure 1 shows the long tailed double triode from which most of the other circuit modules were developed. The input grid can be switched between −10 and +10V while the other grid remains at earth potential. Because the two

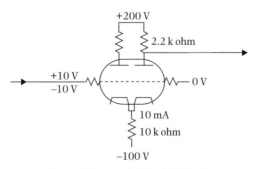

Fig. 1 A long tailed double triode.

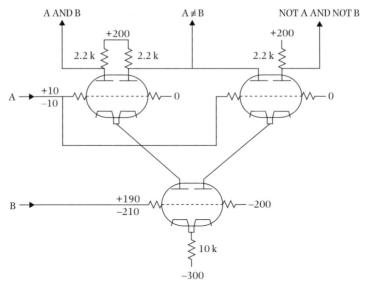

Fig. 2 A half-adder.

cathodes are joined, this steers the current to either anode circuit. With a 2·2k ohm resistance in each anode an output signal of 20V is available at a level of +200V.

The half-adder and adder

One of the circuits developed from the long tailed double triode was the half-adder, shown in fig. 2. This used three double triodes. A complete adder needed two half-adders. This circuit introduces the idea of two layers of valves, one with cathodes at about earth and one with cathodes at about −200V. There is then a need to couple the signal from the anodes at about +200V to the next grids at earth or −200V.

Signal coupling

There are two distinct methods of conveying the signal to a lower level for feeding the grid of another valve, both of which were used in appropriate circumstances. One method was to make use of a wide band DC coupling. This will be described later. The other method was to convey the data in the form of narrow pulses which could be AC coupled by a capacitor and DC restored by a diode at a lower voltage.

It is also important to be able to restore the timing of data signals. In particular, mercury delay lines are temperature dependent and need not only a temperature stabilized enclosure, but also re-timing. A clock pulse gate was used to sample the output of each delay line in order to re-time the signal, reducing each bit to a narrow pulse of 0·25 microsec and then AC couple and DC restore it at a lower level of either earth or −200V.

The circulation unit

Initially most of the circuit modules were developed for the circulation unit, that is the unit which maintains the information circulating in the mercury delay lines. The Pilot Model had 19 delay lines storing 32, 64, or 1024 bits, so these circulation units amounted to a large proportion of the total machine. Figure 3 shows the major components of a circulation unit.

The following is a list of the main modules and their function in the circulation unit:

1. Signal from the transducer, a crystal at the end of the delay line.
2. Receiver to amplify the signal, a 15 MHz carrier modulated by the data at 1 microsecond intervals.
3. Detector to rectify the amplified signal producing bits 1 microsecond wide.

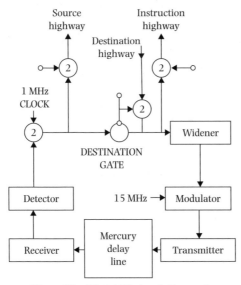

Fig. 3 The Pilot ACE circulation unit.

4. Clock pulse gate for re-timing purposes. The clock pulse width was 0·25 microsecond at 1 MHz.
5. Source gate, which, when selected, sends data to the Source highway leading to the arithmetic unit.
6. Destination gate, interupting the circulation and switching in data from the Destination highway.
7. Instruction Source gate to Instruction highway leading to the control unit.
8. Pulse widener widening the data from clock pulse width to 1 microsecond.
9. Modulator modulating the 15 MHz carrier.
10. Transmitter to a transducer on the mercury delay line.

The data is in the form of 1 microsecond wide pulses from the pulse widener, through the mercury, to the clock pulse gate, and in the form of 0·25 microsecond wide pulses from the clock pulse gate, through the output and input gates and the highways, to the pulse widener.

The modulator, transmitter, and receiver operated at a carrier frequency of 15 MHz. The receiver had three staggered stages and had to have an adjustable and stable gain.

The clock pulse gate

The clock pulse gate (fig. 4) is a double triode gate whose cathodes are supplied with current from the anode of a pentode diode gate. The clock pulses are AC coupled down to the grid of the pentode at about −207V, the cathode being supplied with about 12mA through an 8200 ohm resistance from −300V. The cathode is joined to a diode which stops the cathode falling below −200V, thus cutting off the pentode's current when the clock pulses are down. The left grid of the double triode is fed with the wide signal from the detector and the left anode provides negative output pulses via an AC coupling to one input of the Destination gate.

At the time that this circuit was being developed, germanium diodes became available and were used in the pentode diode gate as well as for DC restoration in some of the AC couplings. Much later they created a reliability problem.

The Destination gate

The Destination gate (fig. 5) consists of three double triodes, two at about earth and one at about −200V. The lower one feeds current to one of the

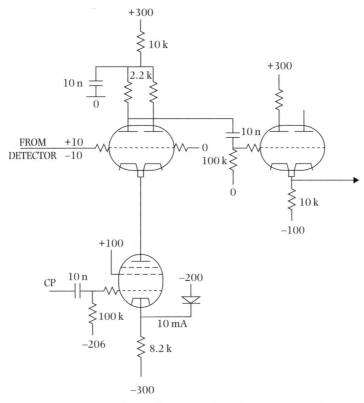

Fig. 4 The clock pulse gate.

upper ones and is controlled by a signal from the central control unit when the circulation is to be interrupted and external data fed in. The left grid of the lower double triode is switched from -190 to -210V to achieve this, the grid of the right valve remaining at -200V. The left grid has to be DC coupled from the anode of a pentode that is in the region at about $+200$V. This introduces the idea of wide band DC coupling (see later).

In the Destination gate the lower double triode feeds current for multiples of 32 microsec to one of the upper double triodes so that either the data from the clock pulse gate or the data from the Destination highway is fed to the pulse widener.

The Source gate

The Source gate (fig. 6) was controlled similarly by the central control unit, but is simpler than the Destination gate, since it is an on–off switch and can operate slowly in a few microseconds.

357

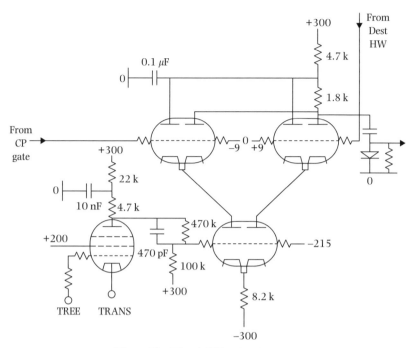

Fig. 5 The Pilot ACE Destination gate.

The pulse widener

The pulse widener consists of two double triodes, each of which is a mono-stable trigger with a period of 0·5 microsec, the second one being fired by a differentiated edge from the first. Their outputs are combined, producing a 1 microsec wide pulse which is used to operate the modulator.

Wide band DC coupling

The circuits for the Source and Destination gates employ the wide band DC coupling patented by E. L. C. White in 1932. As shown in figs 5 and 6, this is used to couple the signal from an anode at about +200V to a grid at about −200V without distortion. At high frequencies the anode load of the pentode is 4·7K ohms and this is coupled down through a capacitor to the left grid of the double triode, producing a step of about 20V. At low frequencies the anode load of the pentode is 22K + 4·7K ohms giving a much larger signal. This signal is conveyed by a voltage divider consisting of two resistances whose values are chosen so that a signal of about 20V arrives at the grid. At intermediate frequencies the ratio of the coupling and decoupling capacitors is

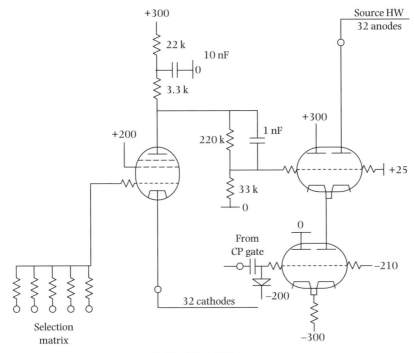

Fig. 6 The Pilot ACE Source gate.

chosen to make the time constants equal, so that the step size at the grid is constant at all frequencies, and it is possible to achieve switching speeds of about 0·1 microsecond.

Figure 7 shows an alternative configuration of the wide band DC coupling, as used in the clock pulse gate and trigger of the ACE computer, described below. In this the anode load at high frequencies comprises two resistors in parallel.

The data distribution system and its control

Nearly all calculations were performed by transfering data from a Source to a Destination under the control of the control unit at the centre of operations. The task of the control unit was to receive an instruction from the Instruction highway, staticise it into a set of valve registers and use the output of these to set the input to three resistance trees. These trees ran the length of the machine and controlled which of the 32 Sources and Destinations was to open. All the Destination gates were under the control of a signal called

Transtim (for transfer time) which opened the particular Destination gate selected by the Destination tree for the required time. The control unit then selected the Source of the next instruction, and continued the cycle of events.

The highway amplifier

To convey the signal from the selected Source to the selected Destination required a highway amplifier (fig. 7). The input to this was connected to 32 anodes distributed along the machine. This implied a high stray capacitance and required an amplifier with a low input impedance provided by the cathode of a valve. This signal was then amplified and distributed to 32 grids in Destination gates with associated stray capacitance. This was one of the important applications for the assisted cathode follower shown on the right-hand side of the figure.

The assisted cathode follower

The conventional cathode follower has the property that it can turn on quite a lot of current on a rising edge in order to charge the stray capacitance, but on a falling edge it can only turn off the standing current from the cathode resistance and so falling edges can be slow. The 'assisted cathode follower' uses an additional valve to provide current to pull the output down. This gets its control signal from the anode of the cathode follower, so that when the cathode follower runs out of current, its anode rises, and this signal is AC coupled to the grid of the valve below. The cathode of this valve is decoupled to earth by a capacitor, so that when its grid rises, it turns on a burst of current. Other applications for the assisted cathode follower included the distribution of clock pulses.

The origin of this assisted cathode follower is not clear. It may have come from EMI with Newman or it may have been developed by Newman. It makes use of negative feedback, the subject of one of Blumlein's patents.

The magnetic drum store of the ACE Pilot Model

As larger problems were programmed on the Pilot Model, the need for more storage became apparent and various options were explored. The development of high speed magnetic recording with in-contact heads was considered to be a long term research project beyond our resources. It was decided to

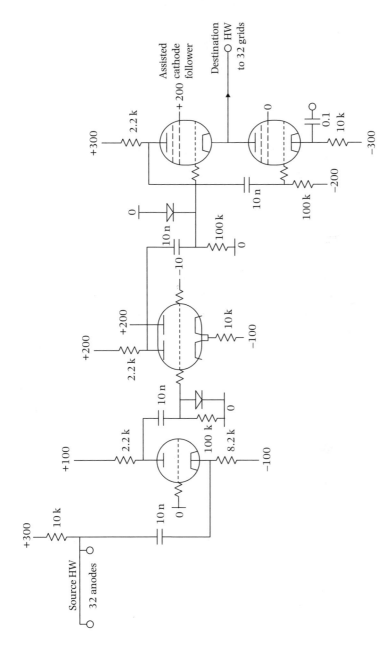

Fig. 7 The highway amplifier.

build a high speed drum with out-of-contact heads and each track storing 1024 bits, the contents of a long delay line.[6] Each transfer copied the whole of the 1024 bits. To achieve this without a buffer store it was necessary to synchronize the drum to the machine, at exactly one-ninth of its speed. This enabled a transfer of 1024 bits in 9 milliseconds.

Synchronization was achieved with the aid of a steel disk having 1024 teeth on its periphery, made by the NPL's Metrology Division workshop, and fixed to the top of the drum. The drum was powered by an integral hysteresis motor driven by six KT66 valves, with negative feedback to maintain synchronization. At the time this drum was designed it was amongst the fastest in the world. It was not intended to be large enough for mass storage, for which a magnetic tape system was anticipated.

Initially the drum had 16 read heads and 16 write heads, which could be shifted to one of two positions, giving access to 512 words in each position. Then an eight position head-shifting mechanism was developed. This was a novel device, known as the 'string of sausages'. It consisted of four steel cylinders, loosely connected in series, with the gaps between them accurately limited to 0·016, 0·032, and 0·064 in. by means of connecting rods and pistons as shown in fig. 8. Each gap was surrounded by a steel magnetic

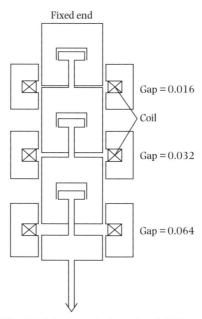

Fig. 8 The Pilot Model magnetic drum head shifting mechanism.

circuit containing a coil. By energizing the three coils independently, it was possible to move the heads to one of eight accurate positions. The device pulled the heads up against a return spring, giving a maximum shift time of 50 milliseconds. This resulted in a total capacity of 4192 words of 32 bits. Although the 'string of sausages' made a noise as of galloping horses, it proved to be remarkably reliable. Special magnetic coatings were developed by Bill Gleed. The recording heads were developed by Lew Page. Fred Osborne was responsible for the precision engineering design.

The magnetic drum of the DEUCE computer worked at the same frequency, but employed a moving coil head shifting mechanism having 16 positions.

The Big ACE

After completing the ACE Pilot Model it was decided to make use of our experience to design and build a larger and faster machine.[7] Although other options were considered, it was decided to continue with mercury delay lines for the main store, since we had experienced no reliability problems with them. It was decided to double the number of long delay lines to 24 and to increase the clock rate by 50 per cent to 1·5 MHz.

As a consequence the word length was increased to 48 bits. This provided enough bits in an instruction to change from a two-address code to a three-address code, with 64 sets of Sources and Destinations, and 64 functions. This was a step towards Turing's original design, which was for a three-address machine with a function box. It was also expected that the longer word length would result in less need to program in floating point, which was somewhat time-consuming on the Pilot Model. Another advance was to eliminate the setup time between transfers, allowing operations at a maximum rate of 30,000 per second.

The Pilot Model, which dissipated about 2 kilowatts, was designed without forced air cooling, and since it had no cabinet, heat was dissipated by convection and radiation. This surprised mathematicians, who were used to cold mechanical desk calculators. The Big ACE had a cooling system (designed by Fred Osborne) and was laid out spaciously in order to avoid temperature problems. Together with the increase in store size, this produced a considerable increase in the dimensions of the ACE.

It was anticipated that this increase in dimensions might lead to timing problems, particularly in view of the increase in clock rate. To alleviate this, the highways were divided into eight parts, each part being handled

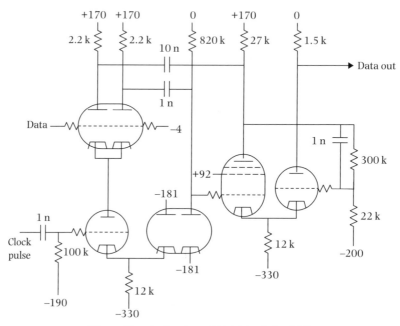

Fig. 9 The clock gate and bistable of the ACE.

by a distribution unit. Variations in transit times were thereby reduced to satisfactory levels and the need to adjust timing on test was eliminated.

The circuit design principles were similar to those adopted for the Pilot Model, with minor changes. The voltages used were increased to +330 and −330V, and a number of specially designed stabilized intermediate voltages were incorporated. The circuits in the circulation units between the detector and the modulator, including the highways, were DC coupled. Figure 9 shows a sample of the circuits: the clock pulse gate and bistable trigger. This figure shows an alternative version of the wide band DC coupling.

During the later stages of designing the ACE, transistors of adequate speed and reliability had become available, and these were used in some emitter coupled circuits for lower speed peripheral applications.

The Big ACE magnetic drum store

This was similar in design to the one on the Pilot Model but considerably faster, and included four drums which could all transfer data simultaneously. Each drum revolved in 5 milliseconds (12,000 rpm), during which 1536 bits could be transferred to or from a delay line. The transfer rate was

3 microseconds per bit. The number of head positions was increased to 16 by using a moving coil mechanism, giving a total of 1024 tracks, each with a capacity of 32K words of 48 bits. This rotation speed required the use of special bearings as used in gyroscopes.

Notes

I am grateful to several ex-colleagues who have provided information and comments, and am particularly grateful to Jack Copeland for his extensive editorial work.

1. Newman, E. A., Clayden, D. O., and Wright, M. A. (1953) 'The mercury-delay-line storage system of the ACE Pilot Model electronic computer', *Proceedings of the IEE*, 100(76), 445–52.
2. For additional information see The Alan Blumlein Homepage <www.AlanTuring.net/blumlein>.
3. Biographical information is from Burns, R. W. (1992) 'A. D. Blumlein—engineer extraordinary', *Engineering, Science and Education Journal of the IEE*, 1, 19–33.
4. Patent number 482740. Blumlein's patents are in the British Library, London.
5. Blumlein's patent 514065 is an interesting application of the long tailed pair. In this patent the two valves share the current equally in the static condition. The two grids are connected to tappings on an LC delay line (a network using discrete inductances and capacitors to produce a delay) so that when a stream of narrow pulses is applied to the beginning of the delay line, the two grids receive the pulses separated by an accurate time interval. The two anodes can be coupled to a bistable trigger, thus producing a pulse which is of an accurate width. The tappings on the delay line can be adjusted to produce a pulse of a specified width.
6. Clayden, D. O., Page, L. J., and Osborne, C. F. (1956) 'The magnetic storage drum on the ACE Pilot Model', *Proceedings of the IEE*, 103, 504–14.
7. Blake, F. M., Clayden, D. O., Davies, D. W., Page, L. J., and Stringer, J. B. (1957) 'Some features of the ACE Computer', Proceedings of a Conference on Data Processing and Automatic Computing Machines at the Weapons Research Establishment, Salisbury, Australia. (A copy is in the Science Museum Library.)

Part V
Technical Reports and Lectures on the ACE
1945-47

20   *Proposed electronic calculator (1945)*

Alan M. Turing

Turing completed his report 'Proposed Electronic Calculator' toward the end of 1945.[1] It was submitted to the Executive Committee of the NPL in February 1946, under the description 'Report by Dr. A.M. Turing on Proposals for the Development of an Automatic Computing Engine (ACE)'.[2] The design set out in 'Proposed Electronic Calculator' was the basis for all the ACE computers.[3]

B. J. C.

[1] Michael Woodger (Turing's assistant at the National Physical Laboratory from 1946) sighted an NPL file showing that 'Proposed Electronic Calculator' was completed in 1945; unfortunately, this file was destroyed in 1952 (Woodger, M., handwritten note (undated), in the Woodger Papers, National Museum of Science and Industry, Kensington, London (catalogue reference M15/78); letter from Woodger to Copeland, 27 November 1999).

[2] Minutes of the NPL Executive Committee, 19 March 1946 (a digital facsimile is in The Turing Archive for the History of Computing <www.AlanTuring.net/npl_minutes_mar1946>).

[3] A copy of the original typewritten report is in the Public Record Office (document reference DSIR 10/385) and it is this which is reproduced here. The diagrams (which are not included in the copy of the report in the PRO) are from the NPL library. The report and diagrams are Crown Copyright and are printed here by permission of the NPL. A number of typographical errors in the original typescript have been corrected. Occasionally a punctuation mark has been added or removed. Material in the text appearing within square brackets has been added by the editor.

PART I.

DESCRIPTIVE ACCOUNT.

PART II.

TECHNICAL PROPOSALS.

PART I.

Descriptive Account.

1. Introductory.

Calculating machinery in the past has been designed to carry out accurately and moderately quickly small parts of calculations which frequently recur. The four processes addition, subtraction, multiplication and division, together perhaps with sorting and interpolation, cover all that could be done until quite recently, if we except machines of the nature of the differential analyser and wind tunnels, etc. which operate by measurement rather than by calculation.

It is intended that the electronic calculator now proposed should be different in that it will tackle whole problems. Instead of repeatedly using human labour for taking material out of the machine and putting it back at the appropriate moment all this will be looked after by the machine itself. This arrangement has very many advantages.

(1) The speed of the machine is no longer limited by the speed of the human operator.

(2) The human element of fallibility is eliminated, although it may to an extent be replaced by mechanical fallibility.

(3) Very much more complicated processes can be carried out than could easily be dealt with by human labour.

Once the human brake is removed the increase in speed is enormous. For example, it is intended that multiplication of two ten figure numbers shall be carried out in 500 μs. This is probably about 20,000 times faster than the normal speed with calculating machines.

It is evident that if the machine is to do all that is done by the normal human operator it must be provided with the analogues of three things, viz. firstly, the computing paper on which the computer writes down his results and his rough workings; secondly, the instructions as to what processes are to be applied; these the computer will normally carry in his head; thirdly, the function tables used by the computer must be available in appropriate form to the machine. These requirements all involve <u>storage of information</u> or <u>mechanical memory</u>. This is not the place for a detailed discussion of the various kinds of storage available* and the considerations which govern their

---

* See § 16.

usefulness and which limit what we can expect. For the present let us only remark that the memory needs to be very large indeed by comparison with standards which prevail in most valve and relay work, and that it is necessary therefore to look for some more economical form of storage.

It is intended that the setting up of the machine for new problems shall be virtually only a matter of paper work. Besides the paper work nothing will have to be done except to prepare a pack of Hollerith cards in accordance with this paper work, and to pass them through a card reader connected with the machine. There will positively be no internal alterations to be made even if we wish suddenly to switch from calculating the energy levels of the neon atom to the enumeration of groups of order 720. It may appear somewhat puzzling that this can be done. How can one expect a machine to do all this multitudinous variety of things? The answer is that we should consider the machine as doing something quite simple, namely carrying out orders given to it in a standard form which it is able to understand.

The actual calculation done by the machine will be carried out in the binary scale. Material will however be put in and taken out in decimal form.

In order to obtain high speeds of calculation the calculator will be entirely electronic. A unit operation (typified by adding one and one) will take 1 microsecond. It is not thought wise to design for higher speeds than this as yet.

The present report gives a fairly complete account of the proposed calculator. It is recommended however that it be read in conjunction with J. von Neumann's 'Report on the EDVAC'.

2. Composition of the Calculator.

We list here the main components of the calculator as at present conceived:-

(1) Erasable memory units of fairly large capacity, to be known as dynamic storage (DS). Probably consisting of between 50 and 500 mercury tanks with a capacity of about 1000 digits each.
(2) Quick reference temporary storage units (TS) probably numbering about 50 and each with a capacity of say 32 binary digits.
(3) Input organ (IO) to transfer instructions and other material into the calculator from the outside world. It will have a mechanical part consisting of a Hollerith card reading unit, and an electronic part which will be internal to the calculator.

(4) Output organ (OO), to transfer results out of the calculator. It will have an external part consisting of a Hollerith card reproducer and an internal electronic part.

(5) The logical control (LC). This is the very heart of the machine. Its purpose is to interpret the instructions and give them effect. To a large extent it merely passes the instructions on to CA. There is no very distinct line between LC and CA.

(6) The central arithmetic part (CA). If we like to consider LC as the analogue of a computer then CA must be considered a desk calculating machine. It carries out the four fundamental arithmetical processes (with possible exception of division, see p. [402]), and various others of the nature of copying, substituting, and the like. To a large extent these processes can be reduced to one another by various roundabout means; judgment is therefore required in choosing an appropriate set of fundamental processes.

(7) Various 'trees' required in connection with LC and CA for the selection of the information required at any moment. These trees require much more valve equipment than LC and CA themselves.

(8) The clock (CL). This provides pulses, probably at a recurrence frequency of a megacycle, which are applied, together with gating signals, to the grids of most of the valves. It provides the synchronisation for the whole calculator.

(9) Temperature control system for the delay lines. This is a somewhat mundane matter, but is important.

(10) Binary to decimal and decimal to binary converters. These will have virtually no outward and visible form. They are mentioned here, lest it be thought they have been forgotten.

(11) Starting device.

(12) Power supply.

3. Storages.

(i) The storage problem. As was explained in § 1 it is necessary for the calculator to have a memory or information storage. Actually this appears to be the main limitation in the design of a calculator, i.e. if the storage problem can be solved all the rest is comparatively straightforward. In the past it has not been possible to store very large quantities of information economically in such a way that the information is readily accessible. There were economical methods such as storage on five-unit tape, but with these the information was not readily accessible, especially if one wishes to jump from point to point. There were also forms with good accessibility, such as storage on relays and valves, but these were quite prohibitively uneconomical. There are now several possibilities for combining economy with accessibility which have been developed, or are being developed. In this section we describe the one which will most probably be used in the calculator.

(ii) <u>Delay line storage</u>. All forms of storage depend on modifying in some way the physical state of some storage medium. In the case of 'delay line storage' the medium consists of mercury, water, or some other liquid in a tube or tank, and we modify its state of compression at various points along the tube. This is done by forcing supersonic waves into the tube from one end. The state of the storage medium is not constant as it would be for instance if the storage medium were paper or magnetic tape. The information moves along the tube with the speed of sound. Unless we take some precautions the sound carrying the information will pass out of the end of the tube and be lost. We can effectively prevent this by detecting the sound in some way (some form of microphone) as it comes out, and amplifying it and putting it back at the beginning. The amplifying device must correct for the attenuation of the tube, and must also correct for any distortion of form caused by the transmission through the tube, otherwise after many passages through the tube the form will be eventually completely lost. We can only restore the form of the signal satisfactorily if the various possible ideal signal forms are quite distinct, for otherwise it will not be possible to distinguish between the undistorted form of one signal and a distorted form of another. The scheme actually proposed only recognizes 2^{1024} distinct states of compression of the water medium, these being sequences of 1024 pulses of two different sizes, one of which will probably be zero. The amplifier at the end of the line always reshapes the signal to bring it back to the nearest ideal signal.

Alternatively we may consider the delay line simply as providing a delay, as its name implies. We may put a signal into the line, and it is returned to us after a certain definite delay. If we wish to make use of the information contained in it when it comes back after being delayed we do so. Otherwise we just delay it again, and repeat until we do require it. This aspect loses sight of the fact that there is still a storage medium of some kind, with a variety of states according to the information stored.

There are, of course, other forms of delay line than those using acoustic waves.

(iii) <u>Technical proposals for delay line</u>. Let us now be more specific. It is proposed to build 'delay line' units consisting of mercury or water tubes about 5′ long and 1″ diameter in contact with a quartz crystal at each end. The velocity of sound in either mercury or water is such that the delay will be 1·024 ms. The information to be stored may be considered to be a sequence of 1024 'digits' (0 or 1), or 'modulation elements' (mark or space). These digits will be represented by a corresponding sequence of pulses. The digit 0 (or space) will be represented by the absence of a pulse at the appropriate time, the digit 1 (or mark) by its presence. This series of pulses is impressed on the end of the line by one piezo-crystal, it is transmitted down the line in the form of supersonic waves, and is reconverted into a varying voltage by the crystal at the far end. This voltage is amplified sufficiently to give an output of the order of 10 volts peak to peak and is used to gate a standard pulse generated by the clock. This pulse may be again fed into the line by means of the transmitting crystal, or we may feed in some altogether different signal. We also have the possibility of leading the gated pulse to some other part of the calculator, if we have need of that information at the time. Making use of the information does not of course preclude keeping it also. The figures above imply of course that the interval between digits is 1 μs.

It is probable that the pulses will be sent down the line as modulation on a carrier, possibly at a frequency of 15 Mc/s.

(iv) <u>Effects of temperature variations</u>. The temperature coefficient of the velocity of sound in mercury is quite small at high frequencies. If we keep the temperatures of the tanks correct to within one degree Fahrenheit it will be sufficient. It is only necessary to keep the tanks nearly at equal temperatures. We do not need to keep them all at a definite temperature: variations in the temperature of the room as a whole may be corrected by altering the clock frequency.

4. <u>Arithmetical Considerations</u>.

(i) <u>Minor cycles</u>. It is intended to divide the information in the storages up into units, probably of 32 digits or thereabouts. Such a storage will be appropriate for carrying a single real number as a binary decimal or for carrying a single instruction. Each sub-storage of this kind is called a <u>minor cycle</u> or <u>word</u>. The longer storages of length about 1000 digits are called <u>major cycles</u>. It will be assumed for definiteness that the length of the minor cycle is 32 and that of the major 1024, although these need not yet be fixed.

(ii) <u>Use of the binary scale</u>. The binary scale seems particularly well suited for electronic computation because of its simplicity and the fact that valve equipment can very easily produce and distinguish two sizes of pulse. Apart from the input and output the binary scale will be used throughout in the calculator.

(iii) <u>Requirements for an arithmetical code</u>. Besides providing a sequence of digits the statement of the value of a real number has to do several other things. All included, (probably) we must:

(a) State the digits themselves, or in other words we must specify an integer in binary form.
(b) We must specify the position of the decimal point.
(c) We must specify the sign.
(d) It would be desirable to give limits of accuracy.
(e) It would be desirable to have some reference describing the significance of the number. This reference might at the same time distinguish between minor cycles which contain numbers and those which contain orders or other information.

None of these except for the first could be said to be absolutely indispensable, but, for instance, it would certainly be inconvenient to manage without a sign reference. The digit requirements for these various purposes are roughly:

(a) 9 decimal digits, i.e. 30 binary,
(b) 9 digits,
(c) 1 digit,
(d) 10 digits,
(e) very flexible.

(iv) <u>A possible arithmetical code</u>. It is convenient to put the digits into one minor cycle and the fussy bits into another. This may perhaps be qualified as far as the sign digit is concerned: by a trick it can be made part of the normal digit series, essentially in the same way as we regard an initial series of figures 9 as indicating a negative number in normal computing. Let us now specify the code without further beating about the bush. We will use two minor cycles whose digits will be called $i_1 \ldots i_{32}$, $j_1 \ldots j_{32}$. Of these $j_{24} \ldots j_{32}$ are available for identification purposes, and the remaining digits make the following statement about the number ξ.

There exist rational numbers β, γ and an integer m such that

$$|\xi - 2^m\beta| < \gamma$$

$$\beta = \sum_{s=1}^{31} 2^{s-1}i_s - 2^{31}i_{32}$$

$$m = \sum_{t=1}^{9} 2^{t-1}j_t - 512$$

$$\gamma = \sum_{u=10}^{17} 2^{u+m-n}j_u$$

$$n = \sum_{v=18}^{23} 2^{v-18}j_v$$

This code allows us to specify numbers from ones which are smaller than 10^{-70} to ones which are larger than 10^{86}, mentioning a value with sufficient figures that a difference of 1 in the last place corresponds to from 2.5 to 5 parts in 10^{10}. An error can be described smaller than a unit in the last place or as large as 30,000 times the quantity itself (or by more if this quantity has its first few 'significant' digits zero).

(v) <u>The operations of CA</u>. The division of the storage into minor cycles is only of value so long as we can conveniently divide the operations to be done into unit operations to be performed on whole minor cycles. When we wish to do more elaborate types of process in which the digits get individual treatment we may find this form of division rather awkward, but we shall still be able to carry these processes out in some roundabout way provided the CA operations are sufficiently inclusive. A list is given below of the operations which will be included. Actually this account is distinctly simplified, and an accurate picture can only be obtained by reading § 12. The account is however quite adequate for an understanding of the main problems involved. The list is certainly theoretically adequate, i.e. given time and instruction tables any required operation can be carried out. The operations are:

(1) Transfers of material between different temporary storages, and between temporary storages and dynamic storage.

(2) Transfers of material from the DS to cards and from cards to DS.

(3) The various arithmetical operations, addition, subtraction, and multiplication (division being omitted), also 'short multiplication' by numbers less than 16, which will be much quicker than long multiplication.

(4) To perform the various logical operations digit by digit. It will be sufficient to be able to do 'and', 'or', 'not', 'if and only if', 'never' (in symbols $A \& B$, $A \vee B$, $\sim A$, $A \equiv B$, F). In other words we arrange to do the processes corresponding to xy, $x + y + xy$, $1 + x$, $1 + (x + y)^2$, 0 digit by digit, modulo 2, where x and y are two corresponding digits from two particular TS (actually TS 9 and TS 10).

5. Fundamental Circuit Elements.

The electronic part of the calculator will be somewhat elaborate, and it will certainly not be feasible to consider the influence of every component on every other. We shall avoid the necessity of doing this if we can arrange that each component only has an appreciable influence on a comparatively small number of others. Ideally we would like to be able to consider the circuit as built up from a number of circuit elements, each of which has an output which depends only on its inputs, and not at all on the circuit into which it is working. Besides this we would probably like the output to depend only on certain special characteristics of the inputs. In addition we would often be glad for the output to appear simultaneously with the inputs.

These requirements can usually be satisfied, to a fairly high accuracy, with electronic equipment working at comparatively low frequencies. At megacycle frequencies however various difficulties tend to arise. The input capacities of valves prevent us from ignoring the nature of the circuit into which we are working; limiting circuits do not work very satisfactorily: capacities and transit times are bound to cause delays between input and output. These difficulties may be best resolved by bending before the storm. The delays may be tolerated by accepting them and working out a time table which takes them into account. Indefiniteness in output may be tolerated by thinking in terms of 'classes of outputs'. Thus instead of saying 'The inputs A and B give rise to the output C', we shall say 'Inputs belonging to classes P and Q give rise to an output in class R'. The various classes must be quite distinct and must be far from overlapping, i.e. topologically speaking we might say that they must be a finite distance apart. If we do this we shall have made a very definite division of labour between the mathematicians and the engineers, which will enable both parties to carry on without serious doubts as to whether their assumptions are in agreement with those of the other party.

For the present we shall merely ignore the difficulties because we wish to illustrate the principles. We shall assume the circuit elements to have all the most agreeable properties. It may be added that this will only affect our circuits in so far as we assume instantaneous response, and that not very seriously. The questions of stable output only involve the mathematician to the extent of a few definitions.

In the present section we shall only be concerned with what the circuit elements do. A discussion of how these effects can be obtained will be given in § 15. The circuit elements will be divided into valve-elements and delay elements.

(i) <u>Delay line, with amplifier and clock gate</u>. This is shown as a rectangle with an input and output lead

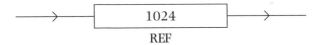

The arrow at the input end faces towards the rectangle and at the output end faces away. The name of the delay line, if any, will be written outside and the delay in pulse periods inside.

This circuit element delays the input by the appropriate number of pulse periods and also standardizes it, i.e. converts it into the nearest standard form by correcting amplitude shape and time.

(ii) <u>The unit delay</u>. This is represented by a triangle, thought of as a modified form of arrow

The input to output direction is indicated by the arrow.

This delay element ideally provides a delay of one pulse period.

(iii) <u>Limiting amplifier</u>. Ideally this valve-element is intended to give no output for inputs of less than a certain standard value, and to give a standard pulse as output when the input exceeds a second standard value. Intermediate input values are supposed not to occur. If we combine this with a resistance network in which a number of input signals are combined the condition takes the form that if the input signals are $s_1\ s_2\ \ldots\ s_n$ there will be zero output unless $\alpha_1\ s_1 + \cdots + \alpha_n s_n \geqslant \beta_1$ and a standard or unit output if $\alpha_1\ s_1 + \cdots + \alpha_n s_n > \beta_2$. This may be simplified by assuming that the inputs $s_1 \ldots s_n$ are always either 0 or 1 and the coefficients $\alpha_1 \ldots \alpha_n$ either 1 or $-\infty$ and also by requiring the integral parts of $\beta_1\ \beta_2$ to be the same. We represent the valve element by a circle, and the inputs with a line and an arrow facing towards it, the outputs with lines and arrows facing away (Fig. 1).[1] A coefficient $-\infty$ (<u>inhibitory</u> coupling) is shown with a small circle cutting a large circle (Fig. 2). The smallest total for which an output is obtained (i.e. integral part of β_1 or β_2 plus 1) is shown inside the circle, but is omitted if it is 1. This number we may call the <u>threshold</u>.

When we require coefficients α larger than 1 we may show more than one connection from one source. Negative coefficients may effectively be shown by means of the negation circuit —|—→— which interchanges 0 and 1. Thus in the circuit of Fig. 3 the valve element D will be stimulated (i.e. emit a standard pulse) if either A is stimulated or both B and C are not.

(iv) <u>Trigger circuits</u>. A trigger circuit, which is shown as an ellipse, differs from a limiting amplifier circuit in that once the inputs have reached the threshold so that it emits one pulse, it will continue to emit pulses until it receives an inhibitory stimulus. It is in fact equivalent to a limiting amplifier with a number of excitatory connections from itself with a delay of one unit. Thus for instance the two circuits shown in Fig. 4 are equivalent. We show the trigger circuits with a different notation partly to simplify the drawing and partly because they will in fact be made up from different circuits. There is also another practical difference. The output from a trigger circuit will be a D.C. voltage, so long as it is not disturbed one way or the other, whereas the output from a limiting amplifier with feedback is more or less pulsiform.

(v) <u>Differentiator circuit and change circuit</u>. We sometimes wish to indicate an output from a trigger circuit either at the beginning or the end of its stimulation. This would in fact be done with a capacity resistance 'differentiator' circuit. Such a circuit designed to produce a positive (excitatory) pulse at the beginning will be denoted by —(B)— and one at the end by —(E)—. These

---

[1] Editor's note. The figures begin on page 429.

are understood to be respectively equivalent to the two circuits of Fig. 5. We may also occasionally wish to make connection to a trigger circuit in such a way that stimulus always changes the condition of the trigger circuit, either from stimulation to non-stimulation or vice-versa. This is indicated by a small square at the connection point thus

and is equivalent to Fig. 6.

(vi) <u>The trigger limiter</u>. Sometimes we wish a continuously varying voltage to initiate a train of pulses, the pulses to be synchronous with the clock and to start approximately when the continuous voltage reaches a certain value. All of the pulses that occur must be of the standard or unit size. There must definitely be no half-size pulses possible. The train of pulses may be stopped by pulses from some other source.

This valve element is indicated by a somewhat squat rectangle containing the letters TL. The continuous voltage input is shown as in an excitatory connection and the stopping pulse as an inhibitory connection, as in Fig.7.

(vii) <u>The adder and other examples</u>. We may now illustrate the use of these circuit elements by means of some simple examples.

The simplest circuit perhaps is that for the logical 'or' (cf. p. [395]). In the circuit of Fig. 8 there is an output pulse from the unnamed element if there is one from any one of A, B, C. We shall find it convenient in such cases to describe this element as A v B v C. The circuits of Fig. 9 are self explanatory in view of our treatment of A v B v C.

An adder network is shown in Fig. 10. It will add two numbers which enter along the leads shown on the left in binary from, with the least significant digit first, the output appearing on the right. An input signal from the top will inhibit any output. The method of operation is as follows. The three valve elements on the left all have stimulation from the same three sources, viz. the two inputs and one corresponding to the carry digit from the last figure, which was formed by the element with threshold 2. We can distinguish the four different possible totals 0, 1, 2, 3 according to which of the valve elements are stimulated. We wish to get an output pulse if the total is 1 or 3. This may be expressed as a pulse if the total is 3 or if it is 1 and not 2 or more. If we write T_n to mean 'the total is n or more' the condition is T_3 v $(T_1$ & $\sim T_2)$.

Using our standard networks for A v B and for A & ~B and observing that the three valve elements on the left of the adder are stimulated respectively in the cases T_1, T_2, T_3 we finally obtain the circuit given.

The adder will be shown as a single block as in Fig. II. The input with the inhibiting circle being of course that shown at the top in the complete diagram.

6. Outline of Logical Control.

A simple form of logical control would be a list of operations to be carried out in the order in which they are given. Such a scheme can be made to cover quite a number of jobs, e.g. calculations from explicit formulae, and has been used in more than one machine. However it lacks flexibility. We wish to be able to arrange that the sequence of orders can divide at various points, continuing in different ways according to the outcome of the calculations to date. We also wish to be able to arrange for the splitting up of operations into subsidiary operations. This should be done in such a way that once we have written down how an operation is to be done we can use it as a subsidiary to any other operation.

These requirements can largely be met by having the instructions on a form of erasable memory, such as the delay lines. This gives the machine the possibility of constructing its own orders; i.e. there is always the possibility of taking a particular minor cycle out of storage and treating it as an order to be carried out. This can be very powerful. Besides this we need to be able to take the instructions in an order different from their natural order if we are to have the flexibility we desire. This is sufficient.

It is convenient to divide the instructions into two types A and B. An instruction of type A requires the central arithmetic part CA to carry out certain operations. Such an instruction, translated from its symbolic form into English might run:-

Instruction 491. A. Multiply the content of TS 23 by the content of TS 24 and store the result in TS 25. Then proceed to carry out the next instruction (i.e. No. 492).

Instructions of type [B] merely specify the number of the next instruction.

Instruction 492. B. Proceed with instruction 301.

We must now explain in more detail how it comes about that we can branch the sequence of instructions and arrange for subsidiary operations.

Let us take branching first. Suppose we wish to arrange that at a certain point instruction 33 will be applied if a certain digit is 0 but instruction 50 if it is 1. Then we may copy down these two instructions and then do a little calculation involving these two instructions and the digit D in question. One form the calculation can take is to pretend that the instructions were really numbers and calculate

$$D \times \text{Instruction } 50 + (1 - D) \times \text{Instruction } 33.$$

The result may then be stored away, let us say in a box which is permanently labelled 'Instruction 1'. We are then given an order of type B saying that instruction 1 is to be followed, and the result is that we carry out instruction 33 or 50 according to the value of D.

When we wish to start on a subsidiary operation we need only make a note of where we left off the major operation and then apply the first instruction of the subsidiary. When the subsidiary is over we look up the note and continue with the major operation. Each subsidiary operation can end with instructions for this recovery of the note. How is the burying and disinterring of the note to be done? There are of course many ways. One is to keep a list of these notes in one or more standard size delay lines (1024), with the most recent last. The position of the most recent of these will be kept in a fixed TS, and this reference will be modified every time a subsidiary is started or finished. The burying and disinterring processes are fairly elaborate, but there is fortunately no need to repeat the instructions involved each time, the burying being done through a standard instruction table BURY, and the disinterring by the table UNBURY.

7. External Organs.

(i) General. It might appear that it would be difficult to put information into the calculator and to take it out, on account of the high speeds associated with the calculator, and the slow speeds associated with mechanical devices; but this difficulty is not a real one. Let us consider for instance the output organ. We will allow the mechanical part of the output organ to work at whatever pace suits it, to take its own time in fact. However we will require it to give out signals stating when it is ready to accept information. This signal provides a gate for the feeding of the information out to the output organ, and also signifies to the calculator that it may note that information as recorded and proceed to feed out some more. The preparation for feeding the information out consists merely in transferring it from dynamic storages onto trigger circuits.

In the case of the output arrangements we have the full power of the calculator behind us, i.e. we can do the conversion of the information into the required form as an ITO.* In the case of the input organ we must go more warily. If we are putting the instruction tables into delay lines, then when the power has been turned off all memory will have been effaced, including the instruction tables. We cannot use instruction tables to get the information back, because the instruction tables are not there. We are able to get over this difficulty as will be seen below.

(ii) <u>Output Organ</u>. The output will go on to 32 columns of some Hollerith cards. All the 12 rows may be used. On the receipt of a signal from the calculator a card will begin to pass through a punch or 'reproducer'. Shortly before each row comes into position for punching a signal is sent back to the calculator and trigger circuits controlling the punches are set up. After the punching another signal is sent to the calculator and the trigger circuits are cleared. The reproducer punch also gives a signal on the final exit of the card. The circuit is shown in connection with CA (Fig. 26).

(iii) <u>Input Organ</u>. Let us first describe the action of this without worrying about the difficulty concerning absence of instruction tables. It is very similar to the output organ in many ways. The input is from 32 columns and 12 rows of a Hollerith card. When the calculator is ready a card release signal goes out to the card reader and a card begins to pass through. As each row comes into position for reading a signal is sent back to the calculator, which then prepares to accept the output from the reader at the moment appropriate for sending it to its destination in the delay line. It is assumed that this destination is already decided by the calculator. A signal is sent back to the calculator on the final exit of the card.

Now let us consider what is done right at the beginning. Arrangements are made for setting into CI and CD a certain invariable initial order and IN. These state that the card is to be transferred into a particular delay line, and that the next order is to be taken from a particular spot, which will actually be in this same delay line. The information in this delay line can contain sufficient orders to 'get us started'. The first few orders obeyed will probably be to take in a few more cards. The information on these will later be sorted to its final

---

* ITO = Instruction Table Operation.

destination. When the final instructions are in place it will be as well to 'read them back'.

Actually it has been arranged that the special initial order consists of 0 throughout so that there is no need to set it up.

(iv) <u>Binary-decimal conversion</u>. It is proposed to do binary-decimal and decimal-binary conversion as ITO. This will be appreciably assisted by the fact that short multiplication is a CAO.*

(v) <u>Instruction-table cards</u>. It was explained in connection with the input organ that the instructions would be on cards, of whose columns all but 32 were available for external use. A proposed use of the 80 columns is suggested below, without proper explanation; the explanation comes later.

	Columns
Genuine input	41–72
Repeat of destination	26–40
Popular name of group	1–8
Detail figure (popular)	9–11
Instruction (popular)	12–25
Job number	73–77
Spare	78–80

Of these the genuine input has already been spoken of to some extent, and will be spoken of again further. The job number and the spare columns do not require explanation. The popular data describe the instruction in letters and figures in a manner appropriate for the operator to appreciate quickly if for instance the cards are listed. In this respect we might say that the popular data is like a telephone number Mol 1380 whereas the genuine input is like the pulses used in dialling: indeed we shall probably carry the analogy further and really only distinguish 10 different letters, as is done on automatic exchanges. The popular data have also another important function, which only appears when we consider that the same instructions will be used on quite different jobs. If we were just to number the instructions serially throughout all the instructions ever used on any job, then, in the set of instructions actually used in any particular job there would be large gaps in the numbering. Suppose now that these instructions were stored in the DS with positions according

---

* CAO = Central Arithmetic Operation.

to their numbers there would be a lot of wasted space, and we should need elaborate arrangements for making use of this space. Instead, when a new job appears we take the complete set of cards involved and make a new copy of each of them; these we sort into the order of popular group name and detail figure. We then renumber them consecutively in the binary scale. This number goes into the columns described as 'repeat of destination'. The renumbering may be done either with a relay counter attached to a collator, or by interleaving a set of master cards with the binary numbers in serial order. To complete the process we have to fill in other instruction numbers in binary form into the genuine input, e.g. if an instruction in popular form were "... and carry out instruction Potpan 15" the genuine input will have to be of form " ... and carry out instruction 001101...1" where 001101...1 is the new number given to Potpan 15 in this particular job. This is a straightforward sorting and collating process.

It would be theoretically possible to do this rearrangement of orders within the machine. It is thought however that this would be unwise in the earlier stages of the use of the machine, as it would not be easy to identify the orders in machine form and popular form. In effect it would be necessary to take an output from the calculator of every order in both forms.

8. Scope of the Machine.

The class of problems capable of solution by the machine can be defined fairly specifically. They are those problems which can be solved by human clerical labour, working to fixed rules, and without understanding, provided that

(a) The amount of written material which need be kept at any one stage is limited to the equivalent of 5,000 real numbers (say), i.e. about what can conveniently be written on 50 sheets of paper.
(b) That the human operator, doing his arithmetic without mechanical aid, would not take more than a hundred thousand times the time available on the calculator, this figure representing the ratio of the speeds of calculation by the two methods.
(c) It should be possible to describe the instructions to the operator in ordinary language within the space of an ordinary novel. These instructions will not be quite the same as the instructions which are normally given to a computer, and which give him credit for intelligence. The instructions must cover every possible eventuality.

Let us now give real examples of problems that do and problems that do not satisfy these conditions.

Problem 1. Construction of range tables. The complete process of range-table construction could be carried out as a single job. This would involve calculation of trajectories by small arcs, for various different quadrant elevations and muzzle velocities. The results at this stage would be checked by differencing with respect to other parameters than time. The figures actually required would then be obtained by interpolation and these would finally be rearranged in the most convenient form. All of this could in theory be done as a single job. In practice we should probably be wiser to do it in several parts in order to throw less responsibility on to the checking arrangements. When we have acquired more practical experience with the machine we will be bolder.

It is estimated that the first job of this kind might take one or two months, most of which would be spent in designing instruction tables. A second job could be run off in a few days.

Problem 2. To find the potential distribution outside a charged conducting cube. This is a problem which could easily be tackled by the machine by a method of successive approximations; a relaxation process would probably be used. In relaxation processes the action to be taken at each major step depends essentially on the results of the steps that have gone before. This would normally be considered a serious hindrance to the mechanisation of a process, but the logical control of the proposed calculator has been designed largely with such cases in view, and will have no difficulty on this score. The problem proposed is one which is well within the scope of the machine, and could be run off in a few minutes, assuming it was done as one of a sequence of similar problems. It is quite outside the scope of hand methods.

Problem 3. The solution of simultaneous linear equations. In this problem we are likely to be limited by the storage capacity of the machine. If the coefficients in the equations are essentially random we shall need to be able to store the whole matrix of coefficients and probably also at least one subsidiary matrix. If we have a storage capacity of 6400 numbers we cannot expect to be able to solve equations in more than about 50 unknowns. In practice, however, the majority of problems have very degenerate matrices and we do not need to store anything like as much. For instance problem (2) above can be transformed into one requiring the solution of linear simultaneous equations if we replace the continuum by a lattice. The coefficients in these equations are very systematic and mostly zero. In this problem we should be limited not by the storage required for the matrix of coefficients, but by that required for the solution or for the approximate solutions.

Problem 4. To calculate the radiation from the open end of a rectangular wave-guide. The complete polar diagram for the radiation could be calculated, together with the reflection coefficient for the end of the guide and interaction coefficients for the various modes; this would be done for any given wavelength and guide dimensions.

Problem 5. Given two matrices of degree less than 30 whose coefficients are polynomials of degree less than 10, the machine could multiply the matrices together, giving a result which is another matrix also having polynomial coefficients. This has important applications in the design of optical instruments.

Problem 6. Given a complicated electrical circuit and the characteristics of its components, the response to given input signals could be calculated. A standard code for the description of the components could easily be devised for this purpose, and also a code for describing connections. There is no need for the characteristics to be linear.

Problem 7. It would not be possible to integrate the area under a curve, as the machine will have no appropriate input.

Problem 8. To count the number of butchers due to be demobilised in June 1946 from cards prepared from the army records. The machine would be quite capable of doing this, but it would not be a suitable job for it. The speed at which it could be done would be limited by the rate at which cards can be read, and the high speed and other valuable characteristics of the calculator would never be brought into play. Such a job can and should be done with standard Hollerith equipment.

Problem 9. A jig-saw puzzle is made up by cutting up a halma-board into pieces each consisting of a number of whole squares. The calculator could be made to find a solution of the jig-saw, and, if they were not too numerous, to list all solutions.

This particular problem is of no great importance, but it is typical of a very large class of non-numerical problems that can be treated by the calculator. Some of these have great military importance, and others are of immense interest to mathematicians.

Problem 10. Given a position in chess the machine could be made to list all the 'winning combinations' to a depth of about three moves on either side. This is not unlike the previous problem, but raises the question 'Can the machine play chess?' It could fairly easily be made to play a rather bad game. It would be bad because chess requires intelligence. We stated at the beginning of this section that the machine should be treated as entirely without intelligence. There are indications however that it is possible to make the machine display intelligence at the risk of its making occasional serious mistakes. By following up this aspect the machine could probably be made to play very good chess.

9. Checking.

It will be almost our most serious problem to make sure that the calculator is doing what it should. We may perhaps distinguish between three kinds of error.

(1) Permanent faults that have developed in the wiring or components, e.g. condensers that have become open circuit.
(2) Temporary errors due to interference, noise reaching unexpected levels, unusual combinations of voltages at some point in the circuit, etc.
(3) Errors due to the use of incorrect instruction tables, or even due to mistaken views as to what the circuit should do.

It will be our intention to install monitoring circuits to detect errors of form (1) fairly soon. The ideal to aim at should be that each conceivable form of failure would give a different indication on the monitor. In practice we should probably simply localise the error to some part, e.g. an adder, which could be changed and then examined at leisure.

Errors of type (2) should not occur when the apparatus is in proper working order, however when a component is beginning to age its deficiencies will often show themselves first in this sort of way. For instance, if the emission of a valve in a Kipp relay circuit is beginning to fail it will eventually not pass on any of the pulses it should, but this will begin with some occasional failures to react. The worst of this can probably be eliminated by frequent test runs in which the conditions of H.T. volts, interference, etc., are all modified in a way calculated to accentuate the deficiencies of the components. Those which are rather down at heel may then be removed, and when the conditions are restored to normal there should be a good margin of safety. We cannot of course rely on this 100%. We need a second string. This will be provided by a variety of checks of the types normally employed in computing, i.e. wherever we can find a simple identity which should be satisfied by the results of our calculations we shall verify it. For instance, if we were multiplying polynomials algebraically we should check by taking a particular value for the variable. If we were calculating the values of an analytic function at equal intervals we should check by differencing. Most of these checks will have to be set up as part of the instruction tables, and the appropriate action to be taken will also be put into them. A few checks will be made part of the circuit. For instance, all multiplications and additions will be checked by repeating them modul[o] 255.

Incorrect instruction tables (3) will often be shown up by the checks which have been put into these same instruction tables. We may also apply a special check whenever we have made up a new instruction table, by comparing the results with the same job done by means of a different table, probably a more straightforward but slower one. This should eliminate all errors on the part of the mathematicians, but would leave the possibility of lost cards, etc., when the table is being used a second time. This may perhaps be corrected by running a test job as soon as the cards have been put into the machine.

There are three chief functions to be performed by the checking. It must eliminate the possibility of error, help to diagnose faults, and inspire confidence. We have not yet spoken at all of this last requirement. It would clearly not be satisfactory if the checking system in fact prevented all errors, but nobody had any confidence in the results. The device would come to no better end than Cassandra. In order to inspire confidence the checking must have some visible manifestations. Certainly whenever a check fails to work out the matter must be reported by the machine. There would not be time for all checks which do work out to be reported, but there could be a facility by which this could be laid on temporarily at moments of shaken confidence. Another facility which should have a good effect on morale is that of the artificial error. By some means the behaviour of the machine is disturbed from outside, and one waits for some error to be reported. This could be managed quite easily. One could arrange to introduce an unwanted pulse at any point in the circuit.

In fact of course we cannot do very much about checking until the machine is made. We cannot really tell what troubles of this kind are in store for us, although one can feel confident that none of them will be insurmountable. We can only prepare against the difficulties we can foresee and hope that they will represent a large percentage of the whole.

10. Time-table, Cost, Nature of Work, Etc.

The work to be done in connection with the machine consists of the following parts:

(1) Development and production of delay lines.
(2) Development and production of other forms of storage.
(3) Design of valve-elements.
(4) Final schematic circuit design of LC and CA.
(5) Production of the electronic part, i.e. LC and CA.
(6) Making up of instruction tables.
(7) External organs.
(8) Building, power supply cables, etc.

(1) Delay lines have been developed for R.D.F. purposes to a degree considerably beyond our requirements in many respects. Designs are available to us, and one such is well suited to mass production. An estimate of £20 per delay line would seem quite high enough.

(2) The present report has only considered the forms of storage which are almost immediately available. It must be recognized however that other forms of storage are possible, and have important advantages over the delay line type. We should be wise to occupy time which falls free due to any kind of hold-up by researching into these possibilities. As soon as any really hopeful scheme emerges some more systematic arrangement must be made.

We must be ready to make a change over from one kind of storage to another, or to use two kinds at once. The possibility of developing a new and better type of storage is a very real one, but is too uncertain, especially as regards time, for us to wait for it; we must make a start with delay lines.

(3) Work on valve element design might occupy four months or more. In view of the fact that some more work needs to be done on schematic circuits such a delay will be tolerable, but it would be as well to start at the earliest possible moment.

(4) Although complete and workable circuits for LC and CA have been described in this report these represent only one of a considerable number of alternatives. It would be advisable to investigate some of these before making a final decision on the circuits. Too much time should not however be spent on this. We shall learn much more quickly how we want to modify the circuits by actually using the machine. Moreover if the electronic part is made of standard units our decisions will not be irrevocable. We should merely have to connect the units up differently if we wanted to try out a new type of LC and CA.

(5) In view of the comparatively small number of valves involved the actual production of LC and CA would not take long; six months would be a generous estimate.

(6) Instruction tables will have to be made up by mathematicians with computing experience and perhaps a certain puzzle-solving ability. There will probably be a great deal of work of this kind to be done, for every known process has got to be translated into instruction table form at some stage. This work will go on whilst the machine is being built, in order to avoid some of the delay between the delivery of the machine and the production of results. Delay there must be, due to the virtually inevitable snags, for up to a point it is better to let the snags be there than to spend such time in design that there are none (how many decades would this course take?). This process of constructing instruction tables should be very fascinating. There need be no real danger of it ever becoming a drudge, for any processes that are quite mechanical may be turned over to the machine itself.

The earlier stages of the making of instruction tables will have serious repercussions on the design of LC and CA. Work on instruction tables will therefore start almost immediately.

(7) Very little need be done about the external organs. They will be essentially standard Hollerith equipment with special mounting.

(8) It is difficult to make suggestions about buildings owing to the great likelihood of the whole scheme expanding greatly in scope. There have been many possibilities that could helpfully have been incorporated, but which have been omitted owing to the necessity of drawing a line somewhere. In a few years time however, when the machine has proved its worth, we shall certainly want to expand and include these other facilities, or more probably to include better ideas which will have been suggested in the working of the first model. This suggests that whatever size of building is decided on we

should leave room for building-on to it. The immediate requirements are:

Room for 200 delay lines. These each require about 6 inches of wall space if they are to be individually accessible, and if this is partly provided by cubicle construction 300 square feet is probably a minimum. To this we might add another 100 square feet for the temperature correction arrangements.

Space for LC and CA. This is difficult to estimate, but 5 eight foot racks might be a reasonable guess and would require another 200 square feet or more. In the same room we would put the input and output organs which might occupy 40 square feet. We should also provide another 100 square feet for operators tables, etc. 400 square feet would not be unreasonable for this room.

Card storage room. We would probably keep a stock of about 100,000 cards, a very insignificant number by normal Hollerith standards. 200 square feet would be quite adequate.

Maintenance workshop. We would do well to be liberal here. 400 square feet.

This total of 1400 square feet does not allow for the planning of operations, which would probably be done in an office building elsewhere, nor for the processing of Hollerith cards which will probably be done on machinery already available to us.

Cost. It appears that the cost of the equipment will not be very great. An estimate of £20 per delay line would be liberal, so that 200 of these would cost us £4000. The valve equipment at £5 per inch of rack space might total £5000. The power supply might cost £200. The Hollerith equipment would be hired, which would be advantageous because of the danger of it going out of date. The capital cost of such Hollerith equipment even if bought would not exceed £2,000. With this included the total is £11,200.

PART II.

Technical Proposals.

11. Details of Logical Control.

In this section we shall describe circuits for the logical control in terms of the circuit elements introduced in § 5. It is assumed that § 5, 6 are well understood.

The main components of LC are as follows:-

(1) A short storage (like a TS) called <u>current data</u> CD. This contains nothing but the appropriate <u>instruction number</u> IN, i.e. the position of the next instruction to be carried out.

(2) A short storage called <u>current instructions</u> CI. This contains the instruction being or about to be carried out.

(3) A tree for the selection of a particular delay line, with a view to finding a particular instruction.

(4) Timing system for the selection of a particular minor cycle from a delay line.

(5) Timing system for the selection of particular pulses from within a minor cycle.

(6) Arrangements for controlling CA, i.e. for passing instructions on to CA.

(7) Arrangements for the continual change of the contents of CD, CI.

(8) Timing arrangements for LC itself.

(9) Starting device.

Let us first describe the starting device. This merely emits pulses synchronously with the clock from a certain point onwards, on the closing of a switch manually. The switch causes a voltage to rise and this eventually operates a trigger limiter. This starting mechanism sets a pulse running round a ring of valve elements providing the timing within a minor cycle (Fig. 12, 13).

In order to check that this circuit is behaving we compare $P32$ with a signal which should coincide with it and which is obtained in another way, stimulating an SOS signal when there is failure. This forms one of the monitoring devices. We are not showing many of them in the present circuits (Fig. 14).

The timing system for the selection of minor cycles is quite simple, consisting chiefly of a 'slow counter' SCA, which counts up to 255 in the scale of 2, keeping the total in a delay line of length 8. The pulses counted are restricted to appearing at intervals which are multiples of eight. As shown (Fig. 15) it is counting the pulses $P10$. The suppression of the outputs at $P9$ prevents undesirable carries from the most significant digit to the least.

The information in CD and CI being in dynamic (time) form is not very convenient for control purposes. We therefore convert this information into static form, i.e. we transfer it on to trigger circuits (Fig. 16).

It will be convenient to make use of a symbolic notation in connection with the valve circuits. We write A & B (or manuscript A + B) to mean 'A and B'. If A and B are thought of as numbers 0 or 1 then A & B is just AB. We write A v B for 'A or B'. With numbers A v B is $1 - (1 - A)(1 - B)$. We also write

−A (manuscript ∼A) for 'not A' or $1 - A$. Other logical symbols will not be used. Where a whole sequence of pulses is involved, it is to be understood that these operations are to be carried out separately pulse by pulse. We shall combine these symbols with the symbol + which refers to the operations of the adder. Thus for example $(A + (P3 \text{ v } P4) \& -P5$ means that we take the signal A and add to it a signal consisting of pulses in positions 3 and 4 and nowhere else (addition in the sense of the adder circuit), and that we then suppress any pulses in position 5, as in Fig. 17. We will also abbreviate such expressions as P5 v P6 v P7 v ... P19 to P5–19, and expressions such as A & P14–18 to A14–18.

In circuit diagrams we have the alternatives of showing the logical combinations by formulae or by circuits. There is little to choose but there may be something to be said for an arrangement by which purely logical combination is not shown in circuit form, in order that the circuits may bring out more clearly the time effects.

We have agreed that there shall be two kinds of instructions, A and B. These are distinguished by CI3. The standard forms for the two types of instructions are:

<u>Type A.</u> Carry out the CA operations given by digits CI5–32, and construct a new CD according to the equation $CD = (CD' + P19) \& -P17$.

<u>Type B.</u> Construct a new CD according to the equation $CD = CI17–32$. Pass the old CD into TS 13.

CD' here represents the old CD. The significance of the formula for CD in case A is this. Normally it is intended that after an operation of type A the next instruction to be followed will be that with the next number, and it might be supposed therefore that the formula $CD = CD' + P17$ would apply. Actually we deviate from this simple arrangement in two ways. Firstly we find it convenient to have a facility by which an instruction may be taken from a TS, viz. TS 6: this has considerable time saving effects. The convention is that a digit 1 in column 17 indicates that the next instruction is to be taken from TS 6. This will involve our having only the digits CI18–32 available to indicate normal positions for instructions and would suggest that the formula should be $CD = CD' + P18$. However if we did this we should always be obliged to have orders of type B in TS 6, for if we had an order of type A we should find that we had to go on repeating that order. If however we have the formula $CD = (CD' + P18) \& -P17$ we can obey an instruction in TS 6 and then revert to the instruction given by CI18–32; a much more convenient arrangement.

It remains to explain why we have P19 rather than P18. This is due to the fact that we wish to avoid the necessity of waiting a long time for our instructions. If the equation were the one with P18 it would mean that the next instruction to be obeyed, after one of type A, is always adjacent to it in time. This would mean that even with the shortest CA operations the next instruction would have gone by before we were ready to apply it; we should always just miss the boat. By putting P19 instead of P18 we give ourselves an extra minor cycle of time which is normally just what we need. In order that the consecutive instructions may be consecutively numbered in spite of this it is best to adopt a slightly unconventional numbering system for the minor cycles (see Fig. 19).

A number of trigger circuits are employed to keep track of the stages which the various processes have reached at any moment. The most important of these are listed below with a short description of the functions of each.

<u>OKCI</u>. This is stimulated when the new instruction has been found and is available at the input of CI, and the CA operations belonging to the last instruction have been carried out. Stimulation begins simultaneously with stimulation of P1, and ends on a P32. The end of OKCI has to wait for the gating of CD, indicating that the new CD is available at its input.

<u>OKCA</u>. Only applies in case A and indicates that the CA operations have been finished.

<u>OKLK</u>.[2] Indicates that we may now begin to look for the next instruction with a view to putting it into CI. It is stimulated when OKCI is extinguished, and is itself extinguished when the new CI has been found.

We may now describe the time cycle of LC. Let us begin at the point where OKLK is stimulated indicating that the search for the new CI may now begin, because we have finished with the old one and information for finding the new one is now available in CD. The new CI is determined by digits 17–32 of CD. Of these digits 23–32 determine the delay line and 18–22 determine the minor cycle within the delay line. A digit 1 in column 17 indicates that the order is to be taken from TS 6 instead of from the longer delay lines. This digit is erased whenever we obey an instruction of type A. Digits 23–32 are set up on trigger circuits and operate via trees as described below. Digits 18–22 determine the time at which we must take the output of the delay line. We compare these digits with the output of the slow counter SCA (Fig. 15) and when they agree we know that the right moment has come. It is convenient

---

[2] Editor's note. Corrected from the mis-typed 'OKCK'.

to arrange that the slow counter is always one minor cycle ahead of time, so as to give us time to organise ourselves before taking the required output. As has been mentioned the order of the digits in CD is arranged rather unconventionally in order to put consecutively numbered minor cycles in alternate positions; this has time saving effects. The required minor cycle now passes into CI and the signal OKSS is given; OKLK is suppressed. When the CA operations belonging to the last instruction have been finished OKCA is stimulated and with it OKCI. We are now able to initiate any new CA operations (case A) and to set up the new CD. When this has been done we have finished with CI and suppress OKCI, which automatically stimulates OKLK beginning the cycle over again (Figs. 22, 22a).

The digits 23–32 determine the delay line required. This amounts to 10 digits and will certainly be adequate for our present programme. Treeing is done in two stages, going first through trees for three or four digits only. These are TRA 000 ... TRA 111, TRB 000 ... TRB 111, TRC 0000 ... TRC 1111. These number 32 valve elements. At the second stage there are 1024 valve elements TREECI 0000000000 ... TREECI 1111111111. The connections are shown for TREECI 1011101101. The connection from CI17 prevents any of the TREECI elements being stimulated when CI17 is stimulated. This is required to deal with the case where the next order is taken from TS 6 and not from the delay lines (Fig. 20).

It is very probable that some other form of tree circuit, not capable of being drawn in terms of our valve elements, will be used, and the same will apply to many parts of the circuit. It is thought worthwhile however to draw these circuits, if only to clarify what it is intended the circuits should do.

We have a similar tree system for the selection of temporary storages.

12. Detailed Description of the Arithmetic Part (CA).

We shall divide the CA operations into a number of types. We shall make provision for 16 types, but for the present will only use nine. The types are distinguished by digits CI5–8.

Type K. Pass the content of TS 6 into a given minor cycle.

Type L. Pass the content of a given minor cycle into TS 6.

Type M. Pass the content of a given TS into TS 6.

Type N. Pass the content of TS 6 into a given TS other than TS 4 or TS 5, or TS 8 or TS 1.

<u>Type O</u>. Pass the content of the first 12 minor cycles of a given DL out onto a card via the reproducer.

<u>Type P</u>. Pass the content of the card at present in the card reader on to a given DL.

<u>Type Q</u>. Pass CI17–32 into TS 6.

<u>Type R</u>. Various logical operations and others yielding results forming one minor cycle, to be performed on the contents of TS 9 and TS 10 and transferred to TS 8.

<u>Type S</u>. Arithmetical operations yielding a result requiring more than one minor cycle for its retention. Results go into TS 4 and TS 5.

<u>Type T</u>. Stimulate a given valve element.

A trigger circuit is associated with each type. With the exception of Q these are all excited for a period consisting of a number of complete minor cycles beginning with a P1 and ending with a P32.

The main components of CA are the 32 temporary storages TS 1–32. Of these TS 1–12 have some special duties.

TS 1 is used to carry the retiring data, i.e. the CD which applied just before the last instruction of type B.

TS 2 and TS 3 contain the arguments for the purely arithmetical operations, or most of them, and for the logical operations.

TS 4 and 5 contain the results of the arithmetical operations. They are frequently connected up in series to form a DL 64. This is because the results of most of the arithmetical operations are sequences of more than 32 but not more than 64 digits.

TS 6 is used as a shunting station for the transfer of information from place to place.

TS 7 is used to carry the digits of a number m when it is proposed to multiply by 2^m.

TS 8 is used to carry the result of logical operations and other operations not requiring more than one minor cycle.

TS 9 and TS 10 are the inputs for the logical operations.

TS 11 will usually be used in connection with error calculations, and accordingly has a special role in the production of multipliers.

TS 12 is used for the timing in 'automatic' multiplication and for the selection of unusual combinations of digits in the multiplier. The word 'automatic' is used because of an analogy from desk machines.

To decide between types K to T we use CI5–8. Digits 5, 6, 7 are treed out to the valve elements TRG 000 ... TRG 111, as in Fig. 23. These tree elements are each associated with two types, which are distinguished by CI8. Thus TRG 000 would be identical with K v L if it were not for timing. For this timing we introduce CATIM which is to be stimulated during the appropriate time in CA operations. K v L is identical with TRG 000 & CATIM (Fig. 24).

In case K we pass the output of TS 6 to COMMIN and hence to the inputs of all the delay lines. We gate the appropriate one of these at the appropriate time, given by TIMCA by comparison of the output of the slow counter SCA with CI.

In case L we do somewhat similarly, passing the appropriate output to COMMOUT and thence to the input of TS 6 at the appropriate time given by TIMCA.

In case M we gate the appropriate output and pass into TS 6.

In case N we pass the output of TS 6 to the inputs of the other TS, only gating the one required.

In case O the first effect is to set the mechanism in motion to pass a card through the reproducer. By means of a commutator arrangement or otherwise the reproducer sends back a series of pulses which indicate the times when the reproducer punches are ready to accept current. In the circuit diagram (Fig. 25) two sets of pulses are shown which are intended to mark the beginnings and ends of these periods. They may be separately provided by the reproducer, or one may be derived from the other by delaying or otherwise. The two sets of pulses each control trigger limiters connected up so as to extinguish one another. (Do not confuse this with the two mutually extinguishing triodes that will normally form part of a trigger circuit or trigger limiter.) One of the trigger limiters TIMOUTCARD stimulates the trigger circuit OUTIM on the first admissible P10. A pulse on the stimulation of OUTIM goes into a slow counter SCB and enables us to keep track of the number of rows of the card that have been punched. The content of SCB is compared with that of SCA and when they agree we know that the minor cycle which we wish to pass out is now available, and TIMCA is accordingly stimulated. TIMCA and OUTIM together permit COMMOUT to pass out to the trigger circuits OUT 1 ... OUT 32 on which it is set up statically and controls the punches.

On the final exit of the card the reproducer sends back a signal to the calculator, which, in combination with O operates a trigger limiter CARDEXOUT. This suppresses CATIM and hence O. CARDEXOUT has feedback to suppress itself, and this will be successful because O will have been suppressed by the time it comes to act.

The behaviour in case P (input) is very similar. The chief difference is that whereas OUTIM was used to gate the output from the calculator INTIM is used to gate the input.

It should be noticed that a completely blank instruction has a definite meaning, viz. to pass the material on the card in the reader into DL 0000000000.

In Fig. 27 TS 01101 typifies any of the TS as regards output connections shown on other diagrams. It is also typical as regards input connections, except as regards TS 4, 5, 8, 1, which have no input connections except those shown on other diagrams.

In the case of operations of type R we shall calculate all of the expressions involved and select them by means of tree elements, digits 18 to 23 being used. The operations so far are:

<u>Digits 000000</u>   TS 8 = TS 9 & TS 10.

<u>Digits 001000</u>   TS 8 = TS 9 v TS 10.

<u>Digits 010000</u>   TS 8 = −TS 10.

<u>Digits 011000</u>   TS 8 = (TS 9 & TS 10) v (−TS 9 & −TS 10).

<u>Digits 100000</u>   TS 8 = o.

As we shall have very much to say about type S we shall make a few remarks first about type T. In order to be able to obtain a rather direct access from the instructions to the valves we shall introduce a number of valve elements which can be stimulated to order. We may have 64 of these, say FLEX 000000 to FLEX 111111. The circuit will be simply as shown in Fig. 31. It is intended that the outputs of these valve elements should be connected in various ways into the circuit when it is desired to try out new circuit arrangements. It is thought that they may often provide means for doing things simply which could be done lengthily as an ITO. To an extent this represents a compromise between the new system of 'control by paper' and the old plugboard and soldering-iron techniques.

We shall also describe the timing arrangements before passing on to type S. We have already mentioned CATIM which determines the timing but we have still to mention what controls CATIM. CATIM is stimulated as soon as the first P1 appears after the signal A, or, in case Q, the first P17. It is extinguished by a variety of means. In cases K and L it is extinguished by the ending of TIMCA indicating that the required minor cycle has just passed through. In cases M, N, R, T, it is only permitted to last for one minor cycle. In case Q it is also only allowed to last for half a minor cycle. In cases O, P the extinguishing signal is CARDEX, which is given by the card reproducer or reader on the final exit of the card, via a trigger-limiter. In case S the signal comes from FINARITH.

The facilities provided under type S are not easily enumerated, because they do not consist of a number of different operations stimulated by different tree valve elements, as for instance applies in the case of the logical processes. Rather they are to be thought of as one process which can be modified in various ways. The standard process always involves converting the content of TS 4 and TS 5 into 'series form', i.e. instead of connecting the outputs of TS 4 and TS 5 to their own inputs they are connected to each other's. When they are so connected their content will be described as the 'partial sum'. Some quantities are then added to or subtracted from the partial sum. If the quantity is to be added then POS is stimulated, otherwise they are subtracted. We may if we wish cancel the original partial sum before adding, in which case we must stimulate CANCEL for a period of two minor cycles. The quantity to be added or subtracted is expressible as the product of a quantity known as the 'multiplicand' and an integer which may be taken to lie in the range -7 to 15, positive values being the more normal. The multiplicand may be taken from TS 3 or from the partial sums register itself. This latter case is convenient for the purpose of multiplying the partial sum by a small integer without a complicated series of previous transfers; if the multiplicand is taken from the partial sums register then SELF is stimulated. The multiplier may also be taken from a variety of sources. It may be taken from TS 2 or from CI or from TS 11, and we accordingly stimulate NOR, GIV or ERR. The multiplier consists of four consecutive digits from whichever source is chosen. The choice of the digits is made by means of a choice of one of the pulses P1 to P32 to enter on a certain line (DIGIT). At present it is suggested that in case NOR this should be P1, resulting in the use of digits $1, 2, 3, 4$, in case [GIV] it should be P23 resulting in the use of digits $23, 24, 25, 26$, in case ERR1 it should be P10, and in case ERR2 it should be P14. In case DIFF these arrangements are to be overridden and the pulse will be stored in TS 12 and taken from there.

In case AUTO the above fundamental process is repeated eight times. In each repetition the multiplicand is taken from TS 3, but it is modified each time by multiplication by 2^4, this effect being obtained by allowing it to circulate in a DL 34 during AUTO. We also wish to take different digits of the multiplier at each repetition of the process; this is done by taking our pulse from TS 12 but allowing it to circulate in a DL 34 also. Facilities are also provided for multiplying the partial sum by a power of 2. Although the circuits are arranged so that this could be combined with other operations, it is not intended that this should be done. The facility consists in enabling the partial sums to be delayed by any time up to 63 and passed through for a period of 2 or 3 minor cycles as desired. The amount of delay is taken from digits 1–5 of TS 7. We stimulate ROTATE 2 or ROTATE 3 according as we wish the rotation to last for 2 or 3 minor cycles.

It may be as well to describe how some rather definite operations are done.

<u>Addition</u>. We do not have a facility for addition of two given numbers so much as for the addition of a given number into the partial sum. To add the content of TS 3 into the partial sum we must stimulate S, POS, GIV, and must also set up the number 1 in columns 24–27. The multiplicand is then TS 3 and the multiplier is 1.

<u>Subtraction</u>. As addition but we do not stimulate POS.

<u>Short multiplication (A)</u>. To multiply TS 3 by 6 (say) proceed as for addition with 0110 in columns 24–27 instead of 1000. We shall very likely also want to cancel the original content of the partial sums register and therefore stimulate CANCEL.

<u>Short multiplication (B)</u>. To multiply the partial sum by 6 we must stimulate S, POS, CANCEL, SELF, GIV, and set up 0110 in CI24–27.

<u>Short multiplication (C)</u>. As B but do not cancel and put 1010 in CI24–27.

<u>Short multiplication with addition</u>. We wish to multiply TS 3 by TS 2 and add into the partial sum. We stimulate POS, NOR, AUTO, DIFF.

<u>Long multiplication with subtraction</u>. If we wish to subtract from the partial sum we do not stimulate POS.

<u>Division</u> is an ITO and will probably be carried out by means of the recurrence relation $u_0 = 3/4$, $u_n + 1 = u_n(2 - au_n)$. The limit of the sequence u_n is a^{-1} provided $1 < a < 2$.

The appropriate instructions for these operations will be found in Fig. 37.

The content of TS 2 or TS 3 is best considered to be a binary integer, i.e. that the least significant digit is in the units position. We must also consider that the most significant digit has reversed sign. The least significant digit appears at time P1 and the most significant at P32. In the partial sums register similarly the least significant digit is to be considered to be in the units position and the most significant to have reversed sign and to appear 63 pulses later. In order to keep track of which part of the partial sum is available at any moment we have a signal ODD which is stimulated during the first minor cycle of the stimulation of S, and thereafter in alternate minor cycles so long as S is stimulated. When the multiplicand is taken from TS 3 we have to make some slight modifications to it before it is in suitable condition for adding into the partial sum. We have to convert the periodic signal with period 32 or 34 into a sequence of 64 digits of which 32 form the original content of TS 3, and the rest is a sort of padding. We may call the 32 digits the genuine digits. Those digits of padding which are less significant than the genuine digits are to be all zero, those which are more significant are to be the same as the most significant genuine digit. It will be seen that this modified multiplicand MUCAND 2 has the same meaning as the original multiplier, but expressed in the code which is appropriate to the partial sum, and multiplied by the power of 2 which is required at the time. It may be necessary to change the sign of this multiplicand, if POS was not stimulated. A simple circuit will do this (Fig. 34).

Owing to the fact that the partial sums register is a closed cycle of 64 there is a danger of carries from the most significant digit on to the least significant. This has to be prevented, and it is done by suppressing the carry in the appropriate adder at the time P32 & −ODD. This is shown by an inhibiting connection on to the adder.

The detailed correctness of the circuits is best verified by working through various particular cases. It is necessary to work several different ones in order to bring out the various different special points involved. In Fig. 35 the preliminaries to a long multiplication have been worked. This shows the setting up of the new CI and the transfer of digits to the valve elements Z1, Z2, Z3, Z4. It brings out the point of adding 2 rather than 1 to the CD in cases A, B, for we are just in time to catch the next instruction. The final stages of the multiplication are shown in Fig. 36. Here it has been assumed that the minor cycle is of length 16, in order to reduce the space occupied by the working.

13. Examples of Instruction Tables.

In this chapter a short account of the paper technique of using the machine will be given. I shall try to give some idea of what the instruction tables for a job will be like and how they are related to the job and to the machine. This account must necessarily be very incomplete and crude because the whole project as yet exists only in imagination.

Each instruction will appear in a number of different forms, probably three or four.

Machine form. When the instruction is expressed in full so as to be understood by the machine it will occupy one minor cycle. This we call machine form.

Permanent form. The same instruction will appear in different machine forms in different jobs, on account of the renumbering technique as described in pp. [385–6]. Each of these machine form instructions arises from the permanent form of the instruction. These permanent forms are on Hollerith cards and are kept in a sort of library.

Popular form. Besides the cards we need some form of the table which can be easily read, i.e. is in the form of print on paper rather than punching. This will be the popular form of the table. It will be much more abbreviated than the machine form or the permanent form, at any rate as regards the descriptions of the CAO. The names of the instructions used will probably be the same as those in the permanent form.

In addition to these we must recognise the 'general description' of a table. This will contain a full description of the process carried out by the machine acting under orders from this table. It will tell us where the quantities or expressions to be operated on are to be stored before the operation begins, where the results are to be found when it is over and what is the relation between them. It will also tell us other important information of a rather dryer kind, such as the storages that must be left vacant before the operation begins, those that will get cleared or otherwise altered in the process, what checks will be made, and how various possible different outcomes of the process are to be distinguished. It is intended that when we are trying to understand a table all the information that is needed about the subsidiaries to it should be obtainable from their general descriptions.

The majority of actual instruction tables will consist almost entirely of the initiation of subsidiary operations and transfers of material. It should be recognised however that the time spent will be in quite different proportions. The three most time consuming operations are multiplication, waiting for material in long delay lines, and transfers of material. In some jobs the input and output of material may also be very time-consuming.

In order to give a fairly complete picture of what the tables are like I am giving examples of two tables, of which one is elementary and does not involve subsidiaries; the other is a more advanced table and consists largely of such orders. Besides these I have added a number of general descriptions of tables.

The fundamental table chosen is INDEXIN, used for finding a minor cycle whose position has been written down in a particular place.

In these tables DL m, n will denote the nth minor cycle of DL m.

INDEXIN (General Description). The minor cycle whose position is described in digits 17–32 of TS 27 is transferred to TS 28. The contents of TS 2, 3, 4, 5, 6, 8, 9, 10 get altered in the process.

Now follows the popular form of the table.

INDEXIN.

1	Q, 0000,0100,0000,0000	2
2	TS 6 – TS 2	3
3	ADD 'A'	4
4	ROTATE 16	5
5	TS 4 – TS 6	6
6	TS 6 – TS 9	7
7	TS 27 – TS 6	8
8	TS 6 – TS 10	9
9	OR	10
10	TS 8 – TS 6	11
11	B,1, INDEXIN 11	
12	TS 6 – TS 28	13
13	B, BURY	

The first column gives the popular form of the name of the instruction, and the last column that of the next instruction to be followed. In most cases this could in theory be omitted because of the instructions being of type A. When the instructions are of type A the middle column describes them in abbreviated form. For instance TS 6 – TS 3 describes the operation of transferring the content of TS 6 into TS 3. Expressions of form Q, . . . mean an instruction of type Q, and the expression after the comma describes what is in columns 17–32. ADD 'A' is to mean 'Add TS 2 into TS 4 cancelling the partial sums', ROTATE 16 means 'Rotate the content of TS 4, TS 5 forwards 16 places', OR is a logical operation.

The expression B, 1, INDEXIN 11 is intended to stand for B in column 3, 1 in column 17 and INDEXIN 11 in columns 17–32.

<u>Outline of operation</u> (INDEXIN). From 1 to 10 we are constructing the instruction which tells us to make the appropriate transfer and putting that instruction into TS 6. The instruction B, 1, INDEXIN 11 requires us to carry out the instruction in TS 6. The new IN formed will be 0, INDEXIN 12 so that we then continue with instruction INDEXIN 12.

The table for INDEXIN is shown in full in Fig. 38.

We use the convention that no digit is shown if the value of the digit is not significant. Both 0 and 1 are shown if either value is possible, and significant.

<u>DISCRIM</u> (General Description). If TS 8 contains any digit 1 then TS 15 is passed into TS 24, otherwise TS 16 is passed into TS 24. The contents of TS 2, TS 3, TS 4, TS 5, TS 8 are altered.

<u>Outline of operation</u>. TS 8 is transferred to TS 2 and then subtracted from zero, passing into the partial sums register TS 4, TS 5. By taking out TS 5 we obtain a minor cycle full of digits 1 or of digits 0 according as there was or was not a digit 1 in TS 8 originally. We then form (TS 5 & TS 15) v (~TS 5 & TS 16) by logical operations and pass it on to TS 24.

This table provides the main means of deciding between two alternative procedures, by setting up one or the other of two instructions, contained in TS 15 or TS 16.

<u>PLUSIND</u> (General Description). 1 is added to the position reference in TS 27, e.g. DL 7, 9 becomes DL 7, 10, but DL 7, 32 becomes DL 8, 1.

<u>TRANS 45</u> (General Description). The following set of transfers is made

$$\text{TS } 22 - \text{TS } 20, \quad \text{TS } 23 - \text{TS } 21.$$

<u>BURY</u> (General Description). The content of TS 1 with 1 added is transferred to the position indicated in TS 31, and 1 is added to the reference in TS 31. We then proceed to carry out the instruction in TS 1.

<u>UNBURY</u> (General Description). The minor cycle whose position is given in TS 31 is taken to be position of the next instruction.

<u>MULTIP</u> (General Description). The number in TS 18, 19 is multiplied by the number in TS 20, 21: the result is brought to standard form by shift of

decimal point. An error is obtained for the product by using the errors in the given numbers and allowing for rounding off. The result is stored in TS 22, 23.

<u>ADD</u> is analogous to MULTIP.

As an example of a more complicated process, I have chosen the calculation of the value of a polynomial.

<u>CALPOL</u> (General Description). The minor cycles of DL 3 taken in pairs contain the coefficients of a polynomial in descending order. Evidently we are restricted to degrees not exceeding 15, and we assume the degree always to be 15, filling up with appropriate zero coefficients. The value of this polynomial will be calculated for the argument in TS 13, TS 14 and the result will be transferred to TS 25, 26. Before starting we require special contents in DL 1, 14 and DL 1, 15. There are

DL 1, 14   0000, 0101, 0000, 0000, 0100, 0110, 0000, 0000

DL 1, 15   0000, 0000, 0000, 0000, 0000, 0100, 0000, 0000

the expression in DL 1, 14 representing the order to transfer DL 3, 1 to TS 6.

CALPOL.

<u>CALPOL</u> 1. Clear TS 22, 23; DL 1, 14 – TS 27; DL 1, 15 – TS 29. CALPOL 8.

<u>CALPOL 8</u>. B, BURY; B, INDEXIN; TS 28 – TS 18; B, BURY; B, PLUSIND; B, BURY; B, INDEXIN; TS 28 – TS 19; B, BURY; B, ADD; B, BURY; B, PLUSIND; TS 27 – TS 2; TS 29 – TS 3; AND; Q, CALPOL 40; TS 6 – TS 15; Q, CALPOL 37; TS 6 – TS 16; B, BURY; B, DISCRIM; B, 1.

<u>CALPOL 37</u>. TS 13 – TS 18; TS 14 – TS 19; B, BURY; B, TRANS 45; B, BURY; B, MULTIP; B, BURY; B, TRANS 45. CALPOL 49.

<u>CALPOL 49</u>. B, CALPOL 8.

<u>CALPOL 50</u>. TS 22 – TS 25; TS 23 – TS 26; B, UNBURY.

The above table for CALPOL has been expressed in a more abbreviated form than the one we gave for INDEXIN, several operations being listed at a time. AND is of course the logical operation and B,1 indicates B with a 1 in column 17.

<u>Outline of operation (CALPOL)</u>. If we denote the polynomial by $a_1 x^{15} +$ $a_2 x^{14} + \cdots$ the calculation proceeds by the equations $b_1 = a_1$, $c_1 = b_1 x$,

407

$b_2 = c_1 + a_2$, $c_2 = b_2 x$, ... After the calculation of each b_r we have to determine whether this is the one required, viz. b_{16} or not. This is done by examining the content of TS 27 which includes the number r and is also, one might say principally, used to describe the position of the next coefficient a_{r+1}. If it is the one required we find ourselves at CALPOL 40 and have to pass b_r out to TS 25, 26. Otherwise we go to CALPOL 31, and after multiplying b_r by x to give c_r we find ourselves back at CALPOL 8 and repeating processes we have done before.

It will be evident that the table CALPOL is somewhat wasteful of space. Each time a subsidiary operation is required we have to repeat B, BURY, and each time we make a transfer we have to do it in two stages, each of which uses a whole minor cycle of which most is wasted. It is possible to avoid this waste of space by keeping the instruction tables in some abbreviated form, and expanding each table whenever we want it. This will require a table EXPAND, and will require each table to include appropriate references to the table EXPAND. These references will however be put in by EXPAND itself (when working under contract to a higher authority), just as EXPAND will put in the references to BURY and UNBURY.

BINDEC (General Description). The number in TS 13, 14 is translated into decimal form of the type $\alpha \times 10^m$ where $1 \leqslant \alpha < 10$, and is transferred into DL 10. The notation of the decimal form is such that the content of DL 10 can be passed out onto a card in the usual way and if the card is then listed the digits of the numbers α, m will then appear on the listing paper in the usual way. Or in other words only the first 10 minor cycles of DL 10 are used, and a decimal digit is represented by the minor cycle in which a pulse occurs, and its significance by the position of it within the minor cycle.

(This account is incomplete as regards signs and some other details.)

14. The Design of Delay Lines.

(i) General. A considerable amount of work has been done on delay lines for R. D. F. purposes. On the whole our problems coincide with the R. D. F. problems but there are a few differences.

(a) Owing to the fact that there will be more than one tank used in the calculator the stability of the delay is of importance. In R. D. F. the delay is allowed to determine the recurrence frequency and the effects of variations in it are thereby eliminated.

(b) In R. D. F. it is required that the delayed signal should not differ from the undelayed by an error signal which is less than 60 dB (say) down on the signal proper. We are less difficult to please in this respect. We only require to be able to distinguish mark from space with a very high probability (e.g. at least $1-10^{-32}$). This requires a high signal to noise ratio, so far as the true random noise and the interference are concerned, but it does not require much as regards hum, frequency distortion and other factors producing unwanted signals of fairly constant amplitude.

Our main concerns then in designing a delay line will be:

(1) To ensure sufficient signal strength that noise does not cause serious effects.
(2) To eliminate or correct frequency and phase distortion sufficiently that we may correctly distinguish mark and space.
(3) To stabilise the delay to within say 0·2 pulse periods.
(4) To eliminate interference.
(5) To provide considerable storage capacity at small cost.
(6) To provide means for setting the crystals sufficiently nearly parallel.

The questions of noise and signal strength are treated in some detail in the following pages. It is found that there is plenty of power available unless either very long lines or very high frequencies are used. The elimination of interference is mainly a matter of shielding and is a very standard radio problem, which in our case is much less serious than usual. Various means have been found by the R. D. F. workers for setting the crystals. Some prefer to machine the whole delay line very accurately, others to provide means for moving the crystals through small angles, e.g. by bending the tank. All are satisfactory.

I list below a number of questions which must be answered in our design of delay lines. In order to fix ideas I have added the most probable answers in brackets after each question.

(1) What liquid should be used in the line? (Either mercury or a water-alcohol mixture.)
(2) Should we use a carrier? If so, of what frequency? (Yes, certainly use a carrier. Frequency should be about 10 Mc/s with water-alcohol mixture, but may be higher if desired when mercury is used.)
(3) What should be the clock-pulse frequency? (1 Mc/s.)

(4) What should be the dimensions of the crystals? (Diameter might be half that of the tank, e.g. 1 cm. Thickness should be such that the first resonances of the two crystals are two or three megacycles on either side of the carrier, if water-alcohol is used. With mercury the thickness is less critical and may be either as with water-alcohol or may have resonance equal to carrier.)

(5) Should the inside of the tank be rough or smooth? (Smooth.)

(6) What should be the dimensions of the tank? (Standard tanks to give a delay of 1 ms. should be about 5′ long whether water-alcohol or mercury. Diameter $1/2''$.[3])

(Keep all the tanks within one degree Fahrenheit in temperature. Correct systematic temperature changes by altering the pulse frequency.)

In order to be able to answer these questions various mathematical problems connected with the delay lines will have to be solved.

(ii) <u>Electromagnetic conversion efficiency</u>. The delay line may best be considered as forming an electrical network of the kind usually (rather misleadingly) described as 'four-pole', i.e. a network which has one input current and one input voltage which together determine an output voltage and current. Such a network is described by three complex numbers at each frequency. In the case where there is little coupling between the output and input, which will apply to our problem, we may take these quantities to be the input and output admittances and the 'transfer admittance'. Strictly speaking we should specify whether the output is open circuit or short circuit when stating the input impedance, but with weak coupling these are effectively the same; similarly for the output impedance. The transfer admittance is the current produced at one end due to unit voltage at the other, and does not depend on which end has the voltage applied to it. In the case of the delay lines the input and output admittances will be effectively the capacities between the crystal electrodes. We need only determine the transfer admittance.

We shall consider the following idealised case. Two crystals of thicknesses d and d′ are immersed in a liquid, with their faces perpendicular to the x-axis. The liquid extends to infinity in both the positive and the negative x-directions, and both liquid and crystals extend to infinity in the y and z directions (Fig. 40). The distance between the near side faces of the crystals is

[3] Editor's note. The typescript has '$1/2''$'.

ℓ. It is assumed that there is considerable attenuation of sound waves over a distance of the order of ℓ but hardly any over a distance of the order of d or d'.

These assumptions are introduced largely with a view to eliminating the possibility of reflections. In practice the reflections would be eliminated by other means. For instance, the infinite liquid on the extreme right and left would be replaced by a short length of liquid in a stub of not very regular shape, so that the reflected waves would not be parallel to the face of the crystal. More likely still, of course, we should have some entirely different medium there.

The physical quantities involved are:

(a) The density ρ. We write ρ for the density of the crystal and ρ_I for that of the liquid. Likewise a suffix I will indicate liquid values throughout.
(b) The pressure p. In the case of the crystal this is understood to mean the xx-component of stress.
(c) The displacement ξ in the x-direction.
(d) The velocity v in the x-direction.
(e) The radian frequency w.
(f) The elasticity η. This is the rate of change of pressure per unit decrease of logarithm of volume due to compression.
(g) The velocity of propagation c.
(h) The mechanical characteristic impedance ζ.
(i) The reciprocal radian wave length β.
(j) The piezo-electric constant ε. This gives the induced pressure due to an electric field strength of unity. This field strength should normally be thought of as in the x-direction, but we shall have to consider the case of a field in the y or z direction briefly also.

These quantities are related by the equations

$$c = \sqrt{\eta/\rho}, \quad \zeta = \sqrt{\eta\rho}, \quad \beta = \frac{w}{c}, \quad v = iw\xi,$$

$$iw\rho v = -\frac{dp}{dx}, \quad p = -\eta\frac{d\zeta}{dx} + E\varepsilon.$$

In what follows we assume that all quantities such as p, v, ξ depend on time according to a factor e^{iwt}, which we omit.

We now consider the 'transmitting crystal', which we suppose extends from x = -a to x = a where d = 2a. The solution of the equations will be of form

$$p = E\varepsilon + B\cos\beta x$$

within the crystal, i.e. for $|x| < a$. Since the pressure is continuous we shall have

$$p = (E\varepsilon + B \cos \beta a)e^{i\beta_I(a-|x|)} \quad \text{if } |x| > a.$$

This gives for the velocity

$$v = \frac{I}{w\rho} \cdot -B\beta \sin \beta x = -iB\zeta^{-I} \sin \beta x \quad \text{if } |x| < a$$

$$v = \zeta_I^{-I}(E\varepsilon + B \cos \beta a)e^{i\beta_I(a-|x|)} \operatorname{sgn} x \quad \text{if } |x| > a.$$

Continuity of velocity now gives

$$B\left(\cos \beta a + \frac{i\zeta_I}{\zeta} \sin \beta a\right) = -E\varepsilon$$

and therefore the velocity at a is

$$\frac{-iB \sin \beta a}{\zeta} = \frac{iE\varepsilon \sin \beta a}{\zeta \cos \beta a + i\zeta_I \sin \beta a}$$

i.e. the velocity at the inside edge of the crystal is

$$\frac{iE\varepsilon}{\zeta} \cdot \frac{I}{\cot \frac{dw}{2c} + iu}$$

where $u = \zeta_I/\zeta$.

Assuming that the exciting voltage is longitudinal we may say that

$$\frac{\text{Velocity}}{\text{Exciting voltage}} = \frac{i\varepsilon}{\zeta d} \cdot \frac{I}{\cot \frac{dw}{2c} + iu}.$$

The effect of the medium between the two crystals we will not consider just yet. Let us simply assume that

$$\frac{\text{Velocity at inside edge of receiving crystal}}{\text{Velocity at inside edge of transmitting crystal}} = \vartheta.$$

We have now to consider the effect of the receiving crystal. Fortunately we can deal with this by the principle of reciprocity. When applied to a mixed electrical and mechanical system this states that the velocity produced at the mechanical end by unit voltage at the electrical end is equal to the current produced at the electrical end by unit force at the mechanical end. Hence

$$\frac{\text{Current at receiving end}}{\text{Force on receiving crystal}} = \frac{i\varepsilon}{d'\zeta} \cdot \frac{I}{\cot \frac{d'w}{2c} + iu}.$$

To these equations we may add that the ratio of force to pressure is the area A' of the receiving crystal, and that the ratio of pressure to velocity is the mechanical characteristic impedance ζ_I. Combining we obtain

$$Y = \text{Transfer admittance} = \vartheta \frac{A' \varepsilon^2 \zeta_I}{dd' \zeta^2} \frac{1}{\left(\cot \frac{dw}{2c} + iu\right)\left(\cot \frac{d'w}{2c} + iu\right)}.$$

Let us now assume that the input to the valve from the receiving crystal consists of a tuned circuit with a fairly low 'Q' as in Fig. 41. Then

$$\text{Voltage attenuation and phase change factor} = \frac{\text{Grid voltage}}{\text{Input voltage}}$$

$$= \frac{Y}{\frac{1}{Liw} + Ciw + \frac{1}{R}}$$

$$= \frac{Y}{Ciw_0} \frac{ww_0}{\left(w + w_0 + \frac{iw_0}{2Q}\right)\left(w - w_0 + \frac{iw_0}{2Q}\right)}$$

where $LCw_0{}^2 \, 1 + \dfrac{1}{4Q^2} = 1, \quad C = C_s + C_x$

$$Q = RCw_0$$

$$= \vartheta \frac{C_x}{C_x + C_s} \cdot \frac{2\pi\varepsilon^2}{\kappa\eta} \cdot R(w)$$

where $\kappa = $ Dielectric constant of crystal.

$\vartheta = $ Attenuation due to viscosity of medium and geometrical causes.

$$R(w) = \frac{u}{\frac{dw_0}{2c}\left(\cot \frac{dw}{2c} + iu\right)\left(\cot \frac{d'w}{2c} + iu\right)} \cdot \frac{ww_0}{\left(w + w_0 + \frac{iw_0}{2Q}\right)\left(w - w_0 - \frac{iw_0}{2Q}\right)}.$$

The quantity $\frac{2\pi\varepsilon^2}{\kappa\eta}$ depends only on the crystal, i.e. on the material of which it is made and its cut and form of excitation. Both ε^2 and η are of the dimensions of a pressure. $4\pi\varepsilon$ is of the dimensions of an electric field, and may be thought of as a constant electric field which has to be added to the varying field in order that the combination should produce the correct pressure variations, somewhat like the permanent magnet field in a telephone receiver. A typical value for $\frac{2\pi\varepsilon^2}{\kappa\eta}$ is 0·004.

Let us now consider the frequency-dependent factor, $R(w)$. The parameter u entering here is the ratio of the characteristic impedances of the crystal and the liquid. It is equal to

$$\frac{\text{Velocity of sound in liquid} \times \text{density of liquid}}{\text{Velocity of sound in crystal} \times \text{density of crystal}}.$$

The velocity of sound in the crystal (X-cut quartz) is 5·72 km/sec. and its density is 2·7. The velocity in water is 1·44 km/sec., and the density 1, hence

u(water) = 0·1 abt.

The velocity in mercury is much the same but the density is 13.5. Hence

u(mercury) = 1·3 abt.

These figures suggest that we consider the two cases where u is small and where u is 1. The latter case may be described by saying that the liquid matches the crystal.

It may be assumed for the moment that our object is to make the minimum value of $|R(w)|$ in a certain given band of frequencies as large as possible. If the width of the band is 2Ω and it is centred on w_0 and if we ignore the variations in ϑ we shall find that the optimum value of u is of the form $N\frac{\Omega}{w_0}$ where N is some numerical constant probably not too far from 1. The value of Q should be as large as possible. With $\Omega = 1$ Mc/s, $w_0 = 10$ Mc/s this seems to suggest that water (u = 0·1) is very suitable. In practice the differences due to the value of ϑ are more serious than those due to u, and there is in any case plenty of power. We would not in practice take Q as large as we could but would rather try to arrange that $|R(w)|$ was fairly constant throughout the band concerned and arg $R(w)$ fairly linear when plotted against w. If water were used one would probably choose the thicknesses of the crystals and the value of Q to give poles of $|R(w)|$ somewhat as shown in Fig. [42]. With this arrangement of the poles the gain corresponding to $|R(w)|$ is 9 dB throughout the range 8 Mc/s and the phase characteristic lies within 5° of the straight line within this range.

With mercury where u is nearly 1 we should put

$$\frac{dw_0}{2c} = \frac{\pi}{2}, \quad \frac{d'w_0}{2c} = \frac{\pi}{2},$$

and then

$$|R(w)| = \frac{2}{\pi}\left(\sin\frac{\pi}{2}\frac{w}{w_0}\right)^2 \left| \frac{ww_0}{\left(w + w_0 + \frac{iw_0}{2Q}\right)\left(w - w_0 + \frac{iw_0}{2Q}\right)} \right|$$

We should probably find it desirable to omit the tuned circuit, in which case R(w) would represent a fairly constant loss of 4 dB. One could use a Q of 2 if one wished, giving a gain of 2 dB instead.

We have assumed above that the crystal is longitudinally excited. If it were transversely excited the figures would be much less satisfactory. At the transmitting end a far larger voltage would have to be applied in order to obtain the same field strength, and at the receiving end the stray capacities will have a more serious effect with transverse electrodes, although if the stray capacity were zero transverse electrodes at the receiving end would actually be more efficient.

(iii) <u>Geometrical attenuation</u>. If a rectangular crystal is crookedly placed in a plane parallel beam, the tilt being such that the one edge of the crystal is advanced in phase by an angle ψ then the attenuation due to the tilt is $\frac{\sin\frac{1}{2}\psi}{\frac{1}{2}\psi}$. With a square crystal whose side is 1 cm. and a frequency of 15 Mc/s this would mean that we get the first zero in the response for a tilt of about $16'$. The setting is probably not really as critical as this owing to curvature of the wave fronts. If the crystals are operating in a free medium without the tube this effect is easily estimable and we find that, for crystals sufficiently far apart the allowable angles of tilt are of the order of the angle subtended at one crystal by the other. It has been found experimentally with tubes operating at 15 Mc/s that tilts of the order of half a degree are admissible.

Now let us consider the loss due to boundary effects. We assume a wave inside the tank of form $p = J_0(\beta'r)e^{-i\beta z + iwt}$ and assume a boundary condition of form $\frac{1}{p}\frac{dp}{dn} = \zeta$ where we do not know ζ nor even whether it is real or complex. The radius of the tank is a, so that the boundary condition becomes $\frac{\beta'aJ_1(\beta'a)}{J_0(\beta'a)} = \zeta a$. Let the solution of $\frac{uJ_1(u)}{J_0(u)} = y$ be u(y). Then we have $\beta^2 + \left(\frac{u(\zeta a)}{a}\right)^2 = \frac{w^2}{c^2}$ and therefore $\mathcal{R}_\beta \mathcal{I}_\beta + \frac{1}{a^2}\mathcal{R}_u \mathcal{I}_u = 0$. But since $\frac{u(\zeta a)}{\beta a}$ is small this means approximately $\mathcal{I}_\beta = \frac{c\mathcal{R}_u\mathcal{I}_u}{a^2 w}$, and the loss in a length ℓ of the tank is $\frac{\ell c}{a^2 w}\mathcal{R}_u\mathcal{I}_u$ nepers. For a given value of ζ there are many solutions of $\frac{uJ_1}{J_0} = \zeta a$ but there is a bounded region of the u plane in which there is always a solution whatever value ζa may have. This means to say that for any boundary condition there is always a mode in which the attenuation does not exceed $\tau\frac{c}{a^2 w}$ where τ is some numerical constant.

The value of τ is about $1 \cdot 9$. It is the largest value of xy such that $(x + iy)J_1(x + iy)/J_0(x + iy)$ is pure imaginary and $y > 0, 0 < x < 2 \cdot 4$.

Taking $\ell c/a^2 w_0 = 0 \cdot 31$ (as p. [418]) the maximum loss in this mode is 6 dB. We should however probably add a certain amount to allow for the fact that not all of the energy will be in this mode. A total loss of 10 dB would probably not be too small.

(iv) <u>Attenuation in the medium</u>. The attenuation coefficient is given by $\frac{2w^2\upsilon}{3c^3}$ where υ is the dynamic coefficient of viscosity, i.e., the ratio of viscosity to density. With water (υ = ·013 c = 1·44 Km/sec.) at a frequency of 10 megacycles and a delay of 1 ms we have a loss of 12 dB. With mercury under the same circumstances the loss is only 1 dB.

These figures suggest that if water is used the frequency should not be much above 10 Mc/s, but that we can go considerably higher with mercury.

(v) <u>Noise</u>. Before leaving the subject of attenuation we should verify how much can be tolerated. The limiting factor is the noise, due to thermal agitation and to shot effect in the first amplifying valve. The effect of these is equivalent to an unwanted signal on the grid of the first valve, whose component in a narrow band of width f cycles has an R. M. S. value of

$$V_N = 4kTf(R + R_e)$$

where T is the absolute temperature, k is Boltzmann's constant and R is the resistive component of the impedance of the circuit working into the first valve, including the valve capacities. R_e is a constant for the valve and describes the shot effect for the valve. In the case that we use mercury and do not tune the input the value of R will be quite negligible in comparison with R_e, which might typically be 1000 ohms. For a pulse frequency of 1 megacycle we must take $f = 10^6$ (the theoretical figure is $\frac{1}{2}10^6$ but this is only attainable with rather peculiar circuits). At normal temperatures $4kT = 1·6 \times 10^{-20}$ and therefore $V_N = 4 \, \mu V$. In the case that we use water and tune the input, we have $R = \frac{Q}{w(C_x+C_s)}$ at the worst frequency. Assuming $\frac{w}{2\pi Q} = 2$ Mc/s (see Fig. 41) and $C_x + C_s = 20$ pf and ignoring the fact that the effect will not be so bad at other frequencies, we have $V_N = 9 \, \mu V$.

Now suppose that we wish to make sure that the probability of error is less than p, and that the difference in signal voltage between a digit 0 and a digit 1 is V. Then we shall need

$$2 \int_{V/2V_N}^{\infty} e^{-\frac{1}{2}x^2} dx < p.$$

(This follows from the fact that a random noise voltage is normally distributed in all its coordinates.) If we put $p = 10^{-32}$ we find $\frac{V}{V_N} \geqslant 24$, $V \geqslant 0·1$ mV.

(vi) <u>Summary of output power results</u>. Summarising the voltage attenuation and noise questions we have:

(a) There is an attenuation factor depending on the material of the crystal and its cut and for quartz typically giving a loss of 48 dB.

(b) There is a factor R depending on the ratio of band width required to carrier frequency, and the matching factor u between crystal and liquid. In practical cases this amounts to gains of 10 dB with water and 2 dB with mercury.

(c) There is a loss factor $C_x/C_x + C_s$ due to stray capacity C_s across the receiving crystal. This might represent a loss of 6 dB.

(d) There is a loss due to the viscosity of the medium. For a water tank with a delay of 1 ms. and a carrier of 10 Mc/s the loss may be 12 dB: with mercury and a carrier of 20 Mc/s it may be 4 dB.

(e) Losses in the walls of the tank. Apparently this should not exceed 10 dB.

(f) The noise voltage may be 4×10^{-6} volts RMS (mercury) or 9×10^{-6} volts RMS (water).

(g) The signal voltage (peak to peak) should exceed the noise voltage (RMS) by a factor of 24 for safety.

These figures require input voltages (peak to peak) of 0·2 volts or 4·5 volts with mercury and water respectively. We could quite conveniently put 200 volts on, so that we have 60 dB (or 53 dB) to spare. There is no danger of breaking the crystals when they are operated with so much damping.

(vii) <u>Phase distortion due to reflections from the walls</u>. We cannot easily treat this problem quantitatively because of lack of information about the boundary conditions and because the ratio of diameter of crystal to diameter of tank is significant. Let us however try to estimate the order of magnitude by assuming the pressure zero on the boundary and considering the gravest mode. In this case the pressure is of form $J_0\left(\frac{k_1 r}{a}\right)e^{-i\beta z + iwt}$ where $2a$ is the diameter of the tank and $k_1 = 2\cdot4$ is the smallest zero of J_0, and $\beta^2 + \frac{k_1^2}{a^2} = \frac{w^2}{c^2}$. In this case the change of phase down the length ℓ of tank is $\phi = \ell\frac{w^2}{c^2} - \frac{k_1^2}{a^2}$. If we are using carrier working we are chiefly interested in $\frac{d^2\phi}{dw^2}$ which turns out to be $-\frac{k_1^2 c\ell}{w_0^3 a^2}$ where w_0 is the carrier frequency. If we suppose that the band width involved is 2Ω, then the greatest phase error which is introduced is $\frac{k_1^2 \Omega^2 c\ell}{2w_0^3 a^2}$. Let us suppose that the greatest admissible error is 0·2 radians, then we must have

$$\frac{\ell c}{a^2 w_0} \leqslant \frac{0\cdot4}{k_1^2}\left(\frac{w_0}{\Omega}\right)^2.$$

Taking $w_0 = 10 \,\text{Mc/s}$
$\Omega = 1 \,\text{Mc/s}$
$c = 1 \cdot 4 \times 10^5 \,\text{cm/sec.}$
$\ell = 1 \cdot 4 \times 10^2 \,\text{cm.}$
$a = 1 \,\text{cm.}$

Then $\frac{c}{w_0} = 2 \cdot 2 \times 10^{-3} \,\text{cm.}$

$$\frac{\ell c}{a^2 w_0} = 0 \cdot 31$$

$$\frac{0 \cdot 4}{k_1{}^2}\left(\frac{w_0}{\Omega}\right)^2 = 6 \cdot 95$$

The situation is thus entirely satisfactory. The carrier frequency could even be halved.

(viii) <u>The choice of medium</u>. In choosing the medium we have to take into account

(a) That a medium with a small characteristic impedance such as water has a slight advantage as regards the factor R(w).
(b) That water is more attenuative than mercury.
(c) That mercury gives wide band widths more easily than water because of closer matching, but that adequate band widths are nevertheless possible with water.
(d) That a water-alcohol mixture can be made to have a zero temperature coefficient of velocity at ordinary temperatures.

On the whole the advantages seem to be slightly on the side of mercury.

(ix) <u>Long lines</u>. The idea of using delay lines with a long delay, e.g. of the order of 0·1 second, is attractive because of the very large storage capacity that such a line would have. Although the long delay would make these unsuitable for general purposes they would be very suitable for cases where very large amounts of information were to be stored: in the majority of such cases the material is used in a fairly definite order and the long delay does not matter.

However such long lines do not really seem to be very hopeful. In order to reduce the attenuation to reasonable proportions it would be necessary to abandon carrier working, or else to use mercury. In either case we should probably be obliged to make the tank in the form of a bath rather than a tube; in the former case in order to avoid the phase distortion arising from reflections from the walls, and in the latter to economise mercury, using a system of mirrors in the bath. In any case the technique would involve much development work.

We propose therefore to use only tanks with a delay of 1 ms.

(x) <u>Choice of parameters</u>. Considerations affecting the carrier frequency are:

(a) The higher the carrier frequency the greater the possible band width.
(b) The difficulty of cutting thin crystals, somewhat modified by the absence of necessity of frequency stability.
(c) The attenuation at high frequencies of the sound wave in the liquid.
(d) The difficulty of setting the crystals up sufficiently nearly parallel if the wavelength is short.
(e) The difficulty of amplification at high frequencies.

Of these (a) and (c) are the most important. A reasonable arrangement seems to be to choose a frequency at which the attenuation in the medium is about 15 dB.

With the comparatively low frequencies and with wide tanks the setting up difficulty will not be serious. With long lines we should probably not attempt to do temperature correction, but would rephase the output.

Considerations affecting the pulse frequency are:

(a) The limitation of the pulse frequency to a comparatively small fraction of the carrier frequency if water is the transmission medium, and the limitation of this carrier frequency.
(b) The finite reaction times of the valves.
(c) The greater capacity of a line if the frequency is high.
(d) Greater speed of operation of the whole machine if the pulse frequency is high.
(e) Cowardly and irrational doubts as to the feasibility of high frequency working.

If we can ignore (e) the other considerations appear to point to a pulse frequency of about 3 megacycles or even higher. We are however somewhat alarmed by the prospect of even working at 1 megacycle since the difficulty (b) might turn out to be more serious than anticipated.

Considerations affecting the diameter of the tank are:

(a) That the crystals are most conveniently adjusted to be parallel by bending the tanks and that the diameter should therefore not be too large.

(b) That the diameter should be at least large enough to accommodate the crystal.

(c) That small diameters give phase distortion (p. [417]).

(d) That with mercury small diameters are economical. At a price of £1 sterling per 1 lb. avoirdupois of mercury a 1 ms. tank of diameter 1″ would contain mercury to the value of about £2–2–6.

A diameter of 1″ or rather less is usual in R. D. F. tanks and appears reasonable in view of these conditions.

(xi) <u>Temperature control system</u>. The temperature coefficient of the velocity of propagation in mercury is quite small at 15 Mc/s, being only 0·0003/degree centigrade. This means that if the length of a 1 ms. line is to be correct to within 0·2 ms. then the temperature must be correct to within two-thirds of a degree centigrade.

15. <u>The Design of Valve-elements</u>.

(i) <u>Outline of the problem</u>. To design valve-elements with properties as described in § 5 and to work at a frequency of say 30 or 100 kilocycles would be very straightforward. When the pulse recurrence frequency is as high as a megacycle we shall have to be more careful about the design, but we need not fear any real difficulties of principle about working at these frequencies, and with such band widths. The successful working of television equipment gives us every encouragement in this respect. A word of warning might perhaps be in order at this point. One is tempted to try and carry the argument further and try to infer something from the success of R. D. F. at frequencies of several thousands of megacycles. Such an analogy would however not be in order for although these very high frequencies are used the bandwidth of intelligence which can be transmitted is still comparatively small, and it is not easy to see how the band width could be greatly increased.

In this chapter I shall discuss the limitations inherent in the problem, and shall also show very tentative circuit diagrams by way of illustration. These circuits have not yet been tried out, and I have too much experience of electronic circuits to believe that they will work well just as they stand. (This does not represent a superstitious belief in the cussedness of circuits and the inapplicability of mathematics thereto. Rather it means that normally the amount of mathematical argument required to get a reliable prophecy of the behaviour of a circuit is out of proportion to the small trouble required to try it out, at any rate if one is in an electrical laboratory. In practice one compromises with a rough mathematical argument and then follows up with experiment. The apparent 'cussedness' of electronic circuits is due to the fact that it is necessary to make rather a lot of simplifying assumptions in these arguments, and that one is very liable to make the wrong ones, by false analogy with other circuits one has dealt with on previous occasions. The cussedness lies more in the minds dealing with the problem than in the electronic circuits themselves.)

(ii) <u>Sources of delay</u>. There are two main reasons why vacuum tubes should cause delays, viz. the input capacity and the transit time. Of these perhaps the first is in practice the more serious, the second the more theoretically unavoidable.

The delay due to the input capacity, when the valves are driven to saturation or some other limiting arrangement is used, is of the order of C/g_m, where C is the input capacity and g_m is the mutual conductance of the valve. We may, for instance consider the idealised circuit Fig. 44. (Coupling with a battery is of course not practical politics, but it produces essentially the same effects as more practical circuits, and is more easily understood.) If I is the saturation current then the grid swing required to produce it is I/g_m and the charge which must flow into the grid to produce this voltage is CI/g_m. If the whole saturation current is available the time required is C/g_m. This argument is only approximate, and omits some small purely numerical factors. However it illustrates the more important points. In particular we can see that Miller effect is not a very serious matter because of the limiting, which reduces the effective amplification factor to 1. On the other hand, if one valve is used to serve several inputs the delay will be correspondingly increased because the capacity has become multiplied by the number of grids served.

This connecting of several grids to one anode, and a number of other practical points will tend to make the actual delay due to input capacity several times greater than C/g_m, e.g. 10 C/g_m.

The delay due to transit time may be calculated, in the case of a plane structure, to be $3d(m/2eV)^{\frac{1}{2}}$ where m, e are respectively the mass and charge of the electron, V is the voltage of the grid referred to cut-off and d is the grid-cathode spacing. In other words the transit time may be calculated on the assumption that the average velocity of the electrons between cathode and grid is one-third of the velocity when passing the grid. This time may be compared with C/g_m which, if C is calculated statically, has the value $\frac{3}{2}d(m/2eV)^{\frac{1}{2}}$, i.e. half of the transit time. That there should be some such relation between C/g_m and transit time can be seen by calculating $C/(g \times \text{Transit time})$, where C is the grid-cathode capacity and g is the actual conductance, i.e., the ratio of current to V.

$$\frac{C}{g \times \text{Transit time}} = \frac{CV}{I \times \text{Transit time}}$$

$$= \frac{\text{Charge on grid}}{\text{Charge in transit}}.$$

Let us now calculate actual values. The voltage V by which the grid exceeds cut-off might be 10 volts which corresponds to a velocity about 1/300 of velocity of light (Note: annihilation energy of electron is half a million volts) or one metre per microsecond. If d is 0·2 cm. the transit time is 0·006 μs. A typical value for C/g_m is 0·002 μs.

The relation between C/g_m and transit time brings up an important point, viz. that these two phenomena of time delay are really inseparable. The input capacity of the tube when 'hot' really consists largely of a capacity to the electrons. When the motion of the electrons is taken into account the capacity is found to become largely resistive (Ferris effect).

Before proceeding further I should try to explain the way I am using the word 'delay'. When I say that there is a delay of so many microseconds in a circuit I do not mean to say that the output differs from the input only in appearing that much later. I wish I did. What I mean is something much less definite, and also less agreeable. Strictly speaking I should specify very much more than a single time. I should specify the waveform of the output for every input waveform, and even this would be incomplete unless it referred both to voltages and currents. We have not space to consider these questions, nor is it really necessary. I should however give some idea of what kind of distortion of output these 'delays' really involve. In the case of the input capacity the distortion may be taken to be of the form that an ideal input pulse of unit area is converted into a pulse of unit area with sharp leading edge and exponentially decaying trailing edge, the time constant of the delay being the 'delay', thus Fig. 44a. In the case of the transit time the curve is probably more nearly of the 'ideal' form (Fig. 44b).

To give the word 'delay' a definite meaning, at any rate for networks, I shall understand it to mean the delay for low frequency sine waves. This is equal to the displacement in time of the centre of gravity in the case of pulses.

In order to give an idea of the effect of these delays we have shown in Fig. 45 a pulse of width 0·2 μs and the same pulse delayed, after the manner of Fig. 44a, by 0·03 μs, this representing our calculated value of 0·003 multiplied by 10 to allow for numerous grids, etc. etc. It will be seen that the effect is by no means to be ignored, but nevertheless of a controllable magnitude.

(iii) <u>Use of cathode followers</u>. In order to try and separate stages from one another as far as possible we shall make considerable use of cathode followers. This is a form of circuit which gives no amplification, and indeed a small attenuation (e.g. 0·5 dB); but has a very large input impedance and a very low output impedance. This means chiefly that we can load a valve with many connections into cathode followers without its output being seriously affected.

Fig. 46 shows a design of cathode follower in which the input capacity effect has been reduced by arranging that the anode is screened from the grid and that the screen voltage as well as that of the cathode moves with the grid. If one could ignore transit time effects this would have virtually zero input capacity.

(iv) <u>The 'limiting amplifier' circuit</u>. When low frequencies are used the limiter circuit can conveniently be nothing more nor less than an amplifier, the limiting effect appearing at cut-off and when grid and cathode voltages are equal. At high frequencies we cannot get a very effective limiting effect at cathode voltage, owing to the fact that the grid must be supplied from a comparatively low impedance source to avoid a large delay arising from input capacity, but on the other hand, in order to get a limiting effect we need a high impedance, high compared with the grid conduction impedance (about 2000 ohms probably).

At high frequencies it is probably better to use a 'Kipp relay' circuit. This is nothing more than a multivibrator in which one leg has been made infinitely long (and then some), i.e. one of the two semi-stable states has been made really stable. An impulse will however make the system occupy the other state for a time and then return, producing a pulse during the period in which it occupies the less stable state. This pulse can be taken in either polarity. It is fairly square in shape and its amplitude is sensibly independent of the amplitude of the tripping pulse, although its time may depend on it slightly. These are all definite advantages.

A suggested circuit is shown in Fig. 47, and the waveforms associated with it at various points in Fig. 48.

(v) <u>Trigger circuit</u>. The trigger circuit need only differ very little from the limiter or Kipp relay. It needs to have two quite stable states, and we therefore return both of the grids of the 6SN7 to -15 volts instead of returning one to ground. Secondly the inhibitory connection is different. In the case of the limiter it simply consists of an opposing or negative voltage on the cathode follower; in the case of the trigger circuit it must trip the valve back, and therefore we need a second cathode follower input connected to the other grid of the 6SN7.

(vi) <u>Unit delay</u>. The essential part of the unit delay is a network, designed to work out of a low impedance and into a high one. The response at the output to a pulse at the input should preferably be of the form indicated in Fig. 50, i.e. there should be a maximum response at time 1 μs after the initiating pulse, and the response should be zero by a time 2 μs after it, and should remain there. It is particularly important that the response should be near to zero at the integral multiples of 1 μs after the initiating pulse (other than 1 μs after it).

A simple circuit to obtain this effect is shown in Fig. 51a. The response is shown in Fig. [51b]. It differs from the ideal mainly in having its maximum too early. It can be improved at the expense of a less good zero at 2 μs by using less damping, i.e. reducing the 500 ohm resistor. It is also possible to obtain altogether better curves with more elaborate circuits.

The 1000 ohm resistors at input and output may of course be partly or wholly absorbed into the input and output circuits. Further the whole impedance scale may be altered at will.

The fact that the pulse has become greatly widened in passing through the delay network does not signify. It will only be used to gate a clock pulse or to assist in tripping a Kipp relay, and therefore will give rise to a properly shaped pulse again.

(vii) <u>Trigger limiter</u>. We can build up a trigger limiter out of the other elements, although we cannot replace it by such a combination in the circuit diagrams because we are not putting a legitimate form of input into all of them. The circuit is (Fig. 52).

The valve P is merely a frequency divider. It can be used to supply all the trigger limiters. The trigger circuit Q should be tripped by the combination of pulse from P and continuous input, and will itself trip R. The arrangement of two trigger circuits prevents any danger of half-pulse outputs, which we are most anxious to avoid. In order that there might be a half-pulse output the trigger circuit Q would have to remain near its unstable state of equilibrium for a period of time of 1 μs. In order that this may happen the magnitude of the continuous input voltage has to be exceedingly finely adjusted; the admissible range is of the form $Ae^{-t} g_m/C$ where A might be say 100 volts (it doesn't matter really) and t is the time between pulses, C and g_m the input capacity and mutual conductance of the valves used in the trigger circuit; C/g_m might be 0.002 μs (we do not need to allow for Miller effect), so that the admissible voltage range is about 10^{-200} volts which is adequately small.

16. Alternative Forms of Storage.

(i) Desiderate for storage systems. A storage system should have a high monetory economy, i.e. we wish to be able to store a large number of digits per pound sterling of outlay: it should also have a high spacial economy. For the majority of purposes we like a form of storage to be erasable, although there are a number of purposes, such as function tables and the greater part of the instruction tables, for which this is not necessary. For the majority of purposes we also like to have a short accessibility time, defining the accessibility time to be the average time which one has to wait in order to find out the value of a stored digit. Normally we shall be interested in the values of a group of digits which are all stored close together, and very often it does not take much longer to obtain the information about the whole group than about the single digit. Let us say that the additional time necessary per digit required is the digit time (reading). We may also define the accessibility and digit times for recording in the obvious analogous way, though they are usually either equal to the reading time or else exceedingly long.

(ii) Survey of available storage methods. The accompanying table [p. 428] gives very rough figures for the various available types of storage and the quantities defined above. This table must not be taken too seriously. Many of the figures are based on definite numerical data, but most are guesses. In spite of the roughness of the figures the table brings out a number of points quite clearly.

(1) All the well established forms of storage (excepting the cerebral cortex) are either very expensive and bulky, or else have a very high accessibility time.

(2) The really economical systems consist of layers packed into the form of a solid. They are read by exposing the layer wanted.

(3) The systems which are both economical and fairly fast have the information arranged in two dimensions. This apparently applies even to the cerebral cortex.

(4) Much the most hopeful scheme, for economy combined with speed, seems to be the 'storage tube' or 'iconoscope' (in J. v. Neumann's terminology).

(5) Some use could probably also be made of magnetic tape and of film for cases where the accessibility time is not very critical.

(iii) <u>Storage tubes</u>. In an iconoscope as used in television a picture of a scene is stored as a charge pattern on a mosaic, and is subsequently read by scanning the pattern with an electron beam. The electron beam brings the charge density back to a standard value and the charge lost by the mosaic registers itself through its capacity to a 'signal plate' behind the mosaic. The information stored in this way on an iconoscope, using a 500 line system, corresponds to a quarter of a million digits.

One might possibly use an actual iconoscope as a method of storage, but there are better arrangements. Instead of putting the charge pattern on to the 'mosaic' with light we can put it on with an electron beam. The density of the charge pattern left by the beam can be varied by modulating either the voltage of the signal plate or the current in the beam. The advantages of this are:

(a) The charge pattern can be set up more quickly with an electron beam than with light.

(b) Less apparatus is required.

(c) The same beam can be used for reading and recording, so that distortion of the pattern does not matter.

It seems probable that a suitable storage system can be developed without involving any new types of tube, using in fact an ordinary cathode ray tube with tin-foil over the screen to act as a signal plate. It will be necessary to furbish up the charge pattern from time to time, as it will tend to become dissipated. The pattern is said to last for days when there is no electron beam, but if we have a beam scanning one part of the target it will send out secondary electrons which will tend to destroy the remainder of the pattern. If we were always scanning the pattern in a regular manner as in television this would raise no serious problems. As it is we shall have to provide fairly elaborate switching arrangements to be applied when we wish to take off a particular piece of information. It will be necessary to stop the beam from scanning in the refurbishing cycle, switch to the point from which the information required is to be taken, do some scanning there, replace the information removed by the scanning, and return to refurbishing from the point left off. Arrangements must also be made to make sure that refurbishing does not get neglected for too long because of more pressing duties. None of this involves any fundamental difficulty, but no doubt it will take time to develop.

	Monetory economy (digits/£)	Spacial economy (digits/litre)	Access. time (reading)	Digit time (reading)	Access. time (recording)	Digit time (recording)	Remarks
Inerasible systems.							
Punched Paper Tape	10^7	$5·10^6$	≥1 min.	·03 sec.	(= reading)	·03 sec.	
Hollerith Cards	10^6	$3·10^5$	≥1 min.	1 ms.	(= reading)	1 ms.	Permutable
Print on Paper	10^8	10^8	30 secs.	10 ms.			Human use. Not very convenient for mechanical or electrical reading.
Film (a) Displayed stationary	10^4	10^4	5 μs	1 μs		1 μs	
(b) Wound on reels	10^9	$3·10^{10}$	≥1 min.	1 μs		1 μs	
Soldered Connections	1000	200	<1 μs	<1 μs	15 mins.	1 min.	
Erasible Systems.							
Plugboards	50	50	<1 μs	<1 μs	30 secs.	10 secs.	
Wheels, etc.	20	$2·10^3$	30 ms.	30 ms.	30 ms.	30 ms.	(Mechanically read.)
Relays	2	2	<1 μs	<1 μs	10 ms.	10 ms.	
Thyratrons	2	2	<1 μs	<1 μs	10 μs	10 μs	
Neons	20	50	<1 μs	<1 μs	30 μs	30 μs	
Trigger Circuits	3	3	<1 μs	<1 μs	1 μs	<1 μs	
Cerebral Cortex	10^5	10^9	5 sec.	30 ms.	30 sec.	5 sec.	Man at £300 p.a. capitalised.
Acoustic delay lines	200	50	1 ms.	1 μs	1 ms.	1 μs	More optimistic estimate than in § 10.
Electric delay lines	100	200	100 μs	<1 μs	100 μs	<1 μs	Circular wave guide with 1 cm. waves. Numerous carriers.
Storage tubes	10^4	10^4	5 μs	1 μs	5 μs	1 μs	Described as 'Iconoscope' by J. v. Neumann.
Magnetic tape	10^8	$3·10^8$	1 min.	10^{-4} sec.	1 min.	10^{-4} sec.	

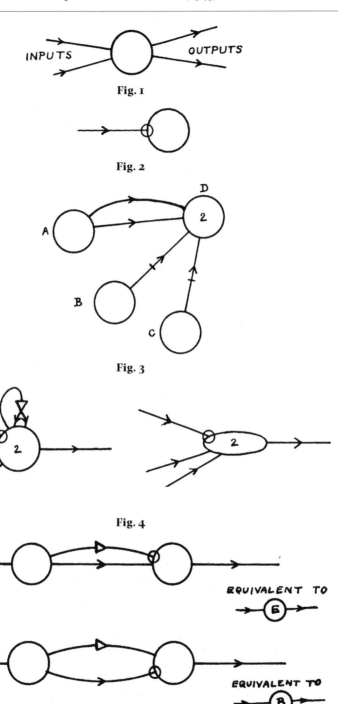

Fig. 1

Fig. 2

Fig. 3

Fig. 4

Fig. 5

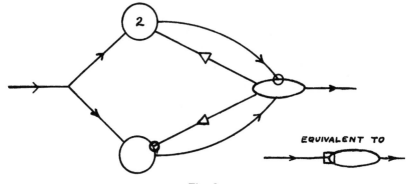

Fig. 6

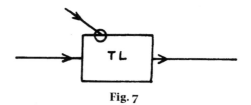

Fig. 7

Fig. 8

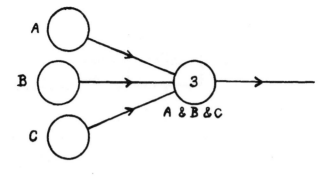

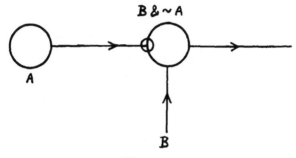

Fig. 9

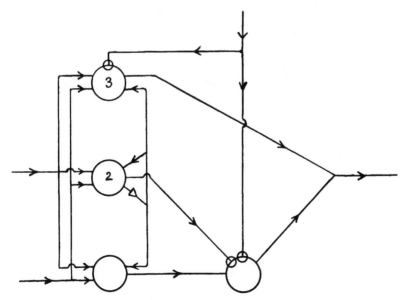

Fig. 10 Adder network.

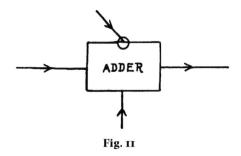

Fig. 11

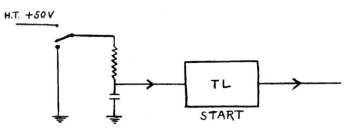

Fig. 12

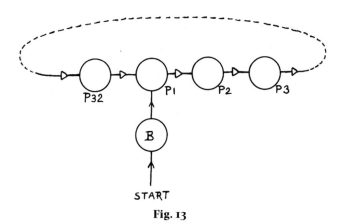

Fig. 13

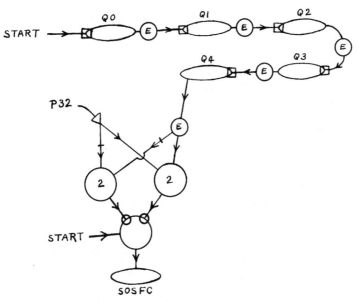

Fig. 14 A checking circuit.

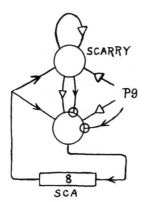

Fig. 15 Slow counter SCA.

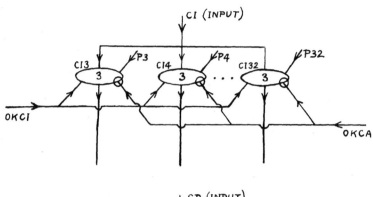

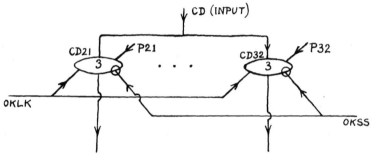

Fig. 16 Staticisers for CI, CD.

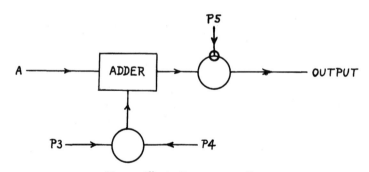

Fig. 17 Illustrating a convention.

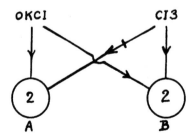

Fig. 18

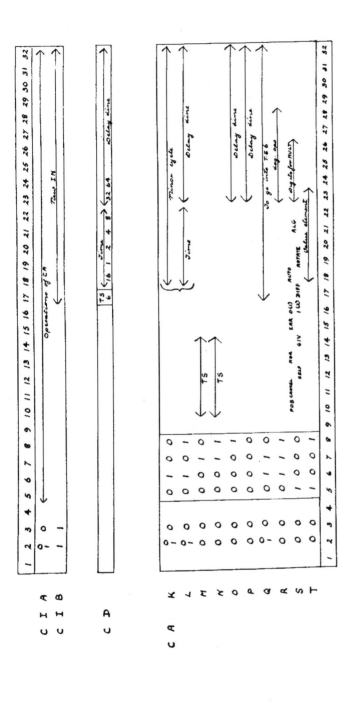

Fig. 19 Meanings of digits in minor cycles.

435

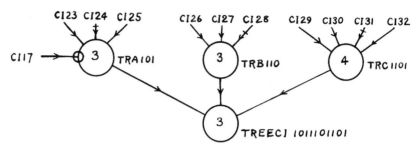

Fig. 20 A tree.

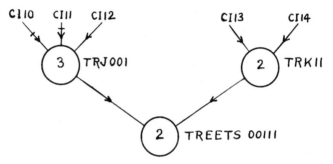

Fig. 21

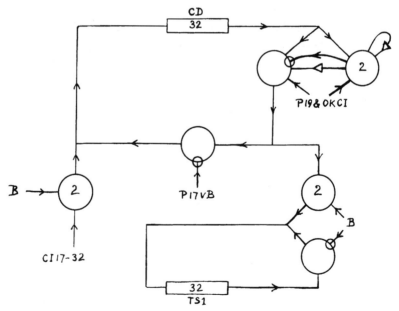

Fig. 22 Circuit for CD [part of LC].

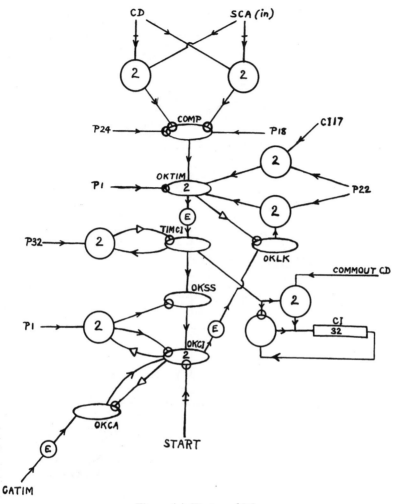

Fig. 22(a) Timing of LC.

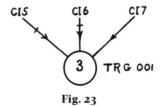

Fig. 23

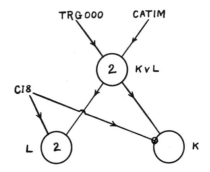

Fig. 24

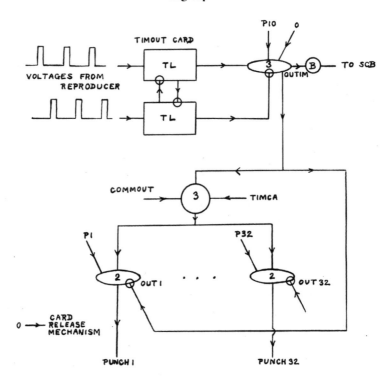

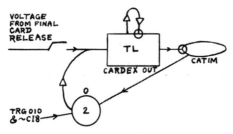

Fig. 25 Output circuit.

438

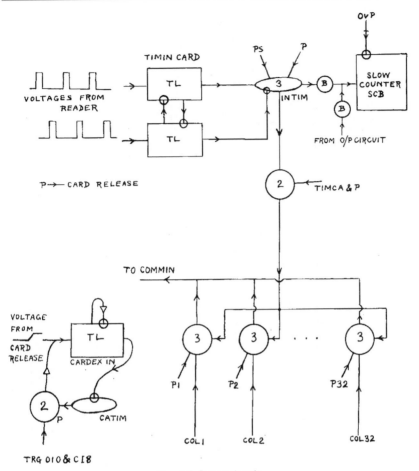

Fig. 26 Input circuit.

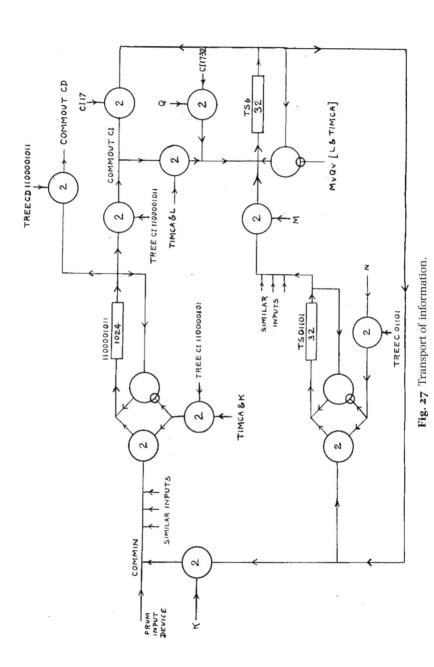

Fig. 27 Transport of information.

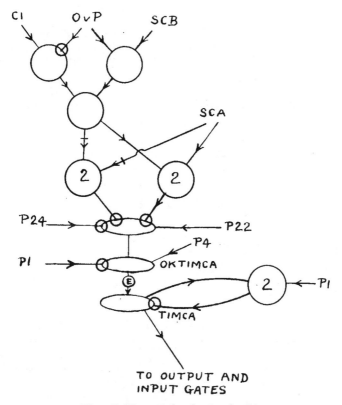

Fig. 28 Minor cycle selection for CA.

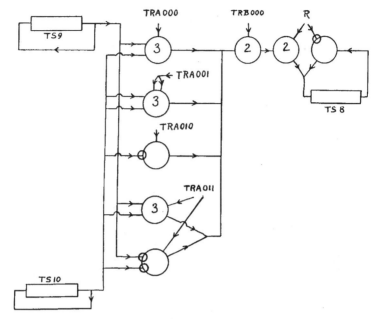

Fig. 29 Logical operations and TS 8.

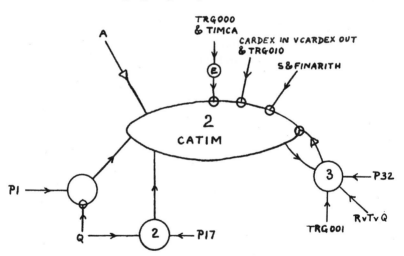

Fig. 30 CATIM.

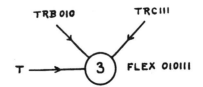

Fig. 31 Type T.

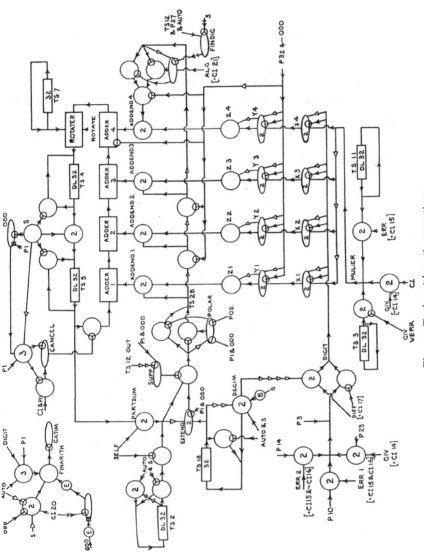

Fig. 32 Truly arithmetic operations.

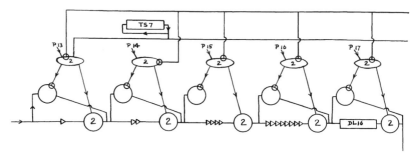

Fig. 33 The rotater.

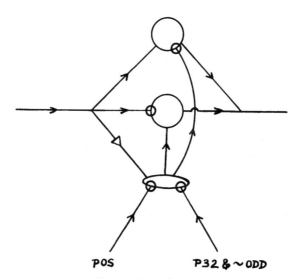

POS   P32 & ~ODD

Fig. 34 Sign changer.

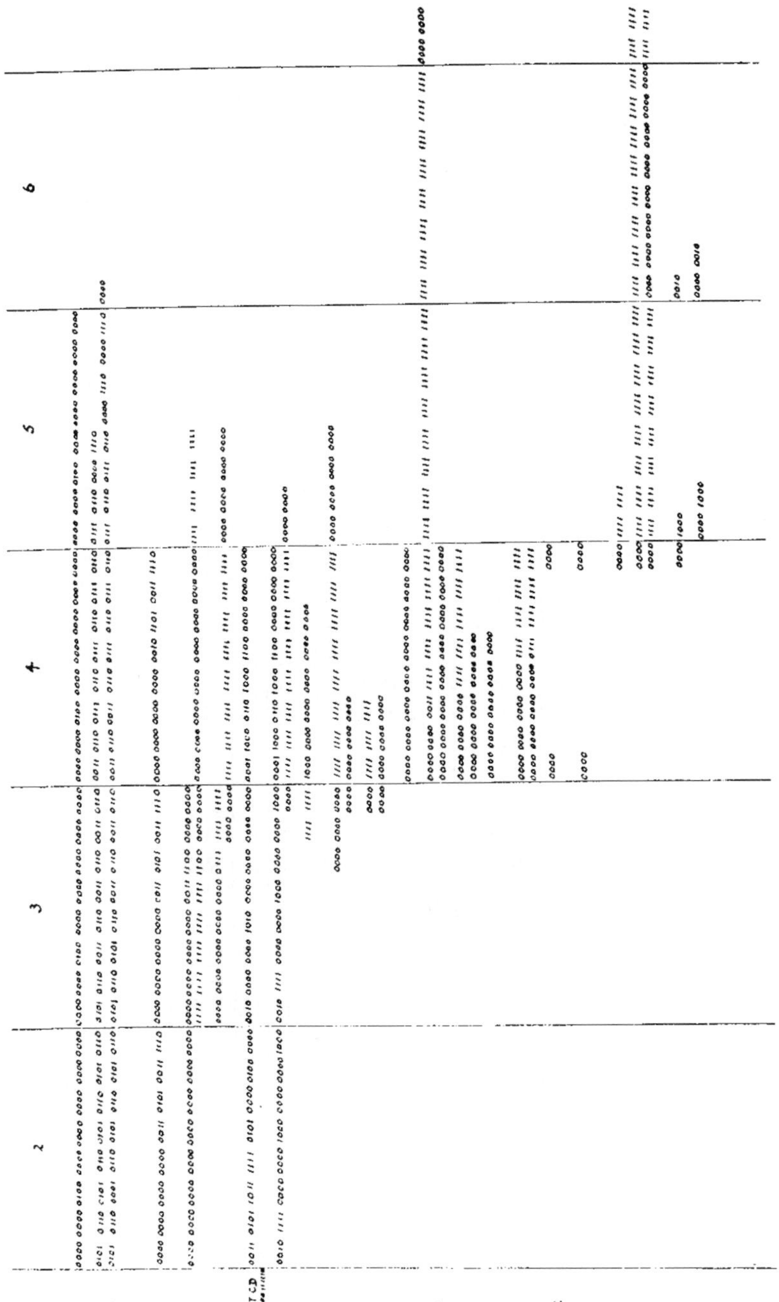

Fig. 35 LC preliminaries to a multiplication.

445

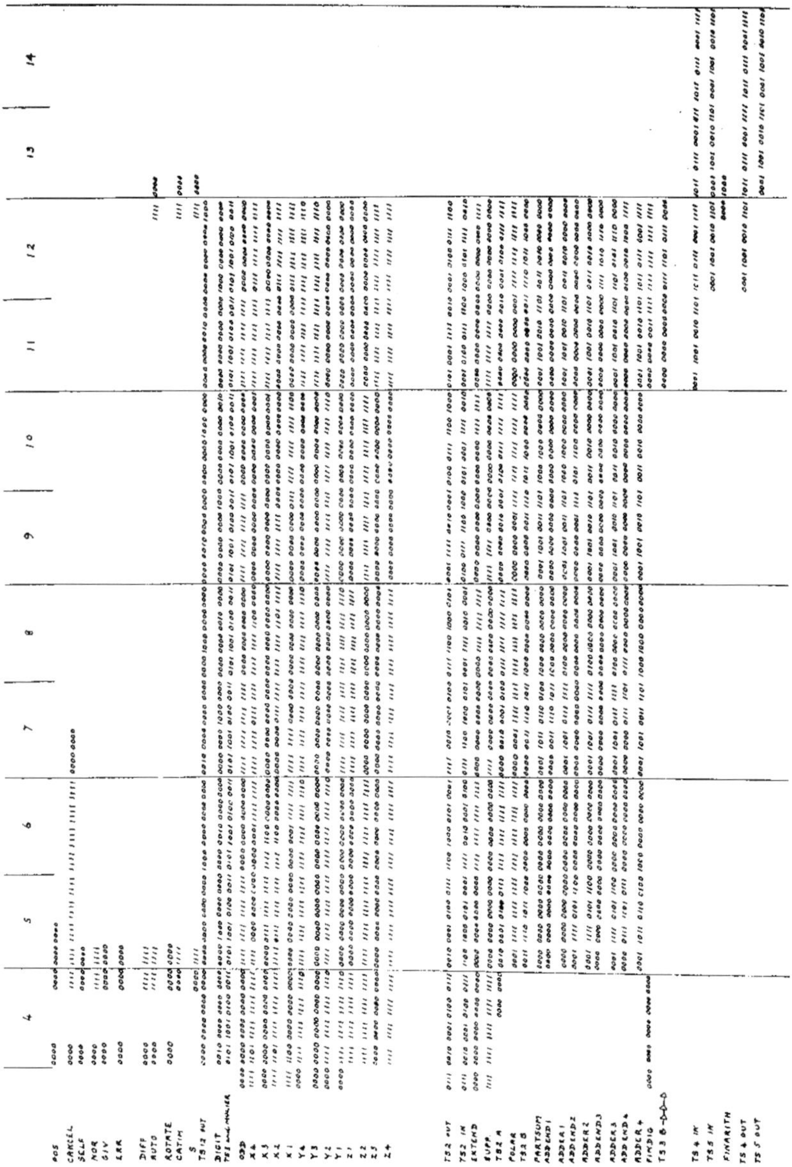

Fig. 36 A multiplication.

	1	2	3	4	5	6	7	8	9	10	11	12	13	14	15	16	17	18	19	20	21	22	23	24	25	26	27	28	29	30	31	32
										POS	CANCEL	SELF	NOR	GIV	ERR	1 / 2	DIFF	AUTO	ROTATE													
Addition			0	1	1	0	0	0		1	0	0	0	1	0		0	0	0					1	0	0	0					
Subtraction			0	1	1	0	0	0		0	0	0	0	1	0		0	0	0					1	0	0	0					
Short Mult. A (×6)			0	1	1	0	0	0		1	0 / 1	0	0	1	0		0	0	0					0	1	1	0					
Short Mult. B (×6)			0	1	1	0	0	0		1	1	1	0	1	0		0	0	0					0	1	1	0					
Short Mult. C (×6)			0	1	1	0	0	0		1	0	1	0	1	0		0	0	0					1	0	1	0					
Long Mult. + add			0	1	1	0	0	0		1	0	0	1	0	0		0	1	0													
Long Mult. + subtract			0	1	1	0	0	0		0	0	0	1	0	0		0	1	0													

Fig. 37 Instructions for certain arithmetical operations.

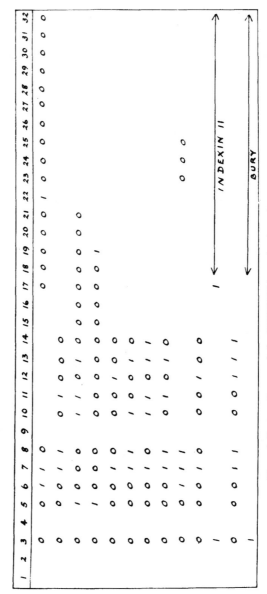

Fig. 38 Instruction cards for INDEXIN (genuine input).

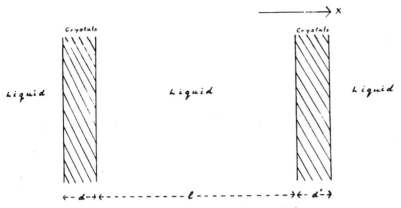

Fig. 40 Ideal arrangement of crystals.[4]

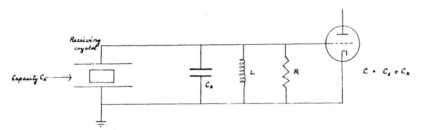

Fig. 41 Receiving crystal circuit.

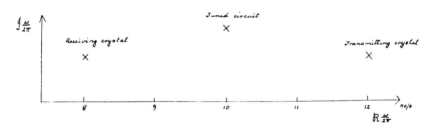

Fig. 42 Suggested arrangement of poles of R(w).

---

[4] Editor's note. Figures 39, 43 and 49 are absent and are not mentioned in the text.

Alan M. Turing

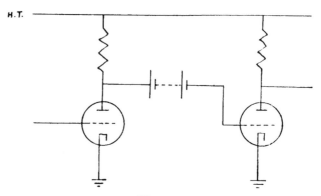

Fig. 44

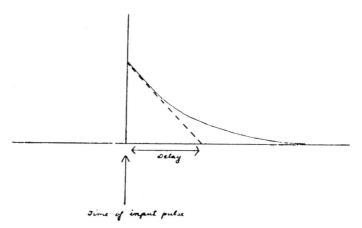

Delay

Time of input pulse

Fig. 44a

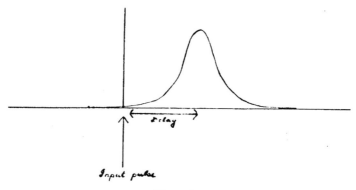

Delay

Input pulse

Fig. 44b

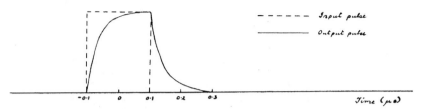

Fig. 45 Probable effect of input capacity on a square pulse such as the clock pulse.

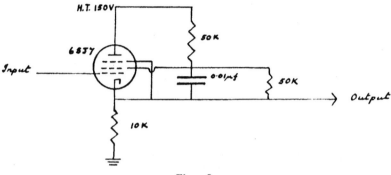

Fig. 46

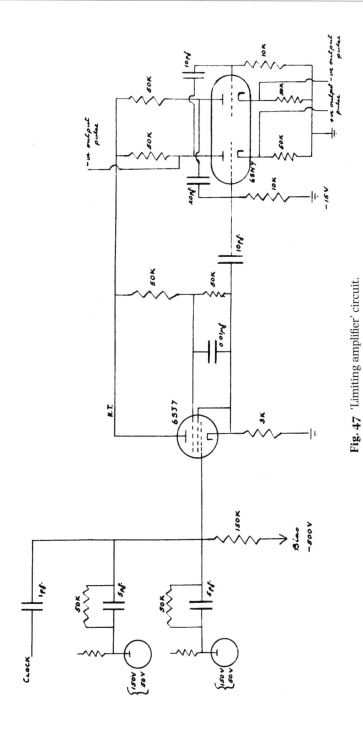

Fig. 47 'Limiting amplifier' circuit.

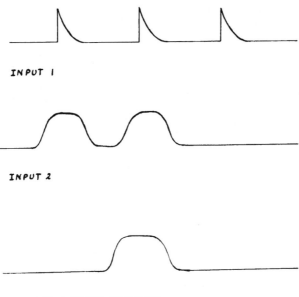

INPUT 1

INPUT 2

INPUT TO CATHODE FOLLOWER

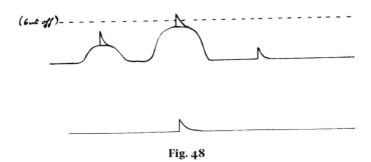

(cut off)

Fig. 48

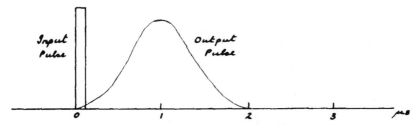

Input
Pulse

Output
Pulse

Fig. 50 Indicial response derivative for unit delay (preferable form).

453

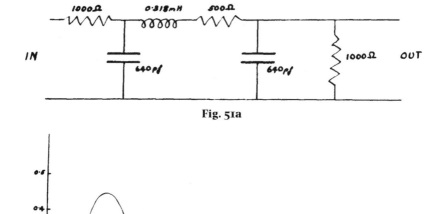

Fig. 51a

Fig. 51b A possible unit delay circuit and corresponding indicial response derivative.

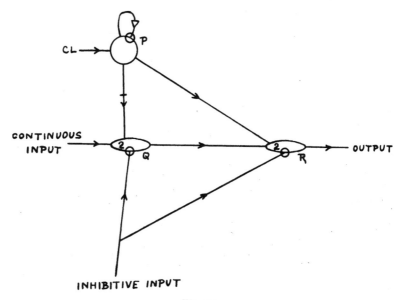

Fig. 52

454

21 Notes on memory (1945)
Alan M. Turing

The following fragments by Turing are from a draft of 'Proposed Electronic Calculator'.[1] The fragments survive only because Turing used the typed sheets as scrap paper, covering the reverse sides with rough notes on circuit design.[2] This material is of interest chiefly because of its remarks concerning the universal machine of Turing's 1936 paper 'On Computable Numbers'.[3] Notoriously, the universal machine received no explicit mention in the final version of 'Proposed Electronic Calculator'. In 1947 Turing described electronic stored-program digital computers as 'practical versions of the universal machine' and it is clear that in designing the ACE, his aim was in effect to replace the paper tape of the universal machine with a practical form of memory for holding instructions and data and to replace the abstract 'scanner' of the universal machine by a 'central pool of electronic equipment'.[4]

B. J. C.

[1] Woodger in interview with Copeland (June 1998); Woodger's comments on item M15/76, typescript, February 2000, in the Woodger papers, National Museum of Science and Industry, Kensington, London. The title 'Notes on Memory' originates with this volume.

[2] Woodger kept the notes, which are in the Woodger Papers (catalogue reference M15/76).

[3] Turing, A. M. (1936) 'On computable numbers, with an application to the Entscheidungs-problem', *Proceedings of the London Mathematical Society*, Series 2, 42 (1936–7), 230–65.

[4] Turing, A. M. (1947) 'Lecture on the Automatic Computing Engine', in B. J. Copeland (ed.) *The Essential Turing*, Oxford: Oxford University Press; the quotations are from p. 383.

General considerations and limitations[5]

<u>Logarithmic accessibility</u>.[6] In 'Computable numbers' it was assumed that all the stored material was arranged linearly, so that in effect the accessibility time was directly proportional to the amount of material stored, being essentially the digit time multiplied by the number of digits stored. This was the essential reason why the arrangement in 'Computable numbers' could not be taken over as it stood to give a practical form of machine. Actually we can make the digits much more accessible than this, but there are two limiting considerations to the accessibility which is possible, of which we describe one in this paragraph. If we have N digits stored then we shall need about $\log_2 N$ digits to describe the place in which a particular digit is stored. This will mean to say that the time required to put in a request for a particular digit will be essentially $\log_2 N \times$ digit time. This may be reduced by using several wires for the transmission of a request, but this might perhaps be considered as effectively decreasing the digit time.

<u>Organisation of large storages. Unit storages</u>.[7] In any very large storage system it is to be expected that the whole is divisible into a large number of comparatively small <u>unit storages</u>. Each of these unit storages may be responsible for the storage of a number of digits which is neither particularly large nor particularly small; 10,000 might be considered a reasonable number to store in a unit storage. Each unit storage is provided with a 'municipal' administration. The purpose of the municipal administration is to accept orders coming along 'telephone' lines from the central administration (or central control CC[8]) requesting particular information, to set going its municipal machinery to obtain this information, and to send it back over the telephone lines. ...

<u>Purely electrical delay lines</u>.[9] Delay lines of this kind, other than wave guides, would be useful for storing a single minor cycle. Two forms suggest

---

[5] Editor's note. This heading is labelled 'C' by Turing.

[6] Editor's note. This paragraph is numbered '(i)' by Turing.

[7] Editor's note. Numbered '(iii)'.

[8] The term 'CC' is from von Neumann's 'First Draft of a Report on the EDVAC'. By the time of the final version of 'Proposed Electronic Calculator' Turing was using the term 'LC' (logical control).

[9] Editor's note. Numbered '(xix)'.

themselves, those made with continuously distributed capacity and induct-
ance and those with lumped constants. In the latter case the number of
sections should be at least equal to the delay in radians for the highest
frequency passed, and should probably be several times this. The delay in
radians is of the order of hundreds, so that we may ignore this possibility.
The distributed type will probably consist of a closely wound coil, with a high
permeability core, and having capacity to a concentric sheath. . . .

22 The Turing–Wilkinson lecture series (1946–7)

Alan M. Turing and James H. Wilkinson

The Lectures[1]

The nine lectures published here were given by Turing and his assistant Jim Wilkinson during the period December 1946 to February 1947.[2] The lectures add substantially to our knowledge of the evolution of the design of the ACE. Turing and Wilkinson describe Versions V, VI, and VII of the design.

In 1980 Wilkinson wrote:

> I joined NPL in May 1946 . . . When I arrived Turing was working on what he called Version V . . . Documentation was not a strong point of Turing's work, and I never saw anything of Versions I to IV. . . . Later in 1946, M. Woodger joined the ACE section, and the three of us worked together. Our main effort was devoted to modifying the logical design of ACE in the light of experience gained in trying to program the basic procedures of numerical analysis. Version V was quickly abandoned and replaced by Versions VI and VII, which were essentially

---

[1] This introduction by Jack Copeland.

[2] The account of the lectures provided by Hodges in his biography of Turing is inaccurate (Hodges, A. (1983) *Alan Turing: The Enigma*. London: Burnett). Hodges stated that the lectures ended in January 1947 (p. 353) and that Turing gave '[o]nly the first two and part of the last' (p. 559). In fact, the series ran until 13 February 1947 and Turing gave half the lectures.

four-address code machines in which each instruction was of the form

A FUNCTION B → C, NEXT INSTRUCTION D.[3]

Version V was simpler than Versions VI and VII, each instruction essentially having the form:

SOURCE, DESTINATION, POSITION OF NEXT INSTRUCTION.

Both Huskey's Test Assembly (also known as Version H) and the Pilot Model ACE were based on Version V.[4] Little is known about Versions I to IV, said by Woodger to post-date Turing's 'Proposed Electronic Calculator'.[5]

The lecture series was proposed in a memo dated 16 November 1946 from Womersley, Superintendent of the Mathematics Division, to Darwin, Director of the NPL:

ACE. Proposal for Lectures by Dr Turing

At an informal meeting yesterday with some Service representatives and Dr Porter of the Military College of Science, a suggestion was made that Dr Turing, rather than explaining his machine to a number of isolated people on many different occasions, should conserve his time by giving a course of lectures intended primarily for those who will be concerned with the technical development of the machine and possibly also with giving advice on matters connected with its components. In view of the fact that the Post Office have a contract for a computing machine for a specific Service application which it is intended shall be constructed on ACE principles, though of course on a very much smaller scale, the future users of the small machine should also be invited. After some discussion the following list of those who should be invited was drawn up.

Post Office	4	Military College of Science	2
The Gramophone Co.	1	A.5	2
British Thomson Houston Co.	1	T.R.E.	2
Standard Telephones	1	A.G.E.	1
Cinema Television Ltd	1	R.R.D.E.	1
		A.S.E.	1

---

[3] Wilkinson, J. H. (1980) 'Turing's work at the National Physical Laboratory and the construction of Pilot ACE, DEUCE, and ACE', in N. Metropolis, J. Howlett, and G. C. Rota (eds) *A History of Computing in the Twentieth Century*. New York: Academic Press; the quotation is from pp. 102–3.

[4] See Wilkinson's chapter 'The Pilot ACE at the National Physical Laboratory'.

[5] Letter from Woodger to Copeland, 25 February 2003.

Additional invitations:

Prof. D. R. Hartree   Dr M. V. Wilkes

Prof. H. S. W. Massey   3 members of the staff of the
                                  Mathematics Div.

It will be noted that no-one is invited from Professor Newman's department in Manchester; this is because they have already had discussions with Dr Turing. The number from the Mathematics Division is limited to 3 because Dr Turing has already given a series of lectures to members of our staff. After some discussion it was agreed that the Ministry of Supply should provide a suitable lecture room in the Adelphi. This has already been arranged. ... The reason for holding these lectures at the Adelphi is that it will be possible in the particular room chosen to provide desk space for each person attending so that he can take copious notes. This will be necessary for those who are to be engaged in the actual design and construction of equipment.[6]

The lectures were held between 2 p.m. and 5 p.m. on successive Thursday afternoons from 12 December 1946 to 13 February 1947 (Boxing Day excepted), in a rather dingy underground room at the Headquarters of the Ministry of Supply, then housed in the Adelphi Hotel, Baker Street, London.[7] Advertised by Womersley as 'lecture–discussions', the lectures resembled tutorials in style.[8] (Womersley's timetable set aside 50 minutes of each weekly session for discussion and criticism by the audience.[9]) It was initially proposed

---

[6] Memo from Womersley to the Director, 16 November 1946 (in the Woodger Papers (National Museum of Science and Industry, Kensington, London; catalogue reference M15); a digital facsimile is in The Turing Archive for the History of Computing <www.AlanTuring.net/womersley_darwin_16nov46>). Womersley's handwritten notes and correspondence concerning the arrangements for the lectures are also in the Woodger Papers (catalogue reference M15); digital facsimiles are in The Turing Archive for the History of Computing <www.AlanTuring.net/womersley_notes_22nov46>. In the end Kilburn attended from the Manchester group; see Chapter 5.

[7] Letter from Wilkes to Copeland, 11 April 1997; Womersley's handwritten notes (see note 6).

[8] Letter from Wilkes to Copeland (see note 7).

[9] Hodges inferred from the fact that Womersley invited '[d]iscussion - in particular, criticism of Dr Turing's technical proposals' (the quotation is from Womersley's memo announcing the lectures) that '[t]hey did not trust him to know what he was talking about' (Hodges, *Alan Turing: The Enigma*, p. 353). This seems far fetched. Critical discussion is, after all, a normal and healthy feature of academic life.

that Wilkes should take an official record of the lectures.[10] Wilkes presumably declined, and the task fell to T. A. H. (Tommy) Marshall, of the Mechanical and Optical Instruments Branch of the Military College of Science, Shrivenham. (Marshall's specialism was servomechanisms.) It is Marshall's notes of the lectures that are published here.[11] Two other sets of lecture notes are known to survive: a complete set of handwritten notes taken by J. G. L. Michel,[12] of the NPL Mathematics Division, and handwritten notes concerning Lectures 6 and 7 taken by Hartree.[13]

Marshall had his notes typed, dividing them into numbered, headed sections. The notes were distributed in booklet form under the title 'The Automatic Computing Engine' (Military College of Science, Shrivenham, February 1947). Comparison of the three sets of notes indicates that Marshall refrained from supplementing the lecture material. He kept the material in more or less its original order. The few exceptions are as follows (section numbers refer to the table of contents below).

1. Marshall placed material concerning delay lines that originally straddled Lectures 1 and 2 into a single appendix, which he located following the text of Lecture 9. This appendix now forms Section 3: 'Supersonic Mercury Delay Lines'.

2. The material in Section 10, 'Sources and Destinations', is drawn mostly from the beginning of Wilkinson's first lecture, but Marshall seems to have prefaced the section with some material drawn from the previous lectures by Turing.

---

[10] 'Notes of a meeting held in the Director's room', 22 November 1946 (in the Woodger Papers (catalogue reference M15); a digital facsimile is in The Turing Archive for the History of Computing <www.AlanTuring.net/womersley_notes_22nov46>).

[11] Marshall's notes are published by permission of the Principal of Cranfield University. (The notes were first published in Copeland, B. J. (ed.) (1999) 'The Turing–Wilkinson lecture series on the Automatic Computing Engine', in K. Furukawa, D. Michie, and S. Muggleton (eds.), *Machine Intelligence* 15. Oxford: Oxford University Press.) A number of typographical errors have been corrected and punctuation marks have been added and removed. The figures have been renumbered. Handwritten corrections marked on Marshall's typescript and initialled 'JHW' have been incorporated.

[12] Michel's notes are in the Woodger Papers (catalogue reference M15/79).

[13] These are among Hartree's papers in the Library of Christ's College, Cambridge. A note by Wilkes that is attached to them wrongly states that the series contained only seven lectures and ran from December 1946 to January 1947. Wilkes himself attended only the first two lectures of the series (letter from Wilkes to Copeland (see note 7)). Hartree attended most or all of the lectures, but the remainder of his notes seem not to have been preserved.

3. Section 14, 'Source and Destination Trees', appears to be part of Lecture 8, and was originally flanked by the material in Sections 24 and 25. Why Marshall moved this section to an earlier position in the exposition is not clear, especially since doing so destroys the otherwise smooth progression in the lectures from Version V through Version VI to Version VII. However, Marshall's arrangement has not been disturbed.

4. Section 15, 'Some Examples of Simple Instructions', came earlier in the exposition, originally lying between the material in Sections 11 and 12.

5. Hartree's lecture notes show that Marshall reversed the order of Sections 23 and 24. The original order has been restored.

6. Sections 27, 28, and 29 are the text of Turing's final lecture (which is titled 'Mechanical Details' in Michel's notes). Marshall relegated this lecture to an appendix, which he called 'Some Possible Circuits'.

Except in respects just noted, the correspondence between the table of contents and the actual lectures is this. Lecture 1 (December 12): Sections 1–4; Lecture 2 (December 19): Sections 5–9; Lectures 3–4 (January 2 and 9): Sections 10, 11, 15, 12, 13 (it is not known where the break between these lectures fell); Lecture 5 (January 16): Sections 16–18; Lecture 6 (January 23): Sections 19–22; Lecture 7 (January 30): Sections 23, 24; Lecture 8 (February 6): Sections 14, 25, 26; Lecture 9 (February 13): Sections 27–29.

As the table of contents shows, 14 of the 30 sections were delivered in whole or part by Wilkinson. Womersley's original intention was that the lectures be given by Turing alone. However, at Darwin's request Turing attended a symposium on digital calculating machinery held at Harvard from 7 to 10 January, and Wilkinson deputized for him during the period of his absence.[14] Wilkinson had joined the NPL only seven months previously, with no prior experience of either electronic engineering or computer design. His official position was that of Turing's half-time assistant.[15] To what extent Wilkinson made use of lecture material provided for him by Turing is a matter for conjecture. Marshall thanks only Turing for 'permission to publish this account in its present form'.

---

[14] Minutes of the NPL Executive Committee, Paper E.910, 15 April 1947 (NPL library; a digital facsimile is in The Turing Archive for the History of Computing <www.AlanTuring.net/npl_minutes_apr1947>).

[15] Wilkinson, in interview with Christopher Evans in 1976 (*The Pioneers of Computing: An Oral History of Computing*, London: Science Museum).

THE AUTOMATIC COMPUTING ENGINE
Table of Contents

1. Introduction.

Digital computing machines of various sorts have been built frequently during the last 300 years with varying degrees of success and have resulted in the various well-known commercial machines of the Brunsviga and Comptometer types. Such machines are invaluable for carrying out simple routine calculations. Their main defect however when used for more complex operations is the amount of labour involved in setting numbers into them and recording intermediate results for use in subsequent parts of the calculations.

The Hollerith machines did much to overcome these difficulties by using high speed electro-mechanical techniques with punched cards as a means of input and output. Even so these machines lacked an effective 'Memory' which is essential for the rapid solution of protracted calculations.

Much work was therefore devoted to the development of analogue type machines, familiar in most Service computing instruments, and the Differential Analyser became a very powerful and useful tool for the mathematician.

The requirements of the ballisticians during the war years brought about the rapid development of digital machines once again and in America the Automatic Sequence Controlled Calculator (ASCC) and the Electronic Numerical Integrator and Calculator (ENIAC) were built. The ENIAC was the most ambitious machine to date and was capable of very high working speeds. Its memory however was still inadequate for many problems.

The machine described here is the Automatic Computing Engine (ACE) which is being designed by the Mathematics Division of the National Physical Laboratory. It is to be a digital computer capable of performing algebraic processes at very high speeds by arithmetical methods. It will be completely flexible and able to cope with a variety of problems and will be fully automatic in operation. Computation will be performed in the binary scale by electronic means and the machine will incorporate a large 'Memory' for the storage of both data and instructions. The normal operating speed of the machine will be one million binary digits per second when all the required data is held within the 'Memory'. Input and output, which is by Hollerith machinery, is of necessity somewhat slower; some 2500 digits per second.

The final version of the ACE will probably contain about 512 'Memory' units capable of storing some 500,000 binary digits and will utilise something like 8000 valves.

2. Representation of Binary Numbers.

In the binary scale only the digits 0 and 1 exist, all numbers being composed of a series of these two digits. Since timing is an essential feature in a machine of this sort and since in graphical representation the time scale is normally from left to right on the paper, and further since in arithmetic processes it is necessary to commence operations with the least significant digit of a number, binary numbers are normally written with the least significant digit first, i.e. to the left. This is the reverse of normal decimal notation. An example will make this clear.

Decimal notation	Binary notation
0	000000......
1	100000......
2	010000......
10	010100......
47	111101......
etc.	

In the machine, the digit 1 is represented by a pulse of 1 μs duration; the digit 0 by the absence of a pulse. There is no separation between pulses. Thus a binary number might be represented by a voltage waveform as below:

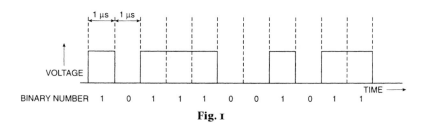

Fig. 1

The above waveform is idealised and is hardly achieved in practice since it would require an infinite bandwidth. In some portions of the machine the pulses are simple voltage pulses and in others are pulses of 15 Mc/s carrier.

Similar pulses are used for timing and sequencing operations throughout the machine.

The position of a pulse with respect to the time scale standard or nominal time may vary in different sections due to a variety of [causes]; similarly due to distortion and attenuation the pulses may be shortened.

Hence waveforms depart considerably from the ideal. Some typical shapes are shown below [Fig. 2].

In addition to representing 1, a pulse indicates 'Truth' in logical operations, and a 0 'Untruth'.

Since binary numbers composed of 32 digits are to be handled, it is convenient to consider 32 pulses (or 'no-pulses') as a group. Such a group is termed a <u>Word</u> and the time period during which it occurs, namely 32 μs, is termed a <u>Minor Cycle</u>. A group of 32 words occurs in 1024 μs and it is convenient to refer to this time period as a <u>Major Cycle</u>.

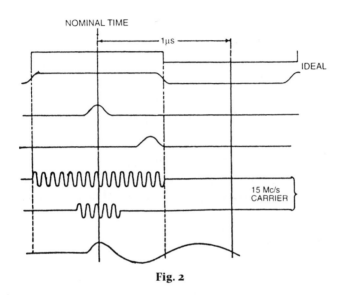

Fig. 2

3. <u>Supersonic Mercury Delay Lines.</u>

The method adopted in the machine for producing a time delay in the transmission of a pulse is an electro-mechanical one. The electrical pulse is converted to a longitudinal pressure wave in a column of mercury, which passes down the columns and is received at the far end after a finite time.

The mercury column is contained in a steel tube of diameter 1 inch. In order to achieve a delay of 1024 μs the tube is approx. 5 feet long.

At either end of the tube is fitted a piezo-electric quartz crystal arranged as shown diagrammatically below.

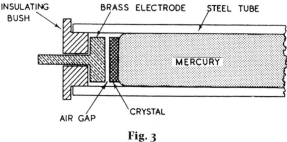

Fig. 3

The mercury forms one electrode and the brass plate the other, there being an air gap between the brass and the crystal to eliminate mechanical coupling.

The pulses are applied to the crystal as pulses of 15 Mc/s carrier of duration 1 μs. Carrier working is adopted for several reasons, chief amongst which is that in non-carrier working the variation of delays of various frequencies is excessive (as much as 4 μs in total non-carrier working for a band pass of 0 to $\frac{1}{2}$ Mc/s). Mercury is chosen as the medium for the propagation of the pressure wave since it has a reasonably constant attenuation of frequencies around 15 Mc/s and the variation of velocity of propagation with temperature at normal air temperatures is small. Further, the acoustic impedance of mercury is more closely equal to that of quartz at normal temperatures and therefore better acoustic matching is achieved than with other possible materials.

A comparison table of the pertinent constants of mercury, water and quartz are given for illustration.

	Mercury	Water	Quartz
Density	13.5 gms/ml	1.00	2.65
Velocity of propagation	1.5 Km/s	1.44	5.71
Temp. coeff. of velocity at 10°C	0.00030 per °C	0.001	—
Acoustic impedance	2.025×10^6	1.44×10^5	1.52×10^6

The conversion efficiency of a quartz-mercury combination is such as to produce a loss of 48 dB.

In order to keep the delay of a mercury line constant to within ±0.5 μs at 10°C (the maximum permissible tolerance) the temperature of the line must be kept within ±1.63°C. This is easily achievable in practice.

Trouble from multiple reflections and standing waves is not present in the long lines since the attenuation is sufficiently great to render the standing wave ratio very small. In the short lines, however, it is necessary to introduce an acoustic diffraction grating to produce artificial attenuation.

The use of lines for storage of pulses.

In order to store pulses in a line it is necessary to receive them as they arrive at the end and feed them back to the beginning. During the feed back the pulses must be amplified and reshaped. Facilities must also be provided to enable pulses to be put into and removed from the line.

The block arrangement is as below.

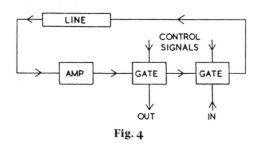

Fig. 4

The precise circuitry is by no means fixed at present, but it is probable that a super-regenerative amplifier will be used which in itself will provide the necessary gating.

4. Symbols Used in Diagrams.

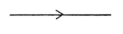

A direct connection. This assumes the instantaneous transmission of pulses between points thus connected. The arrow indicates the direction of flow of signals.

A delay of $1\,\mu s$ ('unit delay'). The triangle is directional.

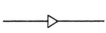

A short delay line of 'length' $32\,\mu s$. It is thus capable of storing a word of 32 binary digits.

A long delay line of 'length' $1024\,\mu s$. (Other 'lengths' may be indicated by the insertion of suitable figures.)

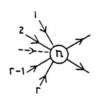

An element which emits a single pulse when it receives on any of its r input lines n or more simultaneous pulses. No pulse is emitted if less than n pulses are simultaneously received. The number n is called the Threshold of the element and elements are referred to by their threshold numbers e.g. a 'Three element'. In the symbol for 'One element' the figure 1 is usually omitted.

This indicates an <u>inhibitory connection</u> to an element. No pulse is emitted by such an element in the microsecond during which it receives a pulse via an inhibitory connection, irrespective of all other inputs.

A <u>Trigger Circuit</u>. Such a circuit of threshold n emits a continuous stream of pulses when it receives n or more simultaneous input pulses. It continues to emit pulses until such time as it receives a pulse on its inhibitory connection, when it stops.

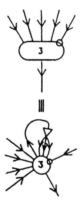

A trigger circuit may be produced from a simple element by the insertion of a unit delay in an output lead and feeding back into the element on n input leads. For example, a trigger circuit of threshold 3 may be produced from a 3-element as shown.

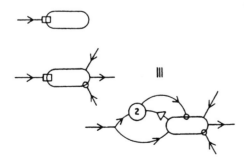

This indicates a connection to a trigger circuit such that when a pulse is received the state of the circuit is changed; i.e., if it was emitting pulses it stops, and vice versa. This connection is equivalent to the additional circuitry on a trigger circuit as shown.

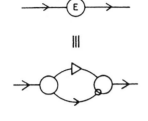

An End element. This gives a pulse out during the microsec after the termination of a series of input pulses.

e.g.

    Input   001110001100

    Output   000001000010

It is made up of two one-elements interconnected as shown.

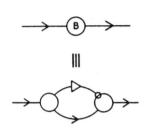

A Beginning element. This gives a pulse out during the first microsec of a series of input pulses.

e.g.

    Input   001110001100

    Output   001000001000

It is made up of two one-elements interconnected as shown.

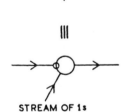

STREAM OF 1s

A polarity changer. Pulses are changed to no-pulses and vice versa.

e.g.

    Input   001110001100

    Output   110001110011

This may be made up as shown.

The above notation covers all the simple units from which the computing and control circuits are constructed. Details of the circuitry of each unit is given in [Sections 27–29].

5. Logical Operations.

It is desirable to have circuits which perform the equivalent of the terms 'and' 'or' and 'not'.

These may be very simply developed from the elements already described as indicated below. In these logical operations a pulse indicates True and no pulse indicates False.

A 'and' B. (A & B)

Suppose it is required to ascertain if two pulses coexist in two channels simultaneously, i.e. A & B.

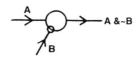

Fig. 5

If each channel is fed to a 2-element, then the element will emit a pulse only when pulses occur on A and B simultaneously, i.e. a pulse on the output indicates 'A & B'.

A 'or' B. (or both) (A ∨ B)

In this case each channel is fed to a 1-element.

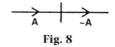

Fig. 6

The element emits a pulse when there occurs a pulse on A or B or both.

A 'and not' B. (A & ~B)

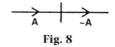

Fig. 7

In this case a pulse is emitted only when a pulse exists on A and no pulse on B.

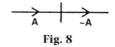

Fig. 8

6. Adding Circuit.

The process of addition is obviously of fundamental importance in a computing machine. It should be borne in mind that the addition of two numbers in reality involves the addition at each stage of three digits, since in general a carry digit has to be added as well as the two addends.

Further, since binary numbers are written with the least significant figure to the left, addition commences at the left and proceeds to the right. An example is given below.

Carry digit	01111100110
A	01011100110 0
B	11100101100 1

Sum	10000111001 1

An adding circuit, operating digit by digit, must therefore be able to distinguish the following combinations of input signals and react in the corresponding manner.

Input			Output	
A	B	Carry	Sum	Carry
0	0	0	0	0
1	0	0 (or permutation)	1	0
1	1	0	0	1
1	1	1	1	1

Since there are three possible input and output signals, ignoring the '0 & carry 0' answer which is implicit, three elements will be required to construct the circuit. It is shown below [Fig. 9].

A single pulse on any one of the A, B, or Carry channels will cause the 1-element to emit a pulse. The output of the 1-element therefore goes straight to the adder output.

Two simultaneous input pulses should, however, produce a 0 in the sum and a carry 1. The 2-element output is therefore used to inhibit the 1-element (which therefore gives 0) and also, when delayed by 1 μs, the next carry digit.

Three simultaneous input pulses produce a carry 1 in the same way and the 1-element is inhibited. The 1 in the sum is produced by the 3-element which is now stimulated.

The adding circuit is usually drawn as below [Fig. 10].

It should be noted that the outputs of the 3-element and the 1-element can be connected together directly, since if the 3-element is emitting the 1-element is always inhibited and vice-versa.

Such a circuit as described will add continuous streams of pulses, digit by digit. However, in the machine it is usually desired to add together words of

32 digits. Since there is no spacing between successive words, a carry digit formed by the addition of the last two digits in the words would be added in to the first pair of the next words. To prevent this, i.e. to break the trains of pulses up into their correct words, the carry digit is suppressed during the first microsec of each word.

	End of Word	Beginning of Word	
Carry	01001	[1]0001.....	([1] suppressed by carry
A	10111	1011.....	suppression)
B	10011	0101.....	
A + B	01101	1110.....	

This is achieved by the modification to the circuit as below [Fig. 11].

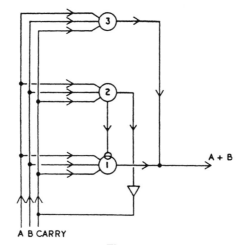

Fig. 9

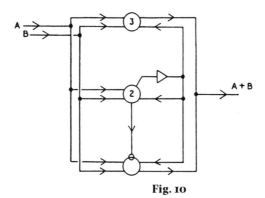

Fig. 10

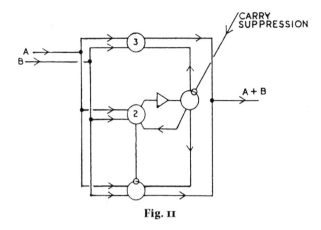

Fig. 11

7. <u>Ring Counter</u>.

It was seen that in the adding circuit it was necessary to suppress the carry digit every 32 μs in order to divide a continuous train of pulses up into words. This calls for some device to count up to 32 μs and then emit a pulse. This may be done by a ring-counter.

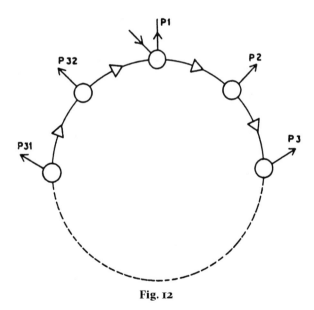

Fig. 12

This is a ring of 32 1-elements connected together through unit delays. When once stimulated, a pulse continues to go round the ring, making the complete circuit in 32 μs.

Timing pulses may be taken from any of the 1-elements and these are termed P1, P2 etc., to P32. P1 corresponds to the first microsec of a word and P32 to the last.

Hence in the adder circuit the carry suppression pulse would be P1.

These timing pulses are used throughout the machine for gating operations.

8. Staticiser.

This is a means of converting a word which is available in a minor cycle onto a set of 32 trigger circuits.

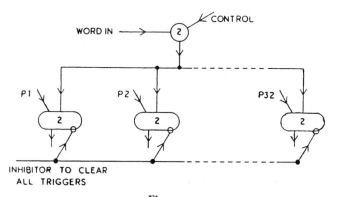

Fig. 13

The 32 trigger circuits each of threshold 2 are connected as shown. When it is required to set up a word on them, the gate (the 2-element) is opened by the control signal and the pulses are applied to all the trigger circuits. To each of the triggers, however, is fed P1 to P32 respectively and therefore only those circuits which received a digit pulse together with a P pulse are turned on, the rest remaining off.

The triggers may be cleared by a pulse on the inhibitor connection which turns all of them off ready for the next word to be staticised.

9. Dynamiciser.

This is a circuit which performs the reverse process to the staticiser in that it converts a set of voltages available on 32 trigger circuits, or from some other source, into a 32 digit word occupying a minor cycle.

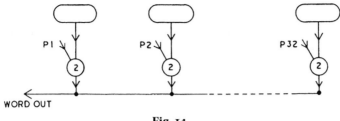

Fig. 14

10. <u>Sources and Destinations</u>.

The 'Memory' of the ACE is composed of a large number of supersonic mercury delay lines. In general, these are either of 'length' 32 μs or 1024 μs, i.e. they delay a pulse by a time corresponding to their length.

If now the output of a line is fed back, after amplification, to the input, a pulse may be caused to travel repeatedly down a line and back through the amplifier for as long as required. Since the lines are either of 32 μs or 1024 μs length they can, with the feedback circuit, be used to store 32 or 1024 pulses each respectively.

<u>Notation</u>.

A short line is designated by a reference number.

A pulse is designated by a subscript at the corresponding position from the output end at the beginning of a minor cycle.

Throughout the machine, pulses are fed from one part to another via the HIGHWAY.

Transfer of numbers is achieved by indicating a SOURCE and a DESTINATION by the control system. Operations are performed on the numbers by coded instructions also from the control system.

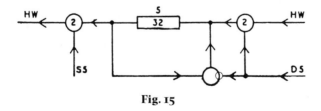

Fig. 15

All D and S pulses have a duration of 1 minor cycle (32 μs). Thus a word on the HW may be put into Temporary Storage Line 5 (TS 5) by a pulse D 5.

This opens the 2-element and admits the word to the line, at the same time inhibiting the feedback loop and therefore clearing any pulses that might have previously been in the line.

D 5 ceases at the end of the minor cycle and the word therefore continues to circulate indefinitely in the line circuit.

To obtain the word from the line a pulse S 5 of duration one minor cycle opens the 2-element and the word passes out to the HW.

This system is adopted throughout the machine.

Certain lines are used exclusively for certain purposes, in Version V.

For example, lines 2 and 3 are always used for <u>addition</u>.

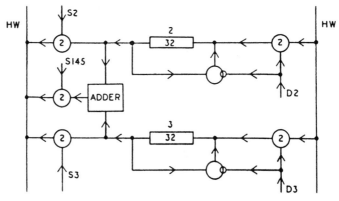

Fig. 16

The arrangement is as above. D 2 admits a word to TS 2 and D 3 to TS 3. The lines feed to an adder and the sum of the words in TS 2 and TS 3 may be passed to HW by a pulse S 145.

The contents of TS 2 and TS 3 may be obtained separately by pulses S 2 and S 3 respectively.

<u>Note.</u> S 145 <u>always</u> gives the sum of whatever is in TS 2 and TS 3 and all addition is carried out here in Version V.

Similarly TS 8 and TS 9 are always used for logical operations, in Version V [Fig. 17].

S 131   gives   TS 8 & TS 9
S 132   gives   TS 8 $\lor$ TS 9
S 133   gives   TS 8 $\not\equiv$ TS 9
S 134   gives $\sim$TS 8

Another special facility is for the multiplication of small numbers by 2^n (where $n = 1, 2, 3, 4$ or 5) in TS 5 [Fig. 18].

The doubling, quadrupling etc., is performed by delaying words 1, 2 μs etc. E.g.

Decimal	Binary
13	1011000.....
26	0101100.....
52	0010110.....
	etc.

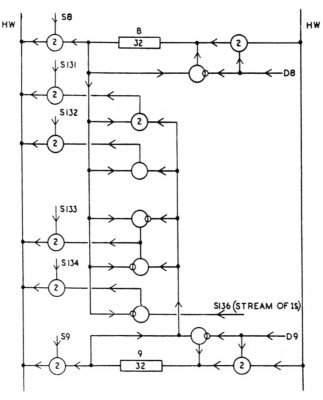

Fig. 17

11. Table of Standard Sources.

This table gives the more commonly used Sources and Destinations.

Note. TS refers to short line (32 μs)

DL refers to long line (1024 μs)

$\left.\begin{matrix} 0 \\ 1 \end{matrix}\right\}$ TS for instructions.

$\left.\begin{matrix} 2 \\ 3 \end{matrix}\right\}$ TS for addition.

$\left.\begin{array}{r}4\\5\\6\\7\end{array}\right\}$ TS for general work.

$\left.\begin{array}{r}8\\9\end{array}\right\}$ TS for logical operations.

$\left.\begin{array}{r}16\\\text{to}\\31\end{array}\right\}$ Ring of 16 TS's.

$\left.\begin{array}{r}64\\65\end{array}\right\}$ Lines of length 64 μs.

66   64 + 65
67   −64 (i.e. 2^{32} − TS 67).

84   64 Δ 1 μs (64 delayed 1 μs), i.e. multiplied by 2.
85   64 Δ 2 μs
to
89   64 Δ 32 μs

96   TC 96 special discriminating trigger circuit.
97   Inhibit TC 96.

$\left.\begin{array}{r}98\\99\end{array}\right\}$ Trigger circuits.

131   TS 8 & TS 9
132   TS 8 ∨ TS 9
133   TS 8 ≢ TS 9
134   ∼ TS 8   (i.e. 2^{32} − 1 − TS 8).
135   0   a stream of 0's.
136   2^{32} − 1   a stream of 1's.
137   ∼ (8 & 9)
145   2 + 3
146   −3   (i.e. 2^{32} − TS 3).
148   Multiplier
149   −2   (i.e. 2^{32} − TS 2).
150
151

152	$5 \Delta 1 \mu s$
to	
156	$5 \Delta 16 \mu s$
158	$260 \Delta 32 \mu s$
160	P32
161	P1 Ring counter
to	
191	P31

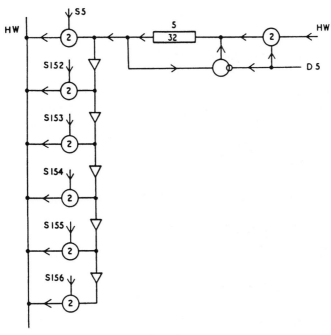

Fig. 18

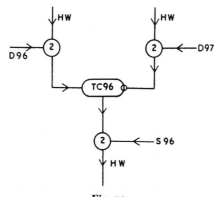

Fig. 19

This [Fig. 19] is a special trigger circuit which is ON if stimulated by D 96 and is put OFF by D 97.

12. Programming (Version V).

An instruction must order the transfers of words to take place and order the operations to be performed on them in correct sequence.

For example, a series of instructions might be 'Take the fourth word in DL 270 and the sixth word in DL 280, add them and put the sum in DL 10'.

This would be ordered as:

1) DL 270.4 routed to TS 2
2) DL 280.6 routed to TS 3 (since TS 2 & TS 3 have adding facilities)
3) S 145 (output of adder) routed to DL 10.

More complicated instructions are built up in this manner and must be followed by the machine in the correct sequence. Since transfer of words and operations take definite time periods to be achieved, the instructions must include information to show the machine how long to allow for each successive stage of the whole series of instructions.

Only one instruction can be obeyed at any one time, in general.

Form of instruction.

Each instruction is in the form of one word of 32 binary digits. The digits form groups as follows.

Digits	1–2	3–12	13–22	23–26	27–32
	Spare	Source	Destination	Characteristic	Timing
	(not used)	Numbers	Numbers		Number
		(0–1023)	(0–1023)		(0–63)

Source number. This indicates from where a word is to be taken, e.g. TS 10.

Destination number. This indicates to where a word is to be routed, e.g. TS 11.

Characteristics.

(1) Digit 23.   0 indicates external operation.

    1 indicates internal operation.

(This use is obsolete. All operations are internal and 23 is therefore always 1.)

(2) Digit 24. This affects the timing number (q.v.).

(3) Digits 25 & 26. These two digits indicate the origin of the next instruction.

All instructions come from one of four sources. The four possible combinations of 0 and 1 for digits 25 and 26 indicate from which source the next instruction is to be drawn as follows.

Digits	Source	
00	TS 0	(Length 32)
10	TS 1	(Length 32)
01	DL 256	(Length 1024) (Usually used)
11	DL 128	(Length 512)

Timing Number.

This may have any value t from 0 to 63 minor cycles and is obeyed in conjunction with digit 24 in the characteristic.

If digit 24 is a 1 the instruction is said to be 'immediate' and carries on for t minor cycles after instruction has been set up.

If digit 24 is a 0 the instruction is said to be 'deferred' and is not obeyed until t minor cycles have elapsed. It is then obeyed in one minor cycle.

13. Control Circuit (Version V).

[Figure 20] shows the control circuit. Its function is to receive an instruction, set it up in a form in which it can be obeyed and time the operation correctly. The main portions are a staticiser in which the instruction is set up, a slow counter for timing and two trigger circuits for bringing about the transfers.

Its action will be considered in parts. An instruction of 32 digits comes in from the 32 μs delay line INST.

TRANSTIM & TIMCI.

These are the names given to two special trigger circuits which initiate the obedience to an instruction [Fig. 21].

Suppose TRANSTIM is initially off. A pulse gated by P1, i.e. at the commencement of a minor cycle, applied to the 'change state' connection will put TRANSTIM on and it will remain on until put off by a second pulse on the same connection. It can therefore only be put on or off at the commencement of a minor cycle.

When TRANSTIM goes off the E-element will send a pulse to TIMCI which will be put on. This is the only way in which TIMCI can be put on. TIMCI will put itself off at the start of the next minor cycle by the 32nd pulse of the previous minor cycle, delayed by 1 μs and gated by P1. TIMCI runs for one minor cycle only, during which time a complete instruction runs into the staticiser.

<u>Note</u>. A transfer in the machine can only take place when TRANSTIM is on. (See [Section 14].)

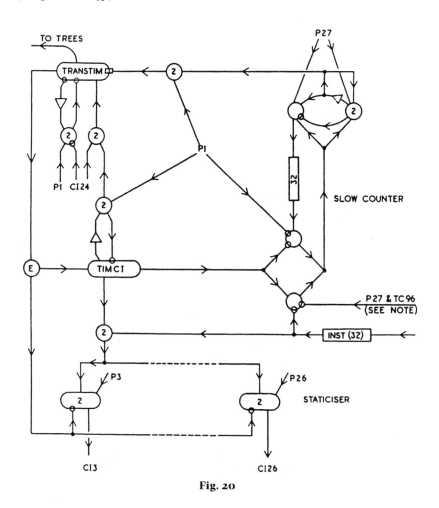

Fig. 20

<u>Slow counter</u>.

This is essentially an adder without the 3-element [Figs 22 and 23].

The slow counter is required to count up to a maximum of 64 minor cycles. Suppose the circuit is as above and initially is clear of all pulses. In each minor cycle a digit is added in the 27th place and the total accumulated. When 64 minor cycles have elapsed there will be a carry digit in the 1st place of the next minor cycle and this passes to TRANSTIM and changes its state. This carry digit is also inhibited from performing a further cycle and so the circuit

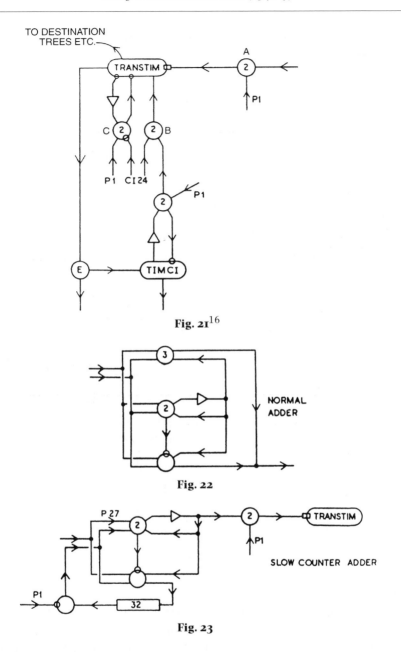

Fig. 21[16]

Fig. 22

Fig. 23

---

[16] Editor's note. It has perhaps not been made sufficiently clear that there are two ways in which TRANSTIM may be put on. (1) by a pulse from A at time P1. (2) By a pulse from B at time P1 when CI 24 = 1. It can be put off in two ways. (1) By a pulse from A at time P1. (2) By a pulse from C at time P1 when CI 24 = 0.

reverts to its state in the first cycle. This is illustrated below.

Position			27	28	29	30	31	32	I	2	
Minor	Cycle	I,	O	O	O	O	O	O	O	O	
"	"	2,	I	O	O	O	O	O	O	O	(i.e., after I mc)
"	"	3,	O	I	O	O	O	O	O	O	
"	"	4,	I	I	O	O	O	O	O	O	
"	"	5,	O	O	I	O	O	O	O	O	
"	"	64,	I	I	I	I	I	I	O	O	
"	"	65,	O	O	O	O	O	O	I*	O	

* This is used to change over TRANSTIM and is inhibited from passing again into the adder. Hence state at 65 is identical with that at I i.e. TRANSTIM will be changed over at intervals of 64 minor cycles.

Now consider the circuit arrangements below.

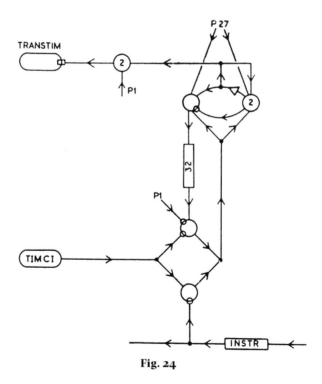

Fig. 24

Suppose that for the first minor cycle TIMCI is delivering a continuous series of pulses and then shuts down. This inhibits the total already in the adder delay line and a complete instruction of 32 digits passing to the inhibitor connection of the lower 1-element goes out of this element as ~INSTR and up to the adder. Here a digit is added in the 27th place and the total traverses the delay line and back to the adder where the process is repeated.

Since a digit is added in the 27th place, only places 27–32 need be considered and these places, it will be remembered, in the instruction are for the timing number. If this timing number is t, the number passed to the adder will be ~t, i.e. $2^6 - 1 - t = 63 - t$. The slow counter runs until a total 64 is accumulated as has already been shown, i.e. it runs for t + 1 minor cycles before a carry digit in the first place operates TRANSTIM.

This is illustrated below. Suppose t is 7.

		Timing Number = 7							
Position		27	28	29	30	31	32	1	2	etc.
Instruction		1	1	1	0	0	0	0	0	...
Passed to adder	1,	0	0	0	1	1	1			
Total at	2,	1	0	0	1	1	1			
	3,	0	1	0	1	1	1			
	8,	1	1	1	1	1	1			
	9,	0	0	0	0	0	0	1		

Thus a pulse goes to TRANSTIM after 8 minor cycles.

Immediate instructions.

Now consider the complete control circuit. Suppose the instruction in the INSTR delay line is an immediate one ordering the transfer of a number from A to B (32 digits).

The timing number will be 1 and instruction will be:

A–B,   11,   1
i.e. Digits: 1, 2   3–12 13–22 23 24 25 26 27 28 29 30 31 32
          Spare   A   B   1   1 0 1 1 0 0 0 0 0

Digit 24 is a 1 since instruction is immediate.

Suppose TIMCI is ON.

The instruction runs out of INSTR and is set up on the staticiser during the first minor cycle.

~INSTR is simultaneously sent up into the slow counter, anything already in the counter delay line being inhibited by TIMCI.

After the first minor cycle TIMCI puts itself off, and since digit 24 was a 1, CI 24 is on, and therefore TRANSTIM is simultaneously started.

During the second minor cycle, the instruction in the staticiser is being obeyed, after which time the slow counter has produced a carry digit in position (1) which puts TRANSTIM off.

TIMCI is started therefore and the staticiser cleared, ready for the next instruction.

Thus it will be seen that in general, with an immediate instruction of timing number t, the instruction will be set up in the staticiser in the first minor cycle and held there while being obeyed for t minor [cycles], i.e. until the end of the $(t + 1)$th minor cycle.

The action is illustrated by the following table for a timing number t.

Minor cycle

1_I TRANSTIM off, TIMCI on. Instruction starts to run in.

2_I TIMCI off, TRANSTIM on. Instruction set up and starts being obeyed.

$(2 + t)_I$ TRANSTIM off, TIMCI on, Staticiser cleared.

Total time for whole operation (t+1) minor cycles.

Deferred instruction.

In this case digit 24 is a 0. As before, the instruction runs into the staticiser during the first minor cycle while TIMCI is ON.

This time, however, TRANSTIM will not come on and allow the transfer to take place until the slow counter stimlates it. TRANSTIM therefore comes on at $(t + 1)_I$ and stays on for 1 minor cycle only before putting itself off.

Minor cycle

1_I TRANSTIM off, TIMCI on. Instruction starts to run in.

2_I TRANSTIM off, TIMCI off.

$(2 + t)_I$ TRANSTIM on. Instruction starts being obeyed.

$(3 + t)_I$ TRANSTIM off, TIMCI on. Staticiser cleared.

Total time for whole operation $(t + 2)$ minor cycles.

Discrimination (use of TC 96).

The inhibitor connection fed with P27 and TC 96 provides the facility for the machine to adopt one of two courses of action according to whether TC 96 is on or off. A deferred instruction, say number n, with timing number 0 is given.

If TC 96 is off, the next instruction to be obeyed will be $(n + 2)$, but if TC 96 is on, a timing number of 1 will be supplied by (TC 96 & P27) and the next instruction will be (n + 3). The examples given later will illustrate this function.

14. Source and Destination Trees.

In order that a combination of 10 signals from a group of CI triggers may be used to tap any source or open any destination a system of inter-connection is arranged, which is known as a TREE.

This may be illustrated by a case of 2 trigger circuits used to select one of $2^2 = 4$ sources say [Fig. 25]. Hence, for example, if CI x is on and CI y is off, a signal will flow to C.

In Versions V and VI the Source and Destination trees control 1024 Sources and Destinations. In Version VII this number is reduced to 512.

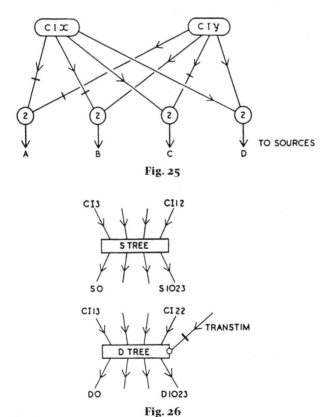

Fig. 25

Fig. 26

Note. The inhibition from TRANSTIM controls when transfer takes place, i.e. TRANSTIM must be on for transfer.

In Version VII there are only 512 Sources and Destinations, each controlled by 9 trigger circuits. In this case the Destination tree is built up as described above with modifications introduced for reasons of available power from trigger circuits.

The 9 triggers are formed into two groups of 4 and 5 each. The 4 group controls 16 4-elements and the 5 group controls 32 5-elements.

The outputs of the 4- and 5-elements are then combined in pairs to select any one of the 512 destinations [Fig. 27].

The Source trees in Version VII are combined with the actual Highways by elements arranged in the manner below [Fig. 28], which illustrates the use of two triggers to control four Sources. In practice, 9 triggers control 512 Sources in this way.

In Versions V and VI there are four lines which can store instructions and feed them as required to INSTR [Fig. 29]. The appropriate one is selected by the four possible combinations of signals from CI 25 and CI 26 as shown.

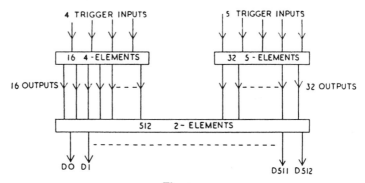

Fig. 27

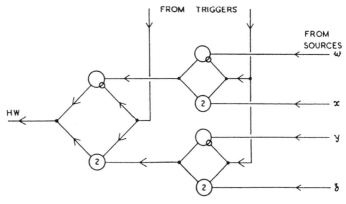

Fig. 28

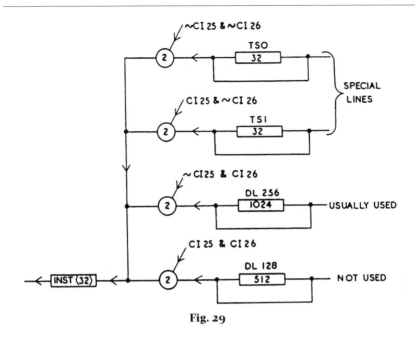

Fig. 29

15. Some Examples of Simple Instructions.

Example 1.

Numbers A and B are in TS 10 and TS 11 respectively. Form A + B and put result in TS 12.

Instruction

(1)	10–2,	11,	1	Immediate instruction, A sent to TS 2.
(3)	11–3,	11,	1	B sent to TS 3.
(5)	145–12,	11,	1	S 145 is TS 2 + TS 3. This is sent to TS 12.

Example 2.

Numbers A and B are in TS 10 and TS 11 respectively. If A = B put A in TS 12. If A ≠ B put (A + B) in TS 12.

Instruction

	(1)	10–8,	11,	1	Puts A into TS 8.
	(3)	11–9,	11,	1	Puts B in TS 9. (8 & 9 have facilities for logical operations.)
	(5)	133–96,	11,	1	S 133 gives a pulse if A $\neq$ B and this will put TC 96 on. If A $=$ B TC 96 will not be put on.
	(7)	136–97,	9,	0	This is to put TC 96 off if it is on. It is ordered as a delayed instruction with zero timing number. There is however an inhibitor connection in the slow counter feed which is operated by (TC 96 & P27). Hence if TC 96 is not on, next instruction will be (9), but if TC 96 is on, a timing number of 1 will be supplied and next instruction will therefore be (10).
If A $=$ B	(9)	10–12,	11,	1	A put into TS 12. Result.
If A $\neq$ B	(10)	10–2,	11,	1	A put into TS 2.
	(12)	11–3,	11,	1	B put into TS 3.
	(14)	145–12,	11,	1	S 145 is (TS 2 + TS 3) and therefore A + B is sent to TS 12. This is the other result.

In this case the next instruction will be (16). If, however, A = B, the last instruction of the table will have been (9), and it is desirable to make the next instruction (16) in this case also. A dummy 'time wasting' instruction is given as (11) with a timing number of 4.

(11) 1023–1023, 11, 4.

The next instruction will now be (16) whichever the result.

16. <u>Multiplier.</u>

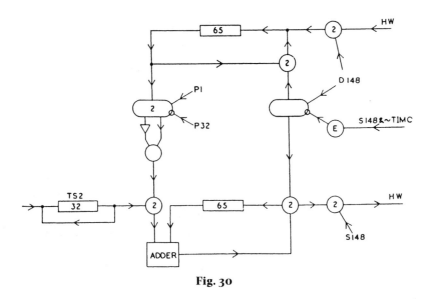

Fig. 30

Multiplication in the binary scale is very simple since it is by a 0 or a 1 only. Consider two numbers A and B each of 32 digits. The product AB may be written:

$$A(B_1 + 2B_2 + 2^2B_3 + \cdots + 2^{31}B_{32})$$

where B_1 to B_{32} are successive digits in the number B.

The multiplier is illustrated above. It consists of two special lines of 65 μs delay each and an adder. The operation is as follows. The number A is first put into TS 2. Number B is then put into the upper delay line. After one minor cycles B occupies the last 32 positions in the line. At the end of the three minor cycles B_{32} is in position one in the line. If B_{32} is a 1 the trigger circuit feeding the adder is stimulated by B_{32} & P1. This opens the 2-element and A runs into the adder. If B_{32} is a 0, the 2-element is not opened and a stream of 0's runs into the adder.

The output of the adder runs into the lower 65 μs delay line. After a further two minor cycles B_{31} is picked out of the upper line, multiplied by A and fed to the adder, where it is added to the total already in the lower line displaced 1 μs. The process is repeated for all the digits in B.

The following table illustrates the action.[17]

$A \times B = C$

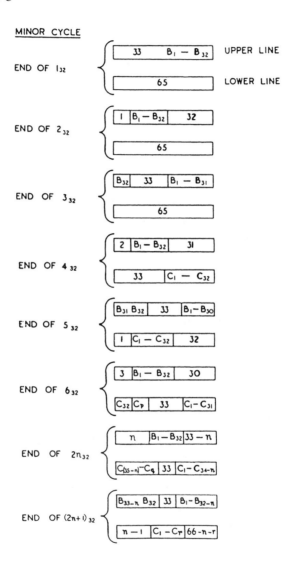

Thus if A and B are 32 digits each, the final total will be running out of the adder during the 66th and 67th minor cycles.

---

[17] Editor's note. Numbers such as the 33 in the top delay line refer to strings of zeroes, e.g. 33 zeroes.

494

An instruction table ordering 32 digit numbers A in 10 and B in 11 to be multiplied together, the product being placed in DL 64, would be as follows.

(1)   10–2,   11, 1   A sent to TS 2.
(3)   11–148,   11, 1   B sent to multiplier upper delay line.
(5)   a–b,   x, y⎫   Other instructions to use up time until (68).
      etc.   ⎬
                         ⎭
(68) 148–64,   11, 2   64 digits product sent to DL 64.
(71)   Next instruction.

It will be seen from the above table that it requires 70 minor cycles to achieve the multiplication of two 32 digit numbers. The time may usefully be filled in by other processes since none of the Highways are in use during this period.

<u>Note</u>. The trigger opening the feedback circuits of the two 65 μs lines is only put off when it receives a pulse from the E-element which is fed with (S 148 & ~TIMCI), i.e. it is put off only when transfer of 64 digit product to a storage line has been effected.

17. Cumulative Adder (D150 and D151).

The function of this device is to count up (to a total consisting of 32 digits) any numbers which are sent to D 150, to store the total and make it available when required.

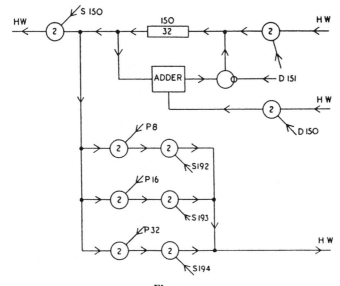

Fig. 31

A number may be put direct into the line by D 151. Numbers coming in on D 150 are added to whatever is already in the line by the adder and put back into the line. The accumulated total may be obtained at S 150 when required.

The facilities offered by S 192, S 193 and S 194 are that the 8th, 16th or 32nd digit of the accumulated total may be examined. This is of use in certain problems.

18. Examples of Repetitive Processes.

Example 1.

Numbers $A_1, A_2 \ldots A_{32}$ are in DL 300. Determine how many are odd and put result in TS 10.

Note. Odd numbers have a 1 in first place. Hence if $(P1 \ \& \ A_r)$ is formed (S 131) it will be a 1 if A_r is odd. The number of such 1's is stored in the cumulative adder D 150.

Instructions

(1)	161–9,	11,	1	Puts P1 into TS 9.
(3)	300–260,	11,	32	Puts all numbers $A_1 \ldots A_{32}$ into DL 260.
(36) = (4)	260–8,	11,	1	Puts A_5 in TS 8.
(6)	131–150,	11,	1	Forms (8 & 9) and sends result to D 150 (cumulative adder) (a 1 if number is odd).
(8)	171–150,	11,	1	P11 sent to D 150. This is done so that after 32 repetitions of the table S 193 will give a pulse.
(10)	158–260,	11,	32	A_1, A_2 etc. now delayed by 1 minor cycle and put back into DL 260. Next number picked out by (4) will be A_6 and so on.
(43) = (11)	193–96,	11,	1	TC 96 put on if table has been repeated 32 times.
(13)	136–97,	9,	0	Discrimination.

Either (15)	1023–1023,	11,	20	Returns to (4) if TC 96 is not on. If TC 96 is on, process is finished but the 1 in 16th place (accumulated from instruction (8)) must be removed from total in 150.
or (16)	150–2,	11,	1	Total of 'odds' in 150 sent to TS 2.
(18)	176–3,	11,	1	P16 sent to TS 3.
(20)	159–10,	11,	1	TS 3 subtracted from TS 2 and result sent to DL 10.
(22)	135–151,	11,	1	Clears cumulative adder for next operation.

Example 2.

To find the larger of two positive numbers.

In order to represent negative numbers, the complementary notation is adopted, i.e. $-A$ is represented by $(2^{32} - A)$ and $|A|$ is restricted to a value $<2^{31}$.

Thus to find the larger of two numbers A and B, both of which are $<2^{31}$, the difference $A - B$ is formed and if negative (i.e. a 1 in 32nd place) $B > A$ and vice versa.

Suppose A is in 10 and B in 11. The larger is to be put into 12.

Instructions

(1)	10–2,	11,	1	A sent to TS 2.
(3)	11–3,	11,	1	B sent to TS 3.
(5)	159–8,	11,	1	(TS 2 − TS 3) sent to TS 8.
(7)	160–9,	11,	1	P32 sent to TS 9.
(9)	131–96,	11,	1	Forms (TS 8 & TS 9). If TS 8 contains a negative number, TC 96 is put on.
(11)	136–97,	9,	0	Discrimination.
Either (13)	10–12,	11,	2	If TC 96 not on, A > B and A sent to DL 12.
or (14)	11–12,	11,	1	If TC 96 is on, B > A and B sent to DL 12.

Note. The above method is applicable also to values of A and B positive or negative provided A and B are then both restricted to the range 2^{30} to -2^{30}. This is necessary since if $A = -B = 2^{31}$ say, the difference $A - B$ would have a 1 in 32 place and it would appear to the machine that the difference was negative, i.e. $B > A$, which is clearly absurd. Hence the further restriction on the values of A and B if either may be negative.

Example 3.

Numbers $A_1, A_2 \ldots A_{32}$ ($2^{30} > A_r > 2^{-30}$) are in DL 300. It is required to find the largest and place it in DL 12.

The method is to compare a pair of numbers, find the larger and compare that with a third and so on until 32 have been examined.

Instructions

(1)	300–260,	11,	32	$A_1, A_2 \ldots A_{32}$ sent to DL 260.
(34) = (2)	260–2,	11,	1	A_3 picked out and sent to TS 2.
(4)	160–8,	11,	1	P32 sent to TS 8.
(6)	163–150,	11,	1	P3 sent to D 150 (cumulative adder). Thus after 32 repetitions of the table S 192 will give a pulse (i.e. in 8th place).
(8)	260–3,	11,	1	A_{10} sent to TS 3.
(10)	159–9,	11,	1	Forms difference ($A_3 - A_9$) and sends to TS 9.
(12)	131–96,	11,	1	Forms (TS 8 & TS 9) and stimulates TC 96 if ($A_3 - A_9$) is negative (i.e. if $A_9 > A_3$).
(14)	136–97,	9,	0	Discrimination.
Either (16)	1023–1023,	11,	2	If TC 96 not on ($A_3 > A_9$) A_3 already in TS 2.
or (17)	3–2,	11,	1	If TC 96 is on ($A_9 > A_3$) A_9 sent to TS 2.
(19)	158–260,	11,	32	All numbers in DL 26 delayed one minor cycle.
(52) = (20)	192–96,	11,	1	If S 192 stimulates TC 96, process has been repeated 32 times.
(22)	136–97,	9,	0	Discrimination.
Either (24)	1023–1023,	11,	13	If TC 96 not on, returns to (6) and repeats table.
or (25)	2–12,	11,	1	If TC 96 is on, process is finished and largest number is sent to DL 12.
(26)	135–151,	11,	1	Clears cumulative adder for next operation.

19. Control and Instruction (Version VI).

A major difficulty in the machine described to date was the permanent association of certain operations with certain delay lines. For example TS 2 and TS 3 were associated with an adder. Thus in order to add two numbers, one had to be put in TS 2, the other in TS 3 and S 145 routed to the required destination.

A table for such a simple process would be, for example:

(1)   10–2,   11,   1
(3)   11–3,   11,   1
(5)   145–12,   11,   1

Three separate instructions are required.

A shortening and simplification of the instruction table would result if two numbers could be taken from <u>any</u> sources A and B, have an operation F performed on them, the result delayed D microseconds and then put in C all by means of one instruction.

Such an operation may now be performed by means of <u>two word instructions</u>.

These two word instructions may be used at any time in addition to the original single word instructions.

Form of two word instruction.

Digits	1, 2	3–12	13–32	23–26	27–32
1st word	Spare	Source A	Destination C	Characteristic	Timing
2nd word	Spare	Source B	Delay D (0–1023 μs)	Operation F	Spare

The first word of a two word instruction is similar to the original one word type except for the characteristic. This must now indicate whether the instruction is a one word instruction or whether it is the first word of a two word instruction.

The significance of digits 23–26 is therefore as follows.

Digit		Meaning	
23	0	External operation	
	1	Internal operation	} as before
24	0	Deferred operation	
	1	Immediate operation	

25, 26,	0	0	Next instruction from TS 0 and present word is a one-word instruction.
	1	0	Next instruction from TS 1 and present word is a one-word instruction.
	0	1	Next instruction from DL 256 and present word is a one-word instruction.
	1	1	Next instruction from DL 256 and present word is 1st word of a two-word instruction.

The second word digits 23–26, i.e. therefore the value of F, determines the nature of the operation to be performed on the numbers from sources A and B by an operation tree.

The code is as follows.

CI 23   CI 26

Op. TREE

F0   F17

Value of F	Operation	
0	A − B	with no Round Carry Suppressions.
1	A − B	with odd R.C.S.
2	A − B	with even R.C.S.
3	A − B	with odd and even R.C.S.
4	A + B	with no R.C.S.
5	A + B	with odd R.C.S.
6	A + B	with even R.C.S.
7	A + B	with odd and even R.C.S.
8	A & B	
9	A ∨ B	
10	A ≢ B	
11 to 17	not yet allocated.	

The ability to add or subtract with various R.C.S. is very useful, since it enables numbers composed of any number of words to be added or subtracted. For example, if the numbers are each 64 digits long, then they must be added with even R.C.S.

Arrangement of secondary control circuit.

A secondary staticiser is arranged in the control circuits as shown [in Fig. 32]. While the first word is running into the primary staticiser (the original one) the second word is running into the secondary staticiser and setting up trigger circuits CJ 3 – CJ 26 [Fig. 33].

Triggers CJ 3 to CJ 12 select the secondary sources SS 0 to SS 1023 by means of a secondary source tree.

Triggers CJ 13 to CJ 26 feed into a delay unit and select the required delay.

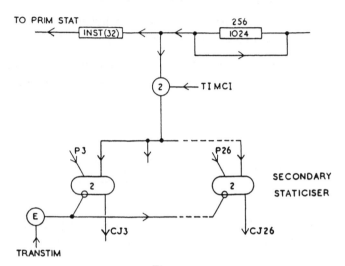

Fig. 32

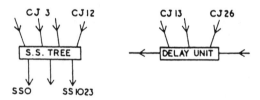

Fig. 33

Delay unit.

This unit is constructed so that any delay from 0–1023 µs may be selected. The arrangement is as below.

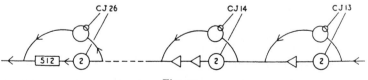

Fig. 34

Thus if the required delay is, for example, 39 μs triggers CJ 13, 14, 15 and 18 will be on and a total delay of 39 μs introduced in the delay unit.

Highway arrangements.

All sources may now be fed into either or both of two Highways. These are known as HW 1 and HW 2 and an S signal puts the source to HW 1 and an SS signal puts the source to HW 2.

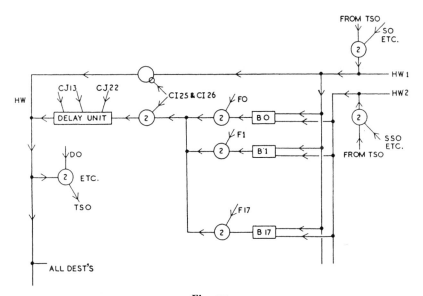

Fig. 35

B 0 to B 17 are the units in which the addition, subtraction and logical operations are performed.

Consider first a one word instruction. It will be set up on the primary staticiser in the normal way and the next one word instruction will be set up on the secondary staticiser. Since the first instruction is of one word only, its digits 25 and 26 will be one of 00, 01, or 10 but not 11. Hence (CI 25 and CI 26) will be 0's and the number on HW 1 will pass directly through to the destination highway.

The source tapped to HW 2 by the instruction in the secondary staticiser will pass to the B units together with the number on HW 1 and will have some operation performed on it, but will not reach the delay unit and destination

HW since the 2-element is not opened by (CI 25 & CI 26). The action is thus quite normal.

If the instruction is the first word of a two word instruction, then digits 25 and 26 will be 11. Thus (CI 25 & CI 26) will be 1's, the direct route from HW 1 to destination HW will be inhibited and the operation selected by F will be performed on the numbers. The result will be delayed D in the delay unit and then passed to the destination highway.

20. Examples of Two-word Instructions.

Example 1.

It is required to add a series of 32 words in DL 300 to a series of words in DL 301, the results to be placed in DL 302 with no delay.

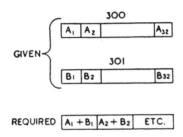

Instructions

       This is 1111

(1)   300–302,   15,   32   (Since 32 pairs are to be added.)
(2)   301–0,   7,   —

Example 2.

There are 32 numbers each of 32 digits, in DL 300. Find how many of them are odd and put result in DL 10.

Note. All odd numbers have a 1 in first place. Hence if a number is, say, A_r and (P1 & A_r) is formed, this will be a 1 if A_r is odd.

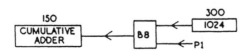

Instruction

(1)   300–150,   15,   32
(2)   161–0,   8,   —
(34) 150–10,   9,   1
=(2)

Compare this simple table with the table required to solve the same problem in the original machine with only one word instruction.

Example 3.

Given 32 numbers of 32 digits each in DL 300 arranged

at the commencement of a major cycle, delay them by 1 minor cycle so that they are arranged

A 32 | A 1 | A 2 | ETC.

at the commencement of a major cycle.

Note. Process is to add SS.135 (a series of 0's), delay result by 32 μs and feed back.

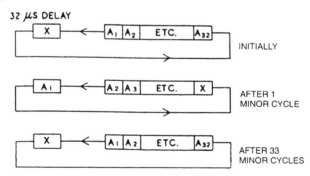

I.e., if timed for 33 and then stopped, arrangement at the commencement of a major cycle has become

A 32 | A 1 | ETC. | A 31

Instruction

(1)   300–300,   15,   33
(2)   135–32,   7,   —

In general.

With n word notation for a series of numbers (i.e. of 32 n digits each) to move up by one complete number (i.e. n minor cycles) instruction is

(1)   300–300,   15, $(32 + n)$
(2)   135–32 n,   7,   —

Example 4.

Given A in TS 10 and B in TS 11 form 8 $(A + B)$ and put in TS 12.

Instruction

(1)   10–12,   15,   1
(2)   11–3,   15,   —

21. A Special Source in the Version VI Control System.

This is S 157. Its function is to enable a particular number of 32 digits to be fed into the machine, for use in the computation, via the instruction table.

The arrangement is as below.

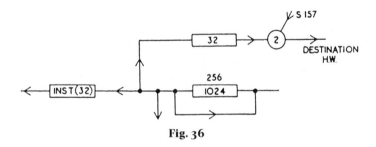

Fig. 36

Instruction

(1)   157–2,   11,   1
(2)   NUMBER
(3)   etc.

(1) & (2) are a two-word instruction which causes instruction (2), which is the number, to be sent via the destination HW to TS 2.

The operation then continues normally.

22. Changing Instruction Tables.

It is frequently necessary that while using a main table, say Table I, at a certain stage another table, say Table II, be brought into use, after which the main table (or a third) be reverted to. This may be effected by link instructions.

Suppose Table I is in DL 301 and Table II in DL 300. A blank is left in the subsidiary Table II at a convenient point at the end for the link.

The link instruction will be supplied by the main table using the facility of S 157.

Main Table I (in DL 301).

(1)	L–M,	11,	1	
	to			Normal instructions.
(9)	N–O,	11,	1	
(11)	157–300,	9,	4	This is a delayed instruction ordering INSTR (16) into the blank in Table II (which is INSTR (17) in Table II). Hence timing number 4.
(16)	301–256,	11,	x	This is the link instruction.
(17)	300–256,	11,	y (47)	This orders Table II into use.

Subsidiary Table II (in DL 300).

(1)	A–B,	11,	1	
	to			Normal instructions.
(15)	C–D,	11,	1	
(17)	LINK (301–256,	11,	x)	Supplied by Table I and brings Table I back into use.

Note. x & y must be chosen to order the new table into use at the correct minor cycle.

Thus $y = 32 + (32 - 17) = 47$

& $x = 32 + (32 - 17) = 47$ also, if both tables commence at (1).

23. Input and Output Units.

The input and output to and from the A.C.E. is by means of Hollerith units and punched cards. The standard Hollerith code is illustrated below. Each card may be punched in 12 rows comprising 80 columns.

The numbering of rows and columns is as shown.

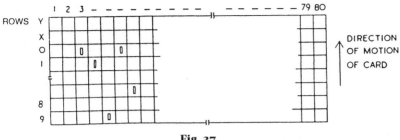

Fig. 37

As normally used in the decimal scale, a ten figure number, say, could be represented by punching holes at positions in the first ten columns corresponding to the value of the digit in each place. Thus a number of 5 digits, for example 62049, would be represented as shown below.

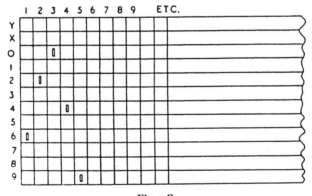

Fig. 38

Cards so punched are used to feed numbers to be used in a computation into the machine. The A.C.E. itself converts such decimal numbers into binary form and stores them.

Cards, however, may be used to record binary numbers directly. Either one or two 32 digit binary numbers may be punched in each row. At the present time only columns 41 to 72 inclusive are being used, with the convention that a 'hole' corresponds to a 1 and 'no-hole' to a 0. Thus a total of twelve (or 24) binary numbers may be punched on each card. The remaining columns are used for indexing and coding of the cards. This method of punching is used for recording output data and also for feeding in instruction tables. In the latter case, the instructions are punched in the normal decimal fashion on cards. The cards are then fed through a Hollerith machine which converts the information to binary form and assembles 12 instructions and punches them on a single card.

Input card reader.

The data is taken from the cards by means of a series of 80 brushes under which the cards pass. As a hole passes under a brush a circuit is completed. The case of card data in binary form only will be considered at present, i.e. instructions.

The cards pass through the reader at a rate of 150 cards/min. Each card has 12 rows and there are 4 'dead' rows between cards, i.e. 25 ms per row.

Each row is actually being read for a period of 8–10 ms out of the 25 ms.

In order to read a card, the necessary instructions must be in the A.C.E. in advance. The table is as follows.

| (1) | 136–138, | 11, | 1 | D 138 controls a knife edge which pushes a card into the reader. |

A signal is now required to indicate when the card is in position for the first row (y) to be read. This is obtained from the reader itself (mechanical contact) and is S 132^+.

(3)	$132^+ - 96,$	11,	1	If a card is in position, TC 96 will be put on and vice versa.
(5)	0–5,	11,	1	Dynamiciser contents sent to TS 5.
(7)	136–97,	9,	26	Discrimination. If TC 96 is off (card not yet ready) brings back to (3). (3) and (7) repeated until TC 96 comes on when (5) is applied.

Wait 15 ms (approx.) and then take next row.

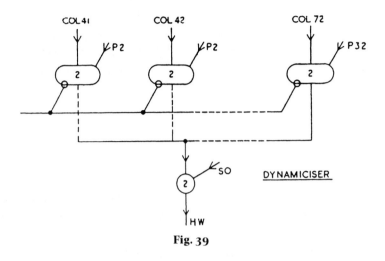

Fig. 39

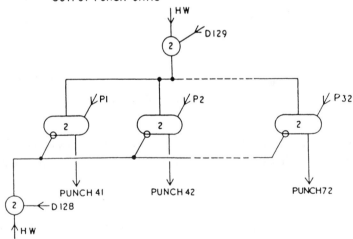

Fig. 40

To record an output number, the number is sent to D 129 to set up the staticiser. These trigger circuits operate the punch relays in the punch unit. Then card is punched (15 ms) and the staticiser cleared.

An alternative output is to a print unit.

24. Arithmetical Operations with Positive or Negative Numbers.

In all the following operations the numbers concerned are composed of a maximum of 32 binary digits, i.e. the modulus of all numbers is less than 2^{32} in magnitude.

It should be remembered that in the A.C.E., as in most digital computers, a negative quantity, say $-A$, is represented by the number $(A \max - A)$ where $(A \max - 1)$ is maximum possible number which can be handled.

Thus in a 32 digit number in the A.C.E. $-A$ would be represented by $2^{32} - A$.

If $|A|$ is restricted to a value $<2^{31}$, the 32nd digit of the word will be a 0 if the number is positive and a 1 if negative. This is the accepted code for negative numbers.

Operation $A + B$.

A and B may be positive or negative numbers.

A two word instruction

(1)   A–C,   15,   1
(2)   B–0,   7,   —

carries out the operation and puts the result in C. Similarly the instruction

(1)   A–C,   15,   1
(2)   B–0,   3,   —
forms $A - B$.

Operation $A + nB$.

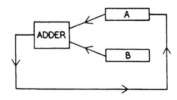

In this case, B is added n times to A.

(1)   A–A,   15,   n
(2)   B–0,   7,   —

The answer accumulates in the A line and method words for positive or negative numbers.

<u>Operation $2^n A$ where $n < 31$ and positive.</u>

This may be achieved by delaying the word for n μs provided that A has less than $31 - n$ digits.

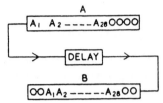

(1)   A–B,   15,   1
(2)   135–2,   7,   —

Method is to add a series of 0's (S 135) and delay result 2 μs and put in B. Works for A positive or negative with above restrictions.

<u>Note</u>. To move a number round n places in a line (and not multiply it by 2^n) e.g., n = z

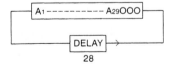

(1)   A–B,   15,   2
(2)   135–2,   7,   —

In this case add zero's and put into B with a delay of 2 μs. The number cannot be fed back into the same line in this case, since this would introduce zero's within the number.

<u>Operation $2^n (A \pm B)$.</u>

(1)   A–C,   15,   1
(2)   B–n,   7,   —
        or 3 for −B.

<u>Operation $2^{-n} A$, A positive.</u>

For example, $2^{-4} A$.

An advance of 4 μs is the same as a delay of 28 μs.

(1)   A–A,   15,   2
(2)   135–28,   7,   —

This works for positive A and number must be returned to original line.

If A is positive or negative, perform a discrimination to determine the sign. If negative, find complement, perform operation, and replace sign afterwards.

Operation A × B (A and B positive or negative).

If A and B are both positive the operation is straightforward. However, if either or both A and B are negative and expressed in the usual convention difficulties arise.

For example, suppose A is negative.

The product formed will be

$$(2^{32} - A)B = 2^{32}B - AB.$$

But the required product is known to be $2^{64} - AB$, in a 64 μs delay line. Hence the quantity $2^{32}(2^{32} - B)$ must be added to the answer to obtain the correct product.

Similarly if B is negative

$$2^{32}(2^{32} - A)$$

must be added.

If both A and B are negative, answer produced will be

$$(2^{32} - A)(2^{32} - B) = 2^{64} - 2^{32}(A + B) + AB.$$

Hence in this case $2^{32}(A + B)$ must be added. (The 2^{64} has no effect.)

An instruction table for signed multiplication will make the operation clear.

A is in line 11 and B in 12. A and B may be positive or negative. The product is to be formed and put in line 64.

(1) 11–2, 11, 1 A put in TS 2.

(3) 12–148, 11, 1 B put in 148.

The multiplication now proceeds normally and the product will be available in minor cycles (3 + 67) and (3 + 68), i.e. in (6) and (7). This time interval is used to determine the corrections, if any, that are required, as follows.

(5) 135–65, 11, 2 Clears out 65 in case it has something
left in it. 65 is used for holding the
correction term.

(8)	11–96,	15,	I ⎫	Determines sign of A by forming
(9)	160–0,	8,	— ⎬	(A & P32) and passing result to TC 96.
			⎭	TC 96 put on if A is negative.
(10)	136–97,	9,	0	Discrimination. Puts off TC 96 if on.

Either

(12)	1023–1023,	11,	2	If A is positive wastes time until (15).

or

(13)	12–65,	11,	I	If A is negative sends B to DL 65.
(15)	160–96,	15,	I ⎫	Determines sign of B by forming
(16)	12–0,	8,	— ⎬	(A & P32) and passing result to TC 96.
			⎭	TC 96 put on if B is negative.
(17)	136–97,	9,	0	Discrimination.

Either

(19)	1023–1023,	11,	3	If B is positive, wastes time until (23).

or

(20)	11–65,	13,	I ⎫	Delayed instruction which adds A to
(21)	65–0,	7,	— ⎬	correct half of DL 65.
				Subtracts number in DL 65 from
(23)	148–64,	15,	48 ⎫	product and puts in DL 64. Timing
(24)	65–0,	1,	— ⎬	number is chosen to obtain result when
				product is coming from the multiplier.

Next instruction will be (72) = (8).

<u>Note</u>. The total time required for signed multiplication is the same as for unsigned.

25. <u>Version VII</u>.

In this version the instructions are in a modified form. Either one- or two-word instructions can be used, the form of the instruction being thus:

Digits	1,	2–10,	11–19,	20–21,	22–26,	27–32,
1st Word	Spare	Source	Destination	Characteristic	Source of next instr.	Timing
2nd Word	Spare	Sec. Source	Delay (21 not used)		Operation (26 not used)	Spare

In this new instructional form, it will be seen that only nine digits are used to describe a source or destination and this limits the total number of sources to 512. It is considered that with the increased flexibility of the machine this number will be adequate.

The Characteristic is of [two] digits and is used to indicate whether the instruction is of one word or two and whether immediate or deferred.

0 = 00   Deferred one-word instruction.

1 = 10   Immediate one-word instruction.

2 = 01   Deferred two-word instruction.

3 = 11   Immediate two-word instruction.

The use of five digits to describe the source of the next instruction facilitates the use of 32 possible sources. These are DL 256 to DL 287 inclusive, and are selected by a small tree in the usual manner:

Thus   0   for digits 22–26 selects DL 256

       1   for digits 22–26 selects DL 257

      n   for digits 22–26 selects DL $(256 + n)$

In the second word of a two-word instruction, the delay required is represented by digits 11–20, 21 not being used, and the operation by digits 22–25, 26 being spare also. The operation code is as before, namely:

Value	Operation			
0	HW1 − HW2	with	NO	R.C.S.[18]
1	HW1 − HW2	with	ODD	R.C.S.
2	HW1 − HW2	with	EVEN	R.C.S.
3	HW1 − HW2	with	ODD & EVEN	R.C.S.
4	HW1 + HW2	with	NO	R.C.S.
5	HW1 + HW2	with	ODD	R.C.S.
6	HW1 + HW2	with	EVEN	R.C.S.
7	HW1 + HW2	with	ODD & EVEN	R.C.S.
8	HW1 & HW2			
9	HW1 ∨ HW2			
10	HW1 $\not\equiv$ HW2			
11	(HW1 & 157) ∨ (HW2 & ~157)			
12–15	Not yet allocated.			

The circuits for the new selector arrangements and new coding are shown below and are self-explanatory [Figs 41 and 42]. The operation of these portions of the circuit is identical with those in the Version VI, though some

---

[18] Editor's note. Round Carry Suppression.

of the units have somewhat different internal construction. Of these, the function unit is noteworthy.

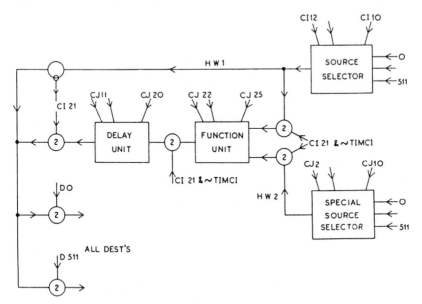

Fig. 41

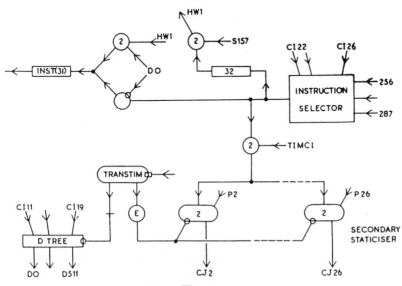

Fig. 42

<u>Function unit.</u>

The operation tree of Version VI has been dispensed with and the circuit is operated direct from trigger circuits CJ 22 and CJ 25 inclusive.
    The circuit is shown below.

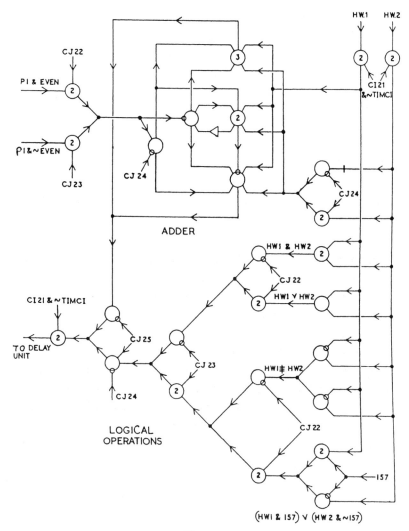

Fig. 43

The upper portion of the circuit is the adder, which is provided with either Odd, Even, Both or No Round Suppression. (Note: EVEN is a multivibrator which is on and off during alternate minor cycles.)

For subtraction (HW1 − HW2), CJ 24 is off and the number on HW2 has its digits inverted, i.e. 0's for 1's and vice versa, by the polarity changer. Thus the number sent to the adder is $2^{32} − 1 − B$, where B is the original 32 digit number on HW2. But the true complement of B is $2^{32} − B$. Hence one must be added to the result in the adder. This is achieved by adding in the carry suppression pulse at the beginning of the minor cycle concerned. It will be seen that the method works for numbers of 32 or 64 digits, but does not for numbers of more digits than 64. If it is essential to correct an answer for a greater number of digits than 64 this must be done separately.

The lower portions of the circuit are for the logical operations and are self explanatory.

The timing circuits of Version VII.

Several changes have been introduced into the timing circuits of the control. The slow counter has been redesigned and the whole circuit simplified as in Fig. [44].

The length of the INSTRUCTION line is now 31 μs. As before, when TIMCI is on, an instruction is allowed to run into the staticiser. Since, however, the INSTR line is of length 31 only, an extra 1 μs delay has to be inserted before the staticiser. TIMCI can be put on only by a pulse from the E-element, i.e. only when TRANSTIM goes off, and puts itself off after one minor cycle by its 32 pulse delayed by 1 μs and gated by P1.

Consider now a single word instruction of the immediate variety with a timing number of 5. The Characteristic (digits 20 & 21) will be 1 and therefore CI 20 will come on. The instruction then runs into the staticiser and simultaneously into the slow counter portion. The BORROW trigger circuit will be on as a result of a P1 pulse at the commencement of the minor cycle and will be put off again during the 26th μs of the minor cycle by P26. While BORROW is off, the 1-element above the counter delay line will be inhibited, and the 2-element will be opened. During the 26th μs, therefore, digit 27 of the instruction will pass to the counter delay line via the polarity changer and the 2-element. Thus if originally a 0 it will enter the line as a 1 and vice versa. The 27th digit will arrive at BORROW during the 27th μs and if a 1 will put BORROW on again. The 28th and subsequent digits of the instruction will therefore pass to the counter delay line unchanged. The action is best illustrated by the following table for an instruction of timing number 5.

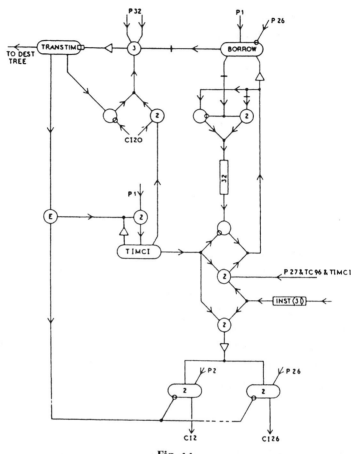

Fig. 44

	Digit numbers					
	27	28	29	30	31	32
Instruction	1	0	1	0	0	0
(timing number 5)						

	Digits passed to counter line						
Minor cycle	(Position in minor cycle)						
	26	27	28	29	30	31	32
End of 1	0	0	1	0	0	0	0

BORROW off during 26th μs and therefore 27th digit of INSTR passed as a 0.

BORROW put on by 27th digit on 27th μs and next passed unchanged.

| End of 2 | 1 | 1 | 0 | 0 | 0 | 0 | 0 |

BORROW off during 26th μs and not put on until 29th μs by the 28th digit delayed 1 μs. Rest of digits pass unchanged.

End of 3	O	I	O	O	O	O	O
End of 4	I	O	O	O	O	O	O
End of 5	O	O	O	O	O	O	O

Now TRANSTIM is not put off until it receives a pulse during the 32nd μs of a minor cycle from BORROW, i.e. until BORROW is off at this time. It will be seen from the example that this will not occur until the end of minor cycle 6 in this case.

Hence this timing circuit produces precisely the same result as the Version VI, i.e. an instruction of timing number t is set up in the first minor cycle and takes place during the next t minor cycles. The next instruction will be set up in minor cycle $(t + 2)$ as before.

The procedure with a deferred instruction is along similar lines. The instruction runs into the staticiser during the first minor cycle. The counter operates as before. CI 20 is off in this case (deferred instruction), and therefore TRANSTIM has put itself off. The transfer does not take place until TRAN-STIM goes on again, i.e. until BORROW is off during the 32nd μs of a minor cycle. This occurs after $(t + 1)$ minor cycles. TRANSTIM then goes on for one minor cycle and puts itself off.

Discrimination.

As in the earlier control circuits discrimination facilities are provided by the use of TC 96. Since in Version VII two-word instructions will largely be used it is desirable that after a discrimination instruction number n the next instruction will be either $(n + 2)$ or $(n + 4)$. This is achieved by the input of (P27 & TC 96 & TIMCI) which, if TC 96 is on, adds a 1 in the 27th place of the number circulating in the counter and hence delays the operation of TRANSTIM by 2 minor cycles.

26. Initial Starting Procedure.

The problem of starting up the A.C.E. is a complex one, since after switching on, all the lines will have a random assortment of spurious pulses in them. The first thing to be done, therefore, is to clear out a line of all spurious pulses.

Three manually operated switches are provided in the control circuit. The first is a switch operating on TRANSTIM and called 'PARALYSIS PERIOD'. This has three settings:

'0'   –   Normal working
'10 ms'   –   10 ms delay after working
'∞'   –   used in conjunction with a SINGLE TRANSFER switch to operate TRANSTIM one at a time.

The third switch is 'P32 PULSES'. It has 3 positions:

'ON'   –   Normal
'Hollerith'   –   P32 pulses gated by card reader and only supplied when a card is in a position to be read.
'OFF'   –   no P32 at all.

With the aid of these switches, the starting up is as follows:

(1) Check that P32 pulses are on.
(2) Set PARALYSIS to ∞ to ensure TRANSTIM off.
(3) Set P32 PULSES to Hollerith.
(4) Set PARALYSIS to 10 ms. TRANSTIM put on only when a card is in position.
(5) Clear the CI staticiser by a manual switch.
(6) Start card feed with a pack of 'Initial Input Cards'.

Initial input cards.

Since the CI staticiser has been cleared, it is set up with a row of zeros, i.e. with an instruction in the Version VII code of 0–0, 0, 0, 0, the last figure the timing number, going to the counter.

The meaning of this instruction is, 'Feed the input dynamiciser to the INSTR line, the instruction being a deferred one-word instruction. The next instruction will come from DL 256 and the timing number of this instruction is 0.'

This means that the 1st row of the first Initial Input Card is fed to the INSTR line. The instructions on these cards then go as shown in the table below.

Minor cycle	Input to INSTR				Origin	Effect
(30)	135–256,	1,	0,	0,	Row 1	A string of 0's pass to 256 for two major cycles (due to 10 ms paralysis of TRANSTIM) and clears the line. Row 2 is wasted.
(31)	0–0,	0,	0,	0,	256	Row 3 to INSTR.
(1)	0–258,	0,	0,	29,	Row 3	Row 4 to DL 258.
(0)	0–0,	0,	0,	0,	256	Row 5 to INSTR.
(2)	0–258,	0,	0,	29,	Row 5	Row 6 to DL 258.
					and so on	

DL 258 can thus be filled with 32 instructions by having odd rows on cards of the required instructions alternating with the instruction 0–258, 0, 0, 29. The first table thus introduced into the machine should be one enabling the more easy assimilation of further instructions.

Synchronisation of EVEN (S 157) with the machine.

This is another starting-up problem. The EVEN multivibrator, which is on and off in even and odd minor cycles respectively, has to be made to synchronise with the arbitrary numbering of the minor cycle. It is found simpler in practice however to achieve the reverse, i.e. to number the minor cycles to correspond to the already functioning multivibrator.

This is done, as soon as sufficient initial instructions have been introduced into the machine, by forming the result $\sim$ (EVEN & P27).

	1	2–10	11–19	20	21	22–26	27	28	29	30	31	32	
P27	0	0–0	0–0	0	0	0–0	1	0	0	0	0	0	
$\sim$(EVEN	1	1–1	1–1	1	1	1–1	0	1	1	1	1	1	If EVEN is on.
& P27)	1	1–1	1–1	1	1	1–1	1	1	1	1	1	1	If EVEN is off.

This result will have two possible values depending on whether EVEN is on or off during the minor cycle in which the operation is performed. If the result is interpreted as an instruction, the two possibilities will be:

	511–511,	3,	31,	62	if EVEN was on
or	511–511,	3,	31,	63	if EVEN was off.

In either case, the instruction is a 'waste-time' one and if obeyed by the machine, will carry it on for either 62 or 63 minor cycles, depending on the original state of EVEN.

The action will be illustrated by reference to a complete instruction table for the process.

Minor cycle

n ⎫	EVEN–8,	I,	o,	I	Sends EVEN to TS 8.
n + 2 ⎪	187–9,	I,	o,	I	P27 to TS 9.
n + 4 ⎪	137–0,	I,	o,	I	Forms ~ (8 & 9) and sends to INSTR.
n + 6 ⎭	511–511,	3,	31,	62 or 63	
⎰ n + 69	Next instruction if n was an even minor cycle.				
⎱ n + 70	Next instruction if n was an odd minor cycle.				

The next instruction obeyed after (n + 6) is always odd, since if n had been odd then that instruction would be (n + 70), i.e. odd also. If n had been even, then the next instruction followed after (n + 6) would be (n + 69), i.e. odd.

Hence, to synchronise the machine with EVEN, a table of the above form is fed through and results in the determination of an odd minor cycle.

27. Polarity Changer.

The following circuits [Figs 45–54] represent possible methods of producing the required effects of the various units. At present it is by no means certain that any of these schemes will be finally adopted. They are intended only to indicate possible solutions.

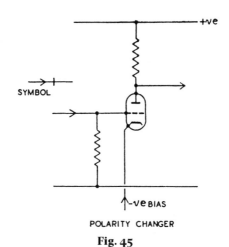

POLARITY CHANGER

Fig. 45

A single valve connected as shown will act as a polarity changer for the conversion of 1's to 0's and vice versa. A 1 is normally represented as a voltage pulse in the positive sense.

The valve is normally biassed to cutoff and the input pulses are large enough to take the valve up to saturation.

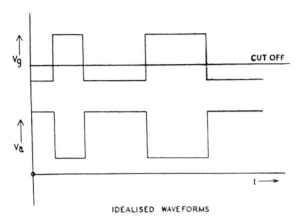

IDEALISED WAVEFORMS

Fig. 46

The circuit is so designed that the input and output pulses are of equal amplitude.

28. <u>N-Elements.</u>[19]

Fig. 47

Such an element has to respond to n or more simultaneous input pulses. Consider the simple network below.

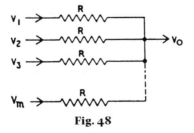

Fig. 48

---

[19] Editor's note. This title added by Copeland.

There are m input voltages applied to the resistors. The output voltage, V_0, is therefore given by

$$V_0 = \frac{1}{m}(V_1+V_2+V_3+\cdots+V_m).$$

If the only possible values of $V_1, V_2, \ldots, V_m$ are either o or v (i.e. no pulse or a pulse) then when the threshold number of pulses n is received

$$V_0 = \frac{nv}{m}.$$

Such an input network connected to a valve can constitute an n-element [Fig. 49].

The valve is so biassed that a voltage $>nv/m$ is required to produce conduction. If the input voltage is $<((n-1)/m)v$, then conduction must not occur [Fig. 50].

The system will work provided v/m is greater than the grid base from cutoff to saturation and is in fact achievable for m = 4 or 5, say, if the pulse amplitude is 50 volts.

Normally, a stage as described above would be followed by an inverter stage in order that positive pulses can be obtained. An inhibitor connection can be conveniently incorporated in this stage [Fig. 51].

With such an inhibitor connection it is necessary to ensure that the pulse it carries is sufficiently large to maintain the second valve in a state of conduction irrespective of the anode potential of the first stage.

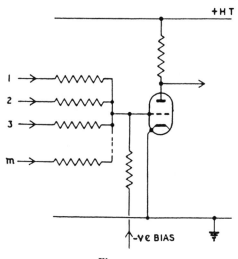

Fig. 49

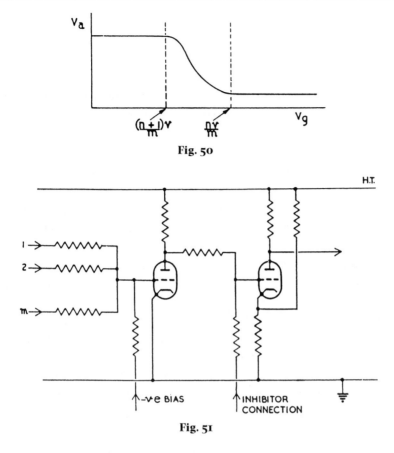

Fig. 50

Fig. 51

So far the use of triodes only has been considered. Pentodes, however, increase the scope of the circuitry, since both control and suppressor grids may be used as inputs. For instance it is convenient to construct a 2-element which has but two input connections using a pentode.

A circuit as [below] is the basis of a very reliable 2-element [Fig. 52].

29. Trigger Circuit.

A trigger circuit must have two stable conditions and must be capable of being changed from one to the other. A suitable circuit is given below [Fig. 53].

In the off condition A is cutoff and B conducting. When the threshold number of pulses is received on the grid of A it goes into conduction and B is cut off. This is the alternative stable condition and corresponds to the on condition, since the anode of B is now near HT potential.

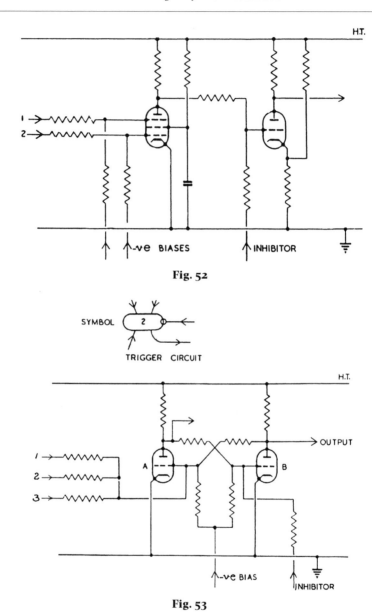

Fig. 52

SYMBOL

TRIGGER CIRCUIT

Fig. 53

The trigger is put off by a pulse on the inhibitor connection to the grid of B. It is necessary to ensure that the inhibitor pulse is large enough to put the trigger off irrespective of the input to the grid of A.

If an output is taken from the anode of A, this consitutes a 'NOT' output.

A 'change-state' connection to a trigger circuit may be achieved by applying the change pulse to both grids simultaneously. As illustration, a possible circuit for TRANSTIM of Version V is given below. Inputs are marked IN and outputs OUT to correspond on symbol and circuit.

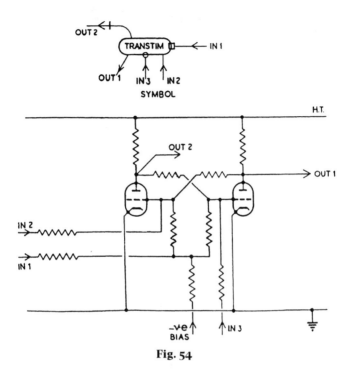

Fig. 54

23 *The state of the art in electronic digital computing in Britain and the United States* (1947)

Harry D. Huskey

Huskey came to the NPL ACE Section in January 1947 from the United States, where he had been involved with the ENIAC and its projected successor the EDVAC. Shortly after Huskey's arrival Womersley suggested that he visit Manchester and Cambridge and prepare a report setting out the status of each of the several computer projects in Britain and the United States.[1]

B. J. C.

Part I. Memory.

1.1. P. O. Research.[2]

After many efforts they finally have long and short [mercury delay] lines working to a certain extent. That is, lines circulated pulses up to periods of

---

[1] The report is in the Woodger Papers (National Museum of Science and Industry, Kensington, London; catalogue reference M12/105). It seems originally to have had no title. The title adopted here is suggested by a phrase of Woodger's in a letter to Huskey about the report on 8 January 1982 (also at M12/105). Some spelling mistakes in the original report have been corrected and in places the punctuation has been altered. Material appearing within square brackets has been added by the editor.

[2] Editor's note. The Post Office Research Station at Dollis Hill, London.

20 minutes or so. On January 17, 1947 their delay line lost pulses when other electrical disturbances occurred in the vicinity such as switching on other equipment. Later this condition was improved, so they said, by added shielding but they still had difficulty when the power shut down started. Their position on March 7th is very much as described above.

They have a short delay line which they asserted worked satisfactorily. The construction details of circuits and delay lines looked very neat.

They expect (with the resumption of power) to continue checking the operation of the circulating circuits and to look into the design of the associated amplifying circuits. They also expect to start design of 32 delay line units and to work on other circuits such as the clock and frequency control circuits.

Their delay line was mounted inside a pipe of about 4″ diameter with screw adjustments at the center to remove sag. No other adjustment (for example, to make the crystals parallel) seems necessary for their one long delay line.

1.2. Cambridge.

In a relatively short time Dr. Wilkes has got a delay into operation and is successful in circulating pulses. In contrast to P. O. Research his delay line is prone to gain pulses instead of losing them. In fact, his method of inserting pulses was to crank a magneto in the room.

This situation (Jan. 29, 1947) led him to favour a screened room in which to place the machine.

Since then I have heard that by raising the levels in the associated amplifier circuits he has overcome this difficulty. This was apparently done at the suggestion of a Mr. Gold.

1.3. Manchester.

A standard cathode ray tube has a [metal] gauze placed in front of the target face and the signal from this [is] amplified and fed through a clocked multivibrator to the control grid of the CRT. If a sweep with gaps (of fixed length) is now placed on the tube and one continues to have it swept the pattern is maintained.

The transit time of the amplifier is accounted for by the fact that there is an accumulation of charge preceding the gap which enables one to anticipate the actual gap. The signal resulting from a particular pattern is plotted on the following amplitude-time graph.

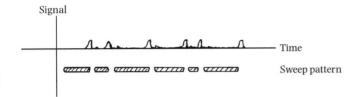

Experiments have been done in which two sweeps are brought together to where they begin to interfere. This experiment seems to indicate that at least 32 lines can be placed on a 5″ scope.

Two identical lines were established and swept alternately and the patterns persisted for some time.

The delay in Prof. Williams present amplifier is such that the sweep time must be 250 microseconds giving an effective pulse frequency of about 125 kilocycles. He has constructed and is testing (March 3rd) a wideband amplifier that should enable him to decrease the sweep time to 100 microsecs, giving an effective frequency of about 330 KC.

The sweep has been triggered at a frequency of 100 cycles per second showing that the persistence of the charge is such that the whole tube can be scanned so as to retain the charge pattern indefinitely.

1.4. Moore School.

People at the Moore School have been successful in circulating pulses in mercury delay lines. There has been a split in the personnel that worked on delay lines, Sheppard who started the work on delay lines has gone with Eckert (The Electronic Control Corporation). The Moore School also tried CRT storages but gave this up as being too difficult.

1.5. Princeton.

Apparently a satisfactory experimental model of the 'selectron' has not yet been made by R. C. A. (Dec. 1946).

1.6. Memory Summary.

	Delay Lines	CRT
Art	Few made. Each works alone.	One connected in circuit. Experiments hopeful but not conclusive.
Volume for 2^{13} 32 digit binary numbers	$6' \times 6' \times 7' = 252$ ft.^{3}	$8' \times 8' \times 2' = 128$ ft.^{3}
Temperature	All lines must be the same temperature (within 1 degree).	Immaterial.
Weight	Heavy.	Relatively light.
Cost	At least £10.	£2 to £3.
Reliability	Unknown, may last indefinitely.	Manufactured in quantity, easily replacible.
Logical Aspects	Numbers are accessible only when they transit the amplifier.	Immediate accessibility.
Restoration of memory	Continuously cycled.	Maximum period unknown. If continuously restored then it is just as good as a delay line.

Part II. Input and output.

2.1. Princeton.

They have been carrying on experiments on magnetic wire and tape. They have built an experimental set-up using two bicycle wheels and a servo-control to take up slack in transfering from one to the other. They seem to be definitely decided in favour of magnetic wire or tape.

2.2. Moore School.

For a year they have been experimenting with wire and tape and low inertia motors to use in the associated servo-controls. Results there indicate that tape or wire can be used to give an input of 30 KC or better.

2.3. N. P. L.

All logical development thus far is based on the use of Hollerith equipment. Experience of IBM and Hollerith is that relay equipment is not as reliable as desired by a factor of, say 10^4. However, proper checking methods can perhaps be devised to cope with these failures to some extent.

2.4. ENIAC.

The ENIAC used two IBM machines (a reader and a gang summary punch) as input and output. These machines accounted for a very considerable portion of the time that the ENIAC was out of operation.

2.5. Manchester.

They intend to have Princeton fabricate their input and output equipment.

2.6. Cambridge.

Dr. Wilkes intends to use Hollerith equipment.

2.7. P. O. Research.

The Ministry of Supply machine will use special paper tape.

2.8. Input and Output Summary.

	Hollerith	Magnetic wire or tape
Development	Well established.	Just beginning.
Speed	1280 binary digits per second (present plans).	At least 30,000 per input unit. (Estimates to 50 KC.)
Sorting	Done on a separate Hollerith machine.	Done internally using erasibility of the tape. 3 or 4 input tapes may be used to to facilitate sorting. Much faster than Hollerith.
Filing subroutines	On groups of cards in drawers. Problems made up by assembling these with any special cards.	On one spool of wire. Probably a separate wire is used for any special instructions.
Assembling subroutines	By hand.	By machine, one spool of wire holds enough to fill a 256 delay line machine 30 times.

Part III. Computer.

All groups generally agree that the computing machine should do addition, subtraction (either directly or by use of complements), and the multiplication of signed numbers as well as some process of discrimination. From this point on there is some difference of opinion. On the other hand, it is generally agreed that operations such as division and square rooting should be left to programming.

3.1. Moore School.

It was agreed that a large machine should have floating decimal point as a built-in feature. However, in the interests of economy of equipment, any small machine would probably be a fixed decimal point machine with what few facilities needed to have a programmed floating decimal point.

3.2. Manchester.

The question of floating decimal point has not been seriously considered here, probably primarily because Prof. Newman is interested in working with whole numbers.

3.3. N. P. L.

So far all plans leave the floating of the decimal point up to the programming.

3.4. Summary.

It is generally agreed that built in floating decimal facilities would be desirable. However, this necessitates considerable extra equipment in a delay line machine, and so seems undesirable in a small machine. In a larger machine the added equipment does not seem out of proportion.

I do not think enough consideration has been given to the CRT type machine to make any decision. The ease of obtaining delays by delayed triggering of the memory sweep may make it much more feasible here.

Part IV. Logical aspects.

4.1. Moore School.

The Moore School developed plans for coded decimal and binary computers with and without floating decimal facilities. As of June, 1946, these had reached their more complicated state and since then the tendency has been to simplify the system as much as possible.

The general tendency at the Moore School was to make, at the price of some extra equipment, one order or instruction do as much as possible. To illustrate, an instruction was of the following form:

a R b = c, t, r.

This order caused at time t the numbers in positions a and b of the memory to be brought out of the memory, the operation R applied and the result placed in position c of the memory. The number r caused this operation to be repeated to the r − 1 consecutive pairs of numbers following a and b (at times t + 1, t + 2, . . ., t + r − 1) and the results went into consecutive places starting with c.

This type of order is particularly convenient with small machines where the whole of the instruction is not taken up with the numbers representing a, b, and c. On the other hand the length of a number or instruction doesn't need to be a power of two long (in contrast to the fact that switching circuits or trees should [have] channels which number a power of 2) so 40-digit numbers and instructions are feasible if needed for this purpose.

4.2. Princeton.

In the von Neumann plan the machine consists essentially of a memory and a static accumulator. All transfers from the memory to the accumulator are essentially additions. The orders used are of the following type:

(1) Clear A	This clears the accumulator.	
(2) x to A	Adds the number in position x in the memory to the contents of the accumulator.	
(3) A to x	Transfers the number in the accumulator to position x of the memory.	
(4) C to x	Transfers control to x; i.e., the next order to be obeyed is in position x in the memory.	
(5) CC to x	Conditional transfer of control; i.e., control is transferred to x in the memory if the number in A is negative. Otherwise, control obeys consecutive orders in the memory.	

Routines are established once and for all (with appropriate blanks in them) and stored on the wire or tape. These are read into the memory as needed.

To assist in preparing a problem they have devised 'flow diagrams' to simplify the actual coding process.

4.3. Manchester.

They are planning to more or less copy the von Neumann scheme. The main difference in their machine is that CRT memory (as against the selectron) leads to a dynamic accumulator (i.e., another CRT tube with a pulse-time adder in series with the restoring circuit).

(It is interesting to note that this method leads to an efficient method of multiplication (using multiplier pulses to trigger the sweep in the memory) wherein a 32×32 binary multiplication takes, on the average, 16 addition operations instead of 32.)

4.4. Cambridge.

Dr. Wilkes has (as of 1-2-47) done no development work on the logical plans. He intends to follow the Moore School pattern very closely.

4.5. N. P. L.

At N. P. L. all efforts have been on the logical developments. Various plans have evolved leading to more and more complicated machines. Certain principles have been established of which one is:

USE THE MINIMUM AMOUNT OF EQUIPMENT,

that is, do everything possible by programming unless it has to be done extremely frequently (the turning on of the machine is not considered a frequent operation). Strangely, this has led to a machine which is a combination of the Moore School and the von Neumann machines. That is, version VIIC (the latest one, 10-3-47) has an accumulative adder (exactly like the von Neumann and Newman-Williams machines) and a function unit (a computer in the Moore School terminology) which comprises an adder (with odd, even, and no round carry suppression), an 'and' unit, an 'or' unit, a 'not-equivalent to' unit, as well as a special device for combining orders to form new orders and a built in multiplier.

This machine has been planned on the premise that switching can be accomplished between pulses (not setting up the resistance network or tree but the final opening and closing of the relevant gates). There <u>are</u> arguments for leaving space between words. For example, in a 256-delay line machine 256 gating circuits (probably 512 tubes) must operate <u>safely</u> inside the distance between consecutive pulses, say .7 microseconds. That is, the gating circuits must operate in .2 microseconds and this will require a considerable amount of power with consequent use of extra valves. By leaving space between words these circuits can be operated more slowly and the only thing lost is the facility of carry over (which can be regained by adding a trigger circuit in the adder) and somewhat inefficient use of the memory (that is, delay lines must be somewhat longer or only 30 digit, say, binary numbers would be used).

This machine has two-word orders (it also uses single word orders for simpler operations) that have many of the aspects of the single word orders of the Moore School plan; that is, they name two sources, an operation, a timing number, and moreover a delay number.

The fact that repetition of subroutines require[s] large numbers of orders has led to the abbreviated code methods whereby not only standard orders are used but special words containing parameters are converted into orders by an interpretation table. The general idea is that these describe the entries to subroutines, the values of certain parameters in the subroutine, how many times the subroutine is to be used, and where to go after the subroutine is finished.

4.6. Summary of Logical Aspects.

This can to some extent be summarized as a 'battle of principles'.

<u>Moore School</u>

1. Simplify set up procedure of the problem to the extent that any mathematician, theoretical engineer, or physicist, can lay out the problem in a form ready to be coded or typed out for the machine. He should be able to do this with the help of a manual of not more than 15 to 20 pages.
2. To assist the operator by having the machine count its own time cycles so that orders do not have to be dovetailed with respect to time.
3. To have as few rules, drills, or manual routines for the operators as possible.

N. P. L.

To minimize equipment. This leads to quite complex manual routines in running the machine. Although this is a stated principle, the operation portions of the machine are larger than any of the other present plans (Princeton, Moore School, Manchester, or Cambridge).

Many people (Prof. Hartree in his lecture before the I. E. E., among others) have expressed the opinion that it will take a huge staff to prepare problems for such an automatic computing machine and keep it going. This idea has been proven by experience with the ENIAC. Several times I have heard Dr. Dederick say that the bottle-neck was in the preparation of problems for the ENIAC. Principles (1) and (2) of the Moore School seem well worthwhile from this point of view.

Part V. Checking.

5.1. Moore School.

The Moore School favors duplicating portions of the apparatus to detect failures and where duplication is not feasible to build in checks which will tend to quickly detect failures (both transient and permanent failures). Use of checking procedures or test problems in the ENIAC has shown the inadequacy of this type of test.

5.2. Princeton.

The von Neumann group has considered building two machines with facilities for comparing between the two.

5.3. Manchester.

They have not given much thought to this problem as yet.

5.4. N. P. L.

Portions of the control should be duplicated for checking purposes (the Moore School intends to duplicate the computing unit). No built-in checking in the memory. Input and output checking is to be programmed as is periodic systematic checks of the memory. Also the operator is expected to produce various identities which can be used to check the numbers resulting from the computations.

Part VI. My general opinion.

1. ENIAC experience has shown that systematic before and after checks do <u>not</u> guarantee correct results. Furthermore, that transient failures leading to such phenomena (successful before and after checking but incorrect results) are extremely hard to find, in fact, some have taken as long as two weeks to find on the ENIAC.

2. Since heater-failure is by far the most frequent type of failure systematic built-in checks should be had to immediately detect these and assist in their being found.

3. Ideally, the machine should have duplicate parts except for the memory. Then when accuracy is desired the memory can be divided into two parts and the computation can be carried on in complete duplicate. When accuracy is not so important the memory is not divided and one has available a much larger capacity machine, or else the comparing circuits can be deenergized and the machine can work on two problems simultaneously.

4. The plan (3) is probably not practical for a small machine but I certainly believe in (2) for any machine. In fact, if space is left between words (with a probable saving of equipment as mentioned above, see 4.5) checking pulses can be inserted in every word space. These will detect not only permanent failures but will probably (with probability ranging from zero to one as the length of the transient failure varies from 0 to one minor cycle) detect transient failures.

5. There are logical facilities in the ACE not needed in most computing problems. The numerical operations of addition, subtraction (with complements), and multiplication, along with a discrimination process seem sufficient. With these other logical operations can be programmed.

6. The computing portions of the ACE can be considerably simplified and certainly should be in any small machine that is built.

7. To facilitate the programming of problems any small machine should count its own time scale (see 4.6).

8. In planning future machines it should be kept in mind that (a) charge storage tube methods are very distinct from delay line methods, (2) that charge storage tube methods will most surely be realized either in the

CRT technique of Williams or in the 'selectrons' of RCA, (c) magnetic input and output methods will (with their increase in speed) make it more necessary to have simple coding methods.

9. The logical development at N. P. L. is far ahead of practical results. In fact, so far ahead that changes in the <u>art</u> may to some extent make it obsolete. Prof. Newman has said that it would seem that N. P. L. must go ahead and build a delay line machine since they have gone so far.

10. I think that the best plan would be to construct a small delay line machine on a plan which is a compromise between the N. P. L. and the Moore School plans; and in the future look toward charge storage tube methods with magnetic input and output.

Index